Crime Scene to Court

The Essentials of Forensic Science

Second Edition

Crime Scene to Court
The Essentials of Forensic Science
Second Edition

Edited by

Peter White
Department of Forensic Science, University of Lincoln, Lincoln, UK

RS•C

advancing the chemical sciences

ISBN 0-85404-656-9

A catalogue record for this book is available from the British Library

Published by The Royal Society of Chemistry,
Thomas Graham House, Science Park, Milton Road, Cambridge CB4 0WF, UK
Registered Charity Number 207890

For further information see our web site at www.rsc.org

Typeset by H Charlesworth & Co Ltd, Huddersfield, UK
Printed in the United Kingdom by Henry Ling Limited, at the Dorset Press, Dorchester, DT1 1HD

Preface

When the group of forensic scientists had their first meeting with the Royal Society of Chemistry to discuss the proposal for a new forensic science textbook, one of the main points of discussion and concern was whether it would attract a significant market. The decision in the end was to continue and provide a book primarily for supporting the teaching of forensic science degree courses in the UK.

At the time of the first meeting and even when the first edition of *Crime Scene to Court* was published in 1998, there was only a handful of universities offering forensic science courses. However, dramatic changes in universities have since lead to the introduction of many forensic science courses and this book has become one of the recommended textbooks for many of the courses. Hence, when the book was published it could not have been more timely and subsequently has exceeded all our market expectations.

It has also been extremely pleasing to note that the book, as originally intended, has appealed to a much wider readership. *Crime Scene to Court* has been used and referred to by the courts, forensic scientists, police and scene of crime officers and read by lay people who just have a fascination for the subject. Interestingly, although originally intended for the UK the book now sits on many bookshelves throughout the world. This can possibly be attributed to the fact that all the authors are recognised experts in their discipline within the UK forensic science profession, which has an international reputation.

As with any scientific subject technology moves on and hence there was always going to be the inevitable question of a revised edition. Since its publication some forensic practices, both scientific and professional, have changed and when approached all the authors agreed the need for a second edition. Furthermore, the authors were also prepared to give up their valuable time to revise their own chapters, for which I am indebted.

Readers can now benefit from these revisions which provide details of current crime scene and laboratory scientific practices but again the original philosophy of producing a relatively non-technical textbook has been adhered to. As indicated earlier there have also been changes in professional requirements. Maintaining a respected and professional forensic science service is crucial and accreditation of laboratories and individuals, forensic science teaching and quality assurance are all issues which have received considerable attention during the past few years. I am delighted that Brian Caddy, with his extensive professional knowledge and involvement in many of these matters agreed to revise and contribute to Chapter 1 to help address these issues.

This second edition also gave the opportunity to consider if any other forensic disciplines should be included. Computer based crimes in civil and criminal cases have risen dramatically within the past decade and special units have been set up to examine such crimes. Hence a new chapter covering this topic has been introduced. Jonathan Henry provides the reader with the benefit of his considerable experience by introducing how different computer based media store information which, when skillfully restored, can provide evidence for courts.

The other new chapter considered very worthy of inclusion covers the subject of Blood Pattern Analysis. Adrian Emes and Christopher Price, both involved as trainers for the Forensic Science Service in this discipline, explain how information regarding location, sequence of events, disturbance of a scene and even which samples should be considered for DNA analysis can all be gleaned from the careful examination of a blood pattern found at a scene or on an item.

As editor I am grateful for these contributions from the new authors and would like to express my thanks to all authors for their support, valuable time and for providing readers with the benefits of their expertise and experiences. I would also like to record my thanks to the Royal Society of Chemistry for its support and Lorraine Stewart for assistance with the typing.

Peter White

Contents

Chapter 4   Marks and Impressions   **82**
         Keith Barnett

Abbreviations

AA	Atomic absorption spectroscopy
ABO	ABO blood groups
ABPI	Association of the British Pharmaceutical Industry
AFR	Automatic fingerprint recognition
ANFO	Ammonium nitrate/fuel oil
BAC	Blood alcohol concentration
BMA	British Medical Association
BMK	Benzyl methyl ketone
BPA	Blood pattern analysis
BrAC	Breath alcohol concentration
CAP	Common approach path
CE	Capillary electrophoresis
CENTREX	Central Police Training and Development Agency
CJD	Criminal Justice Database
CPS	Crown Prosecution Service
CRFP	Council for Registration of Forensic Practitioners
Δ^8-THC	Δ^8-Tetrahydrocannabinol
Δ^9-THC	Δ^9-Tetrahydrocannabinol
EDX	Energy dispersive X-ray analysis
EMIT	Enzyme multiplied immunoassay technique
ESDA	Electrostatic deposition analysis
ESLA	Electrostatic lifting apparatus
FAAS	Flameless atomic absorption spectroscopy
FEL	Forensic Explosives Laboratory
FLP	Fragment length polymorphism
FOA	First Officer Attending
FSS	Forensic Science Service
FTIR	Fourier transform infrared spectroscopy
GC	Gas chromatography
GC-MS	Gas chromatography-mass spectrometry
HLA	Human lymphocyte antigenicity
HMX	Cyclotetramethylene tetranitramine

HOLMES	Home Office Large Major Enquiry System
HPLC	High-performance liquid chromatography
IAFS	International Association of Forensic Scientists
IC	Ion chromatography
ICP	Inductively coupled plasma spectroscopy
IEF	Isoelectric focusing
ILAC	International Laboratory Accreditation Cooperation
IR	Infrared analysis
LC-MS	Liquid chromatography-mass spectrometry
LGC	Laboratory of the Government Chemist
LMG	Leucomalachite green
LSD	Lysergic acid diethylamide
MDA	Methylenedioxyamphetamine
MDMA	Methylenedioxymethylamphetamine
MSP	Microspectophotometry
NAA	Neutron activation analysis
NAFIS	National Automated Fingerprint Identification Scheme
NTCSSCI	National Training Centre for Scientific Support to Crime and Investigation
PCR	Polymerase chain reaction
PETN	Pentaerythritol tetranitrate
PF	Procurator Fiscal
PGC	Pyrolysis gas chromatography
PGM	Phosphoglucomutase polymorphism
POLSA	Police Search Advisor
PSDB	Police Scientific Development Branch
RDX	Cyclotrimethylene trintramine
RFLP	Restriction fragment length polymorphism
RIA	Radioimmunoassay
SEM	Scanning electron microscope
SERRS	Surface enhanced resonance Raman scattering spectroscopy
SGM	Second generation matrix
SIT	Spontaneous ignition temperature
SLP	Single locus probe
SOCO	Scene of Crime Officer
SOP	Standard operating procedure
SSM	Scientific Support Manager
STR	Short tandem repeats
TIAFT	The International Association of Forensic Toxicologists
TLC	Thin-layer chromatography
TNT	2,4,6-trinitrotoluene
VNTR	Variable number tandem repeats

Contributors

R.A. Anderson, *Department of Forensic Medicine and Science, University of Glasgow, Glasgow, G12 8QQ.*

K.G. Barnett, *Forensic Science Service, Birmingham Laboratory, Priory House, Gooch Street North, Birmingham, B5 6QQ.*

V.L. Beavis, *deceased.*

B. Caddy, *5 Kingspark, Torrance, G64 4DX.*

P. Cobb, *deceased.*

M. Cole, *Department of Forensic Science and Chemistry, Anglia Polytechnic University, East Road, Cambridge, CB1 1PT.*

V.J. Emerson, *4 Makins Road, Henley-on-Thames, Oxon, RG9 1PP.*

A. Emes, *The Forensic Science Service, 109 Lambeth Road, London, SE1 7LP.*

A. Gallop, *Forensic Alliance, F5 Culham Science Centre, Abingdon, Oxfordshire OX14 3ED.*

A. Giles, *The Giles Document Laboratory, Manor Lodge, North Road, Amersham, Buckinghamshire, HP6 5NA.*

J. Henry, *Computer Crime Unit, Headquarters Serious Crime Squad, Royal Ulster Constabulary, 29 Knocknagency Road, Belfast, BT4 2PP.*

R.H. Ide, *P.O. Box 5274, Sutton Coldfield, West Midlands, B72 1FF.*

L. Jones, *The Forensic Explosives Laboratory, Defence Science & Technology Laboratory, Fort Halstead, Sevenoaks, Kent, TN14 7BP.*

M. Marshall, *The Forensic Explosives Laboratory, Defence Science & Technology Laboratory, Fort Halstead, Sevenoaks, Kent, TN14 7BP.*

C. Price, *The Forensic Science Service, 109 Lambeth Road, London, SE1 7LP.*

T.J. Rothwell, *Forensic Access, Building F4, Culham Science Centre, Abingdon, Oxfordshire, OX14 3ED.*

R. Stockdale, *Forensic Access, Building F4, Culham Science Centre, Abingdon, Oxfordshire, OX14 3ED.*

J.S. Wallace, *Forensic Science Agency of Northern Ireland, 151 Belfast Road, Carrickfergus, Co. Antrim, BT38 8PL, Northern Ireland.*

N.D. Watson, *University of Strathclyde, Forensic Science Unit, Royal College, 204 George Street, Glasgow, Scotland, G1 1XW.*

N.T. Weston, *Weston Associates, 1 Greenways, Wolsingham, Co. Durham, DL13 3HN.*

CHAPTER 1

Forensic Science

BRIAN CADDY and PETER COBB

1.1 INTRODUCTION

Forensic scientists soon discover when talking to the general public that many people have an extremely limited knowledge of forensic science and the tasks it performs. As conversations continue it becomes apparent that misconstrued ideas often originate from watching television dramas. The material in this chapter is set out to address these misconceptions, by providing a definition of forensic science, a discussion of its origins and how forensic science services have been developed and operate within the United Kingdom. The duties of a forensic scientist and how the high standards of analysis and behaviour that are required are maintained also form an important aspect of this chapter.

1.1.1 Forensic Science – A Definition

If one were to ask one hundred forensic scientists to define forensic science it is possible that one would receive one hundred different definitions but it might be expected that amongst these there would be reference to science and the legal process. A useful working definition therefore is that "forensic science is science used for the purpose of the law". Consequently, any branch of science used in the resolution of legal disputes is forensic science. This broad definition covers criminal prosecutions in the widest sense including consumer and environmental protection, health and safety at work and civil proceedings such as breach of contract and negligence. However, in general usage the term is applied more narrowly to the use of science in the investigation of crime by the police and by the courts as evidence in resolving an issue in any subsequent trial. The narrower definition is implied in the title of

1

this book and the following chapters will discuss the use of science in the investigation of offences such as murder, violent assault, robbery, arson, breaking and entering, fraud, motoring offences, illicit drugs and poisoning. Covering such a range is justification for restricting the definition in a work such as this.

Confusion sometimes exists in the mind of the public between forensic scientists and those involved in forensic medicine (the latter is sometimes referred to as legal medicine). The forensic scientist, as defined above, can be involved in all types of criminal investigation but forensic medicos restrict their activities to criminal and civil cases where a human body is involved. These are nearly always serious cases such as murder and rape *etc*. and will require the participation of pathologists and/or police surgeons.

1.1.2   An Historical Background

The origins of forensic science can be traced back to the 6th century with legal medicine being practised by the Chinese. Within the next ten centuries advances in both medical and scientific knowledge were to contribute to a considerable increase in the use of medical evidence in courts. Other types of scientific evidence did not start to evolve until the 18th and 19th centuries, a period during which much of our modern-day chemistry knowledge was just starting to be developed. Toxicology, the study of poisons, emerged as one of the new forensic disciplines, and was highlighted by the work of Orfila in 1840 with his investigation into the death of a Frenchman, Monsieur Lafarge. Following examination of the internal organs from the exhumed body, Orfila testified on the basis of chemical tests that these contained arsenic, which was not a contamination from his laboratory or the cemetery earth. This evidence resulted subsequently in Madame Lafarge being charged with the murder of her husband, but more importantly raised the problem of contamination, a constant concern for any forensic scientist.

During the latter part of the 19th century there was also considerable interest in trying to identify an individual. One approach, studied by Alphonse Bertillon, was to record and compare facial and limb measurements from individuals. This proved to be unsuccessful due to the difficulties in obtaining accurate measurements. However, this was the first recorded attempt in a criminal investigation to use a classification system based on scientific measurement. Interestingly and in accord with this principle, forensic scientists today use the results from a combination of analytical measurements to discriminate between groups or to compare samples.

A more successful development in personal identification was to come from fingerprint examinations. Although Bertillon is reported to have used latent fingerprints from a crime scene to solve a case, it was Sir William Herschel, a British civil servant in India, and Henry Faulds who were credited with performing most of the early investigations. Faulds, a Scottish physician, is also accredited with establishing the fact that fingerprints remain unchanged throughout the life of an individual. It was not until 1901, however, when Sir Edward Henry devised a fingerprint classification scheme for cataloguing and retrieving prints, that the full potential of personal identification through fingerprint evidence could be used in forensic investigations.

Body fluid samples have also been found to contain information that can help to identify an individual. The progress made in this area has been dramatic, and major advances have occurred within the past decade. Up until 1900 it had been impossible to determine if a blood sample or stain was of human or animal origin, or to classify human blood into four main groups: A, B, AB and O. When tests devised by Paul Uhlenhuth (blood origin) and Karl Landsteiner (blood groups) were used, discrimination between individuals was still poor. The inclusion of the Rhesus test and several different enzyme systems improved discrimination, but it has only been through recent studies of deoxyribonucleic acid (DNA) in human chromosomes that there have been dramatic improvements in the confidence of identifying an individual.

To Edmund Locard (1910) is attributed an important basic principle of forensic science, this in essence being that 'every contact leaves a trace'. Whilst the examination of fingerprints or body fluid, which might be present in only trace amounts, can directly implicate a particular person in a crime, other types of trace evidence e.g., glass, paint, fire accelerants, gunshot residues, etc., can provide links which establish contact between objects and/or people involved with a crime or present at a crime scene.

The ability nowadays to be able to analyse such a variety of materials stems from technological advances that have occurred particularly in the past 50 years. Many of the analytical techniques that have been devised offer unbelievable sensitivity and permit examination of minute quantities (traces) of material which cannot be observed directly by the human eye. To provide some indication of the amount of material being examined in these trace samples consider initially a grain of sugar. This can be seen without any difficulty and weighs about 1 milligram (1 mg; *i.e.*, one thousandth of a gram or 1×10^{-3} g). Now consider one millionth of this quantity which is 1 nanogram (1 ng or 1×10^{-9} g). This amount of sugar cannot be seen, but quantities as small as this can be detected by many analytical techniques. Even lower detection limits can be obtained routinely with some instrumental methods and although beneficial, extreme caution is required at every stage of any investigation and subsequent analysis to ensure a positive result is genuine and not due to contamination or any other artifact.

Rapid developments in computer technology have also played an important role in the advancement of forensic science. Apart from their use in controlling instruments and producing analytical data, computers permit the storage of massive amounts of information that can be searched very quickly. With computers has come the establishment of databases for DNA recovered from body fluids and sometimes tissues and hair, fingerprints and footwear marks *etc.*, the purpose of these particular ones being to help in the identification of an individual or items associated with an individual. These and other databases can save a tremendous amount of time and effort in a case and are beneficial to both the police in following their enquiries and the forensic scientists in providing evidence and information for the courts.

1.1.3 Forensic Science in the United Kingdom

Prior to specialist laboratories being established, the police in many parts of the world relied upon scientific assistance from people who, through their occupations, were able to provide the expertise required. Without a centralised system, knowing whom to approach was a problem and this resulted initially in the formation of formalised institutions, these being established almost invariably as parts of universities or hospitals. Since over time, these were restricted to examining and providing expertise in a limited number of forensic science disciplines, police forces took the step of developing their own forensic science laboratories.

Europe took the lead in this development with the first police forensic laboratory being opened in 1910 in Lyons, France. Thereafter, police laboratories were to appear in Germany (Dresden, 1915), Austria (Vienna, 1923) and other countries including Holland, Finland and Sweden, with these last three all coming into service in 1925. This transition did not occur in the United States of America (USA) until 1923 when the Los Angeles Police Department set up its own forensic science laboratory. The failure to obtain an indictment in a case due to improper handling of evidence prior to laboratory examination was the reason for this change. Many other Police Departments across America followed this lead with the Federal Bureau of Investigation (FBI) laboratory opening in 1932.

Interestingly, the first police forensic science laboratory in the United Kingdom was not established until 1935, when the Metropolitan Police Laboratory sited at Hendon was opened. How this laboratory started is a fascinating history. It all arose from the unofficial efforts of a constable, Cyril Cuthbertson, who was interested in medicine and criminalistics (in the United Kingdom this term is usually associated with the examination of physical evidence such as footwear marks but it has a much wider meaning in the USA and covers most of the activities undertaken in a forensic science laboratory) and became involved in applying scientific tests that helped his police colleagues in their investigations. Following his examination of a document and his attendance at court as a witness, praise for his testimony and skills soon filtered back to Scotland Yard. The Police Commissioner, Lord Trenchard, took a considerable interest in this matter as he could see the benefits of a laboratory dedicated to his police force. As a consequence he, over a period of time, persistantly engaged the Home Office over this matter and this eventually paid off.

The success of this laboratory resulted in the Home Office sanctioning the development of their own forensic laboratories, under the banner of the Home Office Forensic Science Service, to provide regional laboratories for police forces in all areas of England and Wales. These laboratories were all financed from central and local government funds until 1991 when the Forensic Science Service (FSS) became an executive agency of the Home Office. The agency comprises the five operational laboratories of the former Home Office Forensic Science Service (located in Birmingham, Chepstow, Chorley, Huntingdon and Wetherby) and, since 1996, the Metropolitan Police Forensic Science Laboratory in London. Wherever possible, facilities are provided locally but the corporate structure allows the concentration of specialist expertise in particular laboratories so that a comprehensive service is available.

Agency status enables the FSS to charge for the facilities offered on a contract, case by case or item by item basis depending on the circumstances. These facilities are also available to the defence in criminal cases. Where work is performed for both prosecution and defence the work from each is conducted at different laboratories and client confidentiality is maintained. Other agencies which were formerly part of government departments and offer forensic science facilities are the Laboratory of the Government Chemist (LGC), particularly in the area of drugs and documents, and DSTL, formerly the Defence Evaluation and Research Agency, operating under the Forensic Explosives Laboratory (FEL) in respect of explosives. Agency status allows the provision of services to any customer in the United Kingdom or overseas.

Forensic Science Northern Ireland in Belfast is also a government agency and provides forensic services to the province. In Scotland forensic science facilities are still provided by individual police forces with laboratories in Aberdeen, Dundee, Edinburgh, and Glasgow.

Although the laboratories referred to above can generally be regarded as the 'official' laboratories there is a wide range of practitioners and practices throughout the country providing an independent forensic service to clients. These include university departments, public analysts, large and small practices and sole practitioners. Although these may undertake prosecution work, they have a particular role in working with lawyers retained by a defendant in a criminal case to explore the strengths and weaknesses of scientific evidence tendered by the prosecution. Whilst this may include the laboratory examination of original or new material in a case it will usually involve an evaluation of the results obtained by the original scientist and the interpretation offered. The latter may require modification in the light of further information provided by the client or discovered by the retained expert.

Following the change to agency status the 'official' and private laboratories, together with other institutions, are all competing against each other for custom from the police or in offering a service for the defence. Unfortunately, although some of these laboratories have sought quality control and accreditation of their procedures and facilities, as described later in this chapter, there is no system of accreditation or regulation of the forensic science profession. This means that any organisation can now offer and supply forensic science services whether or not they have the technical competence and experience.

There are hidden dangers in a totally 'privatised' forensic science service any country might adopt. For example, commercial pressures and competition could lead to compromised standards. Constraints on budgets could also restrict both the amount of material submitted and the analytical work to be performed. The danger with this scenario is that these restrictions could prevent the forensic scientist from reaching a conclusion which might provide a court with either stronger evidence to support a prosecution, or show that the accused could not have perpetrated a criminal act. Therefore, it is essential that these dangers are identified and appropriate controls are put into place.

There has been one other development in forensic science, this being the introduction of civilians, called Scene of Crimes Officers (SOCOs) or Crime Scene Examiners, into police forces to carry out the searching of crime scenes and collection of evidence. Contrary to belief, it is nowadays quite rare for a forensic scientist to attend a scene and one reason for introducing SOCOs into the forensic system was to reduce the amount of time forensic scientists were being called away from their laboratory work.

Over the years there has become an increasing recognition of the importance of crime scene investigation and the need for collection, packaging and transport of material of potential evidential value. Whilst in earlier days this would have been carried out by a detective or a scientist, it is now usually performed by specialists who have received extensive training in all aspects of crime scene examination including latent fingerprints, evidential traces and photography. This professionalism should ensure that the integrity of items received by the scientist for examination cannot be disputed.

In conclusion the forensic science work performed in the United Kingdom has for many years been regarded very highly throughout the world for its integrity. The implications of any changes implemented, such as the change in status of the forensic laboratories, must be monitored and reviewed and actions taken where necessary to preserve the reputation of the service and more importantly, to ensure that it is not responsible for any miscarriage of justice.

1.2 WHEN IS FORENSIC SCIENCE REQUIRED?

A police officer investigating an incident will seek clarification of three issues:

1. Has a crime been committed?
2. If so, who is responsible?
3. If the responsible person has been traced is there enough evidence to charge the person and support a prosecution?

This clarification is seldom the isolated duty of one officer and the ultimate trial will reveal the involvement of the specialist police officers and civilian staff, lawyers and scientists. Forensic science can be expected to make a contribution to the clarification of all three issues.

1.2.1 Has a Crime Been Committed?

In most cases there may be no doubt that a crime has been committed but there are a number of occasions when only a scientific examination of items can inform the investigator that this is the case. For example, alleged possession of an illicit drug will require identification of the seized material. Similarly, to support an offence of driving under the influence of drink or drugs a blood sample taken from a motorist will require an accurate analysis not only to establish that alcohol or a drug is present but that any alcohol exceeds a permitted level. The presence of semen on a vaginal swab from an under-age girl will be evidence of illegal sexual activity. Similarly the demonstration of toxic levels of a poison in tissues removed at post-mortem from a body of an individual believed to have died from natural causes will be a strong indication of a crime. Doubts as to the authenticity of a document may be resolved by scientific examination and provide evidence of fraud.

1.2.2   Who is Responsible?

If a latent fingerprint is developed and recovered from a crime scene and the criminal's prints are already in a database then the person potentially responsible for that crime may soon be identified. Similarly the existence of a database of DNA profiles may enable identification of an offender who has bled at the scene of violence or who has left other body fluids in a sexual assault. Although specific identification of an offender may not be provided by scientific examination useful leads may be produced which will enable the investigator to reduce the field of enquiry.

1.2.3   Is the Suspect Responsible?

Irrespective of any support received from the scientist, the usual diligent police investigator often produces a suspect and the investigator will look to the scientist to provide corroborative evidence to enable a charge to be made and to assist the court in deciding guilt. The scientific examination will normally be directed towards two aspects:

1. Examination of material left on the victim or at the scene which is characteristic of the suspect.
2. Examination of the clothing and property of the suspect for the presence of material characteristic of the victim or the scene.

1.2.3.1   Materials Characteristic of the Suspect.   The biological, physical or chemical characteristics of materials found on the victim or at the scene can help to confirm the identity of the suspect and/or provide evidence of their involvement or presence at the crime scene. Blood, semen, saliva, fingerprints, hair and teeth are all characteristics of an individual.

Finding fibres from clothing or the characteristic pattern of the soles of shoes worn by a suspect may provide evidence of their involvement, as can any material found that may be associated with their particular occupation. The characteristics of a vehicle, if used in the crime, such as oil drips or tyre marks may indicate where it had been parked or driven over ground near to or at the scene. Paint, glass or plastic from the vehicle after a collision may help to identify the particular vehicle and hence the owner who could become a suspect. Finally, the characteristic marks that may arise from weapons, tools or other items used in committing a crime, for example, a knife in a stabbing, a screwdriver used in forcing an entry or a firearm used in a robbery, especially if found on the suspect, could provide further evidence of a suspect's association with a crime.

Clearly it is very unlikely that all these possibilities will be realised in a single case but knowing the circumstances surrounding a case and taking into account previous experiences the forensic scientist should be in a position to exploit their skills to the benefit of the investigator and the courts.

1.2.3.2 Materials Characteristic of the Scene or Victim. The crime scene could be in a building or outdoors, but any search usually yields materials that are characteristic of the particular location as identified below:

1. Domestic premises – external and internal painting, external and internal glass, furnishings, crockery and glassware, *etc*.
2. Commercial premises – as for domestic plus process materials.
3. External scenes such as gardens, waste ground and fields – soil, vegetation and miscellaneous debris.

Where the scene involves a living or dead victim, biological and clothing characteristics discussed above for the suspect will also apply.

1.3 DUTIES OF THE FORENSIC SCIENTIST

Having established when the service of a forensic scientist could be required their duties can now be identified as follows:

1. Examine material collected or submitted in order to provide information previously unknown or to corroborate information already available.
2. Provide the results of any examination in a report that will enable the investigator to identify an offender or corroborate other evidence in order to facilitate the preparation of a case for presentation to a court.

3. Present written and/or verbal evidence to a court to enable it to reach an appropriate decision as to guilt or innocence.

Under the adversarial system of trial used in the United Kingdom, the United States of America and many other parts of the world the individual forensic scientist may be regarded as, and claimed to be, an independent witness for the court but may not always be so regarded. It is essential therefore for the scientist to be able to demonstrate competence, impartiality and integrity by attention to issues such as the following:

1. The scientist should only give evidence on work carried out personally or under their direct supervision. However, an expert witness can interpret factual evidence given by another witness under oath in the light of scientific findings and knowledge.
2. Where scientific examinations are relied on for legal purposes the methods used should be based on established scientific principles, validated and, preferably, published in reputable scientific literature, so that they can be scrutinised by the scientific community at large.
3. Where the scientific findings require interpretation the basis of any interpretation should be available to the scientific community.

It is important to recognise that the responsibilities of the individual forensic scientist are personal and not corporate. Thus in giving evidence he or she is completely and solely responsible for their own experimental results and for the opinions expressed. However the corporate environment will usually be a supportive structure to provide appropriate training, standardised methods and procedures, evaluation of performance and a quality management system. Attention to the last can be a real source of reassurance to the individual forensic scientist, the criminal justice system and the public at large.

1.4 QUALITY IN FORENSIC SCIENCE

There are many definitions of quality but for our purposes that of the International Organisation for Standardisation (ISO) is appropriate:

Quality: **The totality of features and characteristics of a product or service that bear on its ability to satisfy stated or implied needs.**
(ISO Standard 8402: 1986)

The ultimate 'customer' to be satisfied is the court and it will expect there to be in place a total quality management system that will ensure the integrity of material examined by the scientist, the examination carried out and the testimony given.

1.4.1 Quality at the Scene – Laboratory Chain

The quality control system must clearly extend outside the laboratory environment and places a responsibility on all involved in an investigation to maintain, as often specified, a chain of custody. A more appropriate expression would be a chain of integrity since the court will need to know not only the identity of the links in the chain but also their behaviour as illustrated in Table 1.1.

Even on this short consideration it can be noted that apart from knowing the identity of the 'custodians' of items at each stage, the court will need to be assured of their awareness of the consequences of any deficiency in the processes in which they are involved. For this reason in many investigations most of the process is conducted by specialist personnel with occasional assistance from laboratory based scientists. All involved will need to protect the items from the twin problems of deterioration and contamination. The latter is a vital matter when a suspect has been arrested since all possible contact between items from two sources must be prevented with proof that the appropriate actions have been taken.

Table 1.1 *The scene-laboratory links in the chain of custody to ensure quality in this procedure*

Link	Category	Comment
1	Preservation of the scene	This can be difficult in the early stages, particularly when injury or hazard is involved but the police must establish access control as soon as possible. Thereafter access must be restricted to those who can make a real contribution to the investigation.
2	Search for material of potential evidential value	This must be systematic with careful records kept of the location of all material collected.
3	Packaging and labelling of collected material	This must ensure that the material arrives at the laboratory as far as possible in the condition in which it is collected and that it can be related to the source.
4	Storage and transmission to the laboratory	Again preservation of the condition is a priority *e.g.* refrigeration may be appropriate.

1.4.2   Laboratory Quality Procedures

Clearly if the integrity of the articles received by the laboratory has been maintained the responsibility is then transferred to the scientist. The methods used for the examination of various evidential materials are detailed in subsequent chapters but certain principles apply to all examinations:

1. Prevention of contamination is a prime requirement, particularly as such small amounts of material can be examined and characterised. The scientist must be able to demonstrate that the procedures used have prevented the adventitious transfer of evidential material between two sources.
2. Security of all items must be assured by recording the name of all individuals having contact with them. This is usually achieved by signing an attached label but attention must always be directed towards the avoidance of leaving items unattended in the laboratory.
3. Careful permanent records should be kept at each stage of the laboratory examination to avoid any possibility of confusion by assigning results to the wrong item.
4. All the procedures and methods used by the scientist should be fully documented. These are often refered to as Standard Operational Procedures (SOPs) and Standard Methods (SMs). Forensic science laboratories normally have a comprehensive system detailing procedures to be followed and methods to be used. However, the final source of assurance is the competence and integrity of the individual.

1.5   ACCREDITATION OF FORENSIC SCIENCE FACILITIES

In common with many industrial organisations and scientific laboratories there is an increasing call for third party accreditation of forensic laboratories. In the United States of America this call has been met by the American Society of Crime Laboratory Directors through their Laboratory Accreditation Board. Under the auspices of this board the organisation, staffing and facilities of a laboratory are subjected to evaluation and on-site inspection before accreditation. A full re-inspection is carried out every five years. A good proportion of the many forensic science laboratories in the United States have been accredited by this process and some in other parts of the world especially Asia and Australia.

Given the smaller number of laboratories in the United Kingdom it has proved more convenient to use a well established system of accreditation applicable to all laboratories offering testing services to a client. The United Kingdom Accreditation Service (UKAS) is recognised by the government as the body for accrediting all types of laboratories and in this role has established a number of standards all of which have now been subsumed under two major standards namely ISO/IEC 17025 and ISO 9000:2000. The former is associated with laboratory tasks and some management aspects and the latter mainly with management issues. All these processes are being internationalised through the International Laboratory Accreditation Cooperation (ILAC) organisation. Clearly the wider recognition will be of great value to laboratories engaged in work supporting trade across international boundaries but given the increasing international nature of crime there must be attractions in the concept of all forensic laboratories working to a common high standard.

The accreditation process involves a series of steps. The applicant laboratory submits documentation, including its quality manual, to UKAS who then assign a Technical Officer and a Lead Assessor to be responsible for advising whether the laboratory should be accredited. This advice will be based on a pre-assessment visit for informal discussion and a broad review of the quality system followed by a formal inspection of the laboratory by an assessment team. At the end of the inspection the team will discuss any non-compliance found and agree the appropriate corrective action. If the assessment team are satisfied with the corrective actions UKAS will review all the evidence and decide whether to accredit. Following accreditation the laboratory will be subjected to regular surveillance and re-assessment visits to ensure that standards are being maintained.

A Forensic Science Working Group that took part in formulating standards for forensic science spent much time in defining what was meant by an objective test. Their definition is stated below:

An objective test is one which, having been documented and validated, is under control so that it can be demonstrated that all appropriately trained staff will obtain the same results within defined limits.

Objective tests are controlled by:

1. Documentation of the test.
2. Validation of the test.
3. Training and authorisation of staff.
4. Maintenance of equipment.

and where appropriate by:

1. Calibration of equipment.
2. Use of appropriate reference materials.
3. Provision of guidance for interpretation.
4. Checking of results.
5. Testing staff proficiency.
6. Recording of equipment/test performance.

The forensic scientist is required to tackle a wide variety of problems, many of which have no commercial analogue. This means that widely publicised and used methods such as those of the British Standards Institution may not be an option. The issues raised in the foregoing definition will assist the forensic scientist to develop the necessary degree of objectivity in the method applied to a particular problem.

1.6   PERSONAL ACCOUNTABILITY IN FORENSIC SCIENCE

The ultimate role of the forensic scientist is the presentation of expert testimony to the court trying the issue and in fulfilling this role the witness is completely and solely accountable for the experimental results presented and for the opinions expressed. These must be justified to the court, often in the face of fierce cross-examination and the witness cannot shelter behind the laboratory manual or base an opinion on a consensus or majority vote. This requires the witness to be a professional in the best sense of the word, that is, to have an initial developed competence which is continuously maintained together with a powerful sense of integrity. Clearly an employing organisation will have a responsibility in this respect but the public at large may seek the reassurance of membership of an independent professional body. The latter would seek to provide evidence of competence with a code of conduct and advice on professional behaviour.

This was recognised in recent times by a gathering of professional forensic scientists and although there is still no professional body for forensic scientists the outcome of their deliberations was the establishment, with government support, of a register of competent forensic scientists.

1.6.1   The Council for the Registration of Forensic Practitioners (CRFP)

Registration of forensic scientists with the CRFP is purely voluntary but since the scheme has the support of the government and the judiciary it is anticipated that the courts will expect, as a measure of their competence, that most forensic scientists will register.

In order to become registered a forensic scientist has to be assessed against a set of criteria that are identified below.

1. Knowing the hypothesis or question to be tested.
2. Establishing that items submitted are suitable for the requirements of the case.
3. Confirming that the correct type of examination has been selected.
4. Confirming that the examination has been carried out competently.
5. Recording, summarising and collating the results of the examination.
6. Interpreting the results in accordance with established scientific principles.
7. Considering alternative hypotheses.
8. Preparing a report on the findings.
9. Presenting oral evidence to court and at case conferences.
10. Ensuring that all documentation is fit for purpose.

The process requires that candidates submit brief details of a series of approximately 60 cases that they have investigated over the previous six months. An assessor will then select six cases from this list and request that the candidate submits full details of these cases in an anonymised form. Collectively these cases should enable the assessor to identify compliance with the ten criteria.

If the candidate meets the assessment criteria then he/she will be placed on the register in one of the defined areas as listed below:

1. Drugs – all areas of drug work not associated with toxicological investigations.
2. Firearms – ballistics, comparison microscopy and classification.
3. Human contact traces – DNA, body fluids, blood distribution, hairs and others (*e.g.* serology).
4. Incident reconstruction – fire and explosive investigation, metallurgy and material failures, traffic accident reconstruction, tyre examination, tachograph and non-metallurgic component failures.
5. Marks – tools, footwear, tyre marks, packaging and manufacturing marks and any others.
6. Particulates and other traces – fibres, glass, paint, explosive residues, gunshot residues, plant materials, pollutants, chemical traces and stains, any other particulate materials.

7. Questioned documents – handwriting, documents and other related materials.
8. Toxicology – toxicology, all aspects of alcohol analysis and interpretation.

Candidates are registered for four years before they are required to re-register. Re-registration requires the submission of information on continuous professional development and maintenance of professional competence. All those registered must comply with a code of conduct which is outlined below:

1. Recognise that your overriding duty is to the court and to the administration of justice: it is your duty to present your findings and evidence, whether written or oral, in a fair and impartial manner.
2. Act with honesty, integrity, objectivity and impartiality: you will not discriminate on grounds of race, beliefs, gender, language, sexual orientation, social status, age, lifestyle or political persuasion.
3. Comply with the code of conduct of any professional body of which you are a member.
4. Provide expert advice and evidence only within the limits of your professional competence and only when fit to do so.
5. Inform a suitable person or authority, in confidence where appropriate, if you have good grounds for believing there is a situation which may result in a miscarriage of justice.

In all aspects of your work as a provider of expert advice and evidence you must:

6. Take all reasonable steps to maintain and develop your professional competence, taking account of material research and developments within the relevant field and practising techniques of quality assurance.
7. Declare to your client, patient or employer if you have one, any prior involvement or personal interest which gives, or may give, rise to a conflict of interest, real or perceived; and act in such a case only with their explicit written consent.
8. Take all reasonable steps to ensure access to all available evidential materials which are relevant to the examinations requested; to establish, so far as reasonably practicable, whether any may have been compromised before coming into your possession; and to ensure their integrity and security are maintained whilst in your possession.
9. Accept responsibility for all work done under your supervision, direct or indirect.

10. Conduct all work in accordance with the established principles of your profession, using methods of proven validity and appropriate equipment and materials.

11. Make and retain full, contemporaneous, clear and accurate records of the examinations you conduct, your methods and your results, in sufficient detail for another forensic practitioner competent in the same area of work to review your work independently.

12. Report clearly, comprehensively and impartially, setting out or stating:

    (a) your terms of reference and the source of your instructions;
    (b) the material upon which you based your investigation and conclusions;
    (c) summaries of your and your team's work, results and conclusions;
    (d) any ways in which your investigations or conclusions were limited by external factors; especially if your access to relevant material was restricted; or if you believe unreasonable limitations on your time, or on the human, physical or financial resources available to you, have significantly compromised the quality of your work.
    (e) that you have carried out your work and prepared your report in accordance with this Code.

13. Reconsider and, if necessary, be prepared to change your conclusions, opinions or advice and to reinterpret your findings in the light of new information or new developments in the relevant field; and take the initiative in informing your client or employer promptly of any such change.

14. Preserve confidentiality unless:

    (a) the client or patient explicitly authorises you to disclose something;
    (b) a court or tribunal orders disclosure;
    (c) the law obliges disclosure; or
    (d) your overriding duty to the court and to the administration of justice demand disclosure.

15. Preserve legal professional privilege: only the client may waive this. It protects communications, oral and written, between professional legal advisers and their clients; and between those advisers and expert witnesses in connection with the giving of

legal advice, or in connection with, or in contemplation of, legal proceedings and for the purposes of those proceedings.

The introduction of the CRFP is a major step forward for the forensic science profession in the United Kingdom. However, the criteria used for registration are not standards in the accepted meaning.

1.6.2 Standards of Competence

Another body in the United Kingdom has been involved with drawing up standards of competence for forensic scientists, this being the Forensic Science Sector Committee of the Science Technology and Mathematics Council which is responsible to one of the new government Sector Skills Councils.

The standards are written in a generic form to enable all the different disciplines to be described. They are presented as a series of units with each unit being divided into a set of elements. The Units are listed in Table 1.2 and an example of one of the elements in Figure 1.1.

Having described standards of competence it then becomes necessary to develop a strategy for assessing scientists against such standards. Such assessment strategies are presently being developed.

Finally the United Kingdom has seen in recent years an enormous growth in undergraduate degrees in the forensic sciences and the quality

Table 1.2 *Professional standards of competence in forensic science*

Unit	Element
1. Prepare to Carry out Examination	1.1 Determine case requirements 1.2 Establish the integrity of items and samples 1.3 Inspect items and samples submitted for examination
2. Examine Items and Samples	2.1 Monitor and maintain integrity of items and samples 2.2 Identify and recover potential evidence 2.3 Determine examinations to be undertaken 2.4 Carry out examinations 2.5 Produce laboratory notes and records
3. Undertake Specialist Scene Examination	3.1 Establish the requirements for the investigation 3.2 Prepare to examine the scene of the incident 3.3 Examine the scene of the incident 3.4 Carry out site surveys and tests
4. Interpret Findings	4.1 Collate results of examinations 4.2 Interpret examination findings
5. Report Findings	5.1 Produce report 5.2 Participate in pre-trial consultation 5.3 Present oral evidence to courts and inquiries

You must ensure that you:	You need to know and understand:
a. make laboratory notes and records contemporaneously and that they are fit for purpose, accurate, legible, clear and unambiguous	1. why it is important to record information contemporaneously
	2. why it is important to ensure that notes and records are fit for purpose, accurate, legible, clear and unambiguous
b. order notes and record information in a way which supports validation and interrogation	3. what information you need to record
	4. which recording systems you need to use
c. uniquely classify records and file them securely in a manner which facilitates retrieval	5. when notes and records are complete
	6. the systems you use to order your notes and record information
d. accurately collate laboratory notes on work carried out by others into the overall records	7. the importance of ordering notes and information
	8. the classification systems you use to ensure records are easily retrievable
	9. how the classification system operates
	10. how to file records securely
	11. the importance of collating notes accurately
	12. the identity of others who might wish to use the notes
	13. the ways in which the notes might be used

Figure 1.1 *An example of an element associated with standards of competence (from www.crfp.org.uk)*
UNIT 2: EXAMINE ITEMS AND SAMPLES
Element 2.5: Produce laboratory notes and records

of such degrees has been of concern to many in the profession. For this reason the Forensic Science Society has begun a programme of developing standards for such degrees that will be offered to the Universities as part of an accreditation programme. These are in the early stages of development at present but standards for crime scene investigation, laboratory analysis and interpretation and presentation of forensic science evidence are almost complete at the time of writing.

1.7 CONCLUSION

From what has preceded it is hoped that the reader will have an understanding of the role of the forensic scientist, how he/she achieves a professional status and how he/she interacts with the legal process. Working as a forensic scientist can be physically, emotionally and intellectually demanding but also intellectually rewarding. The succeeding chapters will show why this is so.

1.8   BIBLIOGRAPHY

Murder Under the Microscope, Philip Paul, Macdonald and Co. London, 1990.
Science and the Detective, Brian H.Kaye, VCH, Weinheim, 1995.
A World List of Forensic Science Laboratories and Practices, 8th Edition, The Forensic Science Society, 1997.
Directory of Consulting Practices in Chemistry and Related Subjects, The Royal Society of Chemistry, 1996.
www.ukas.com
www.ilac.com
www.european-accreditation.org
www.crfp.org.uk
www.forensic-science-society.org.uk

CHAPTER 2

The Crime Scene

NORMAN WESTON

2.1 INTRODUCTION

Forensic evidence starts at the scene. If evidence is missed or incorrectly handled at the scene, no amount of laboratory analysis or processing will be able to rectify the problem and the scene usually cannot be revisited to have another attempt at obtaining additional evidence.

The people who bear the responsibility of examining the scene of any crime can include a police officer, a detective, a crime scene examiner, a scientific support officer or a forensic scientist. Historically, the person charged with investigating a crime has been prepared to consult those with specialist knowledge or professional skills who may add to, or account for, observations made at the scene of the crime. Throughout the 19th century, the level of technical support increased across a broad front, ranging from the increasing skills of the chemists in detecting poisons, to the introduction of photography, both for recording purposes and crime detection. This culminated in the first major work to recognise the significance of scientific approaches to crime detection with the publication in 1892 of *Criminal Investigation*, a book by Hans Gross which influences the art of crime detection to this day. Around the turn of the century the ability to both recognise an individual, and their involvement in crime, by the uniqueness of their fingerprints, outstripped all other developments in crime investigation. For the first time, evidence at the scene of a crime in the form of a fingermark, could be compared against a databank of known criminals and provide the investigator with a named individual.

The value of 'trace' or 'contact' evidence was, as previously stated, first recognised in 1910 when Edmund Locard introduced his theory of interchange. It is the finding, recovery, and scientific investigation of

21

these traces which can provide the links in a chain of evidence, which are essential to assist the investigator. Unfortunately, in the early part of the 20th century, the ability to analyse minute traces of evidence by biological, physical or chemical techniques did not significantly exist. The great leap forward in analytical techniques and the electronic revolution in all branches of science has enabled Locard's trace evidence, whether blood, clothing fibres, glass, paint, soil from shoes *etc.*, from the scene of crime, to be matched with a suspect in such a way that it provides increasingly objective and significant evidence to link the suspect with the crime. In the case of DNA analysis from body fluid traces, this evidence approaches the same levels of certainty as fingerprint evidence and can also provide the name of an individual from a database. A further valuable evidence type has been provided by the consumer society's fascination with fashionable footwear and an infinite variety of patterned shoe soles. Particularly when worn or damaged, these can provide evidence of much greater significance than a plain leather sole. In parallel with these improving methods of analysis, many new techniques for developing and recovering evidence have become available, particularly in the recovery of fingerprints and shoe marks.

It is against this background of rapidly increasing technology that the crime scene is now examined. There is, therefore, a clear need that all who are required to deal with a crime scene should be trained to a high state of awareness, knowledge and skill. Increasingly, the solution of many crimes and particularly major crimes, depends on a thorough investigation of the crime scene by specialist crime scene examiners supported, when necessary, by other experts with scientific knowledge or expertise.

In practice, the time, effort or expense involved in crime scene investigation is tempered by considerations of the seriousness of the offence and the likelihood of recovering evidence of value which will identify the perpetrator of the crime. In this chapter, these constraints have been largely ignored and consideration has been given to what can be achieved. Two hypothetical cases have been used to illustrate the general principles of crime scene investigation and the actions of the scientific support personnel involved. Many different types of evidence can be gathered from crime scenes, but for the purpose of this text they are limited to those which are covered in some detail in the succeeding chapters. Furthermore, legal issues which influence the actions of police investigators and their scientific support staff will be referred to only in general terms.

2.2 THE ORGANISATION OF SCIENTIFIC SUPPORT WITHIN THE POLICE SERVICE OF ENGLAND AND WALES

Initially, scientific support in its various forms developed in a haphazard manner shadowing the techniques available. In the early part of the last century, there was a need for police officers with skills in photography, the development of latent fingermarks, the identification of fingermarks, and with some awareness of how to recover and deal with the other forms of forensic evidence likely to be found at the scene of crime. Their work expanded until, from the 1960s onwards, it was clearly necessary to form specialist departments to carry out this work. From the 1980s onwards, this led increasingly to the formation of Scientific Support Departments, which would encompass or have access to all the specialist areas likely to be needed in crime scene investigation and headed by a Scientific Support Manager (SSM).

At the same time, it was realised that as this work became specialised and technical, careers in scientific support would develop in their own right. There has been considerable variation in the way these have developed but the essential elements will be summarised.

2.2.1 The Fingerprint Bureau (Department)

This has grown out of the initial need for police officers to recover and photograph fingermarks at the crime scene, and then bring these back for comparison and identification. Increasingly, it became apparent that these officers could be most effective by specialising in fingerprint identification, leaving others to do the recovery. It was also realised that the skills of a trained police officer were not essential to do this work, and that concentration and pattern recognition skills were the most important. The large majority of Fingerprint Bureaux in the United Kingdom now employ a mainly civilian, *i.e.* non-police officer staff, specifically and intensively trained to identify fingermarks and to give evidence of identification in court, as fingerprint experts.

The introduction of the National Automated Fingerprint Identification System (NAFIS) has enabled fingerprint co-ordinates to be fed into a remote terminal and comparisons made with a centralised database. All on-screen identification comparisons are verified by fingerprint experts.

2.2.2 The Scene of Crime Department

The development of these units followed from the changes in the fingerprint departments. Advantages were seen in training officers who would develop increasing experience of all types of crime scene. Currently, these Scene Examiners assess the scene, control the scene, record the scene by document and photography, examine the scene and recover all types of evidence, interpret the scene from the evidence, collect and control the exhibits, liaise with all who are involved with the case and prepare reports and statements of evidence. The Scene of Crime Department is increasingly seen as able to advise on scientific matters and provide intelligence to link scenes or crimes together. At the scene they are the immediate source of advice on Health and Safety issues.

Again, this is a specialist career in its own right for which academic qualifications and technical skills in relevant areas may be more important than training as a police officer. Consequently, over 60% of Crime Scene Examiners in England and Wales are now non-police officers.

2.2.3   The Photographic Services Department

The duties of this department are less clearly defined. Variable amounts of photography are carried out in fingerprint and scene of crime departments as part of their normal duties. Many Forces employ specialist professional photographers to provide a service which can produce photographic and video images of the highest quality, especially when non-routine techniques are required. In addition, processing facilities are now required to provide a service in many other areas of police work, apart from the scientific support department's requirement. There is increasing use of photography in areas such as traffic, surveillance and public order offences. Photographers often form part of the scientific support team at major incidents and increasingly video recordings are made both for evidential purposes and as aids to "briefings". The increasing use of (electronic) imaging equipment and enhancement techniques is developing and changing the work of this department.

2.2.4   In-Force Laboratories and Scientific Services

In the early years, latent fingerprints were either photographed as they appeared, or enhanced by 'dusting' with powders and then photographed. Since that time, a wide range of techniques for the detection and enhancement of fingerprints have been developed world-wide. The problem of selecting the sequence of processes which are likely to be the most effective for the surface being examined has been solved largely by the work of the Home Office Police Scientific Development Branch (PSDB). This organisation has produced manuals and handbooks which give guidance using 'flow charts' on the best process for each circumstance. The pioneering work of the Serious Crime Unit (SCU) of the former Metropolitan Police Forensic Science Laboratory in London, with a shrewd combination of technical skills and serendipity, has developed the use of light sources including quasi lasers, lasers and photographic techniques to supplement chemical processes. The SCU has had notable success both in the laboratory and at scenes, as has the Specialist Fingerprint Unit (SFU) of the Birmingham Forensic Science Laboratory.

The success of these units has led to some police forces setting up their own laboratories to provide many of the more frequently used techniques both within their Force and at serious crime scenes. There is, however, still a need for the SCU and the SFU of the Forensic Science Service to support the Forces who lack these techniques, as well as providing the more sophisticated, or 'state of the art', techniques which can only be justified as a central service. Scientific Support Departments also examine shoe marks, tool marks and other 'physical' evidence which can be used to link offences or connect offenders. Increasingly,

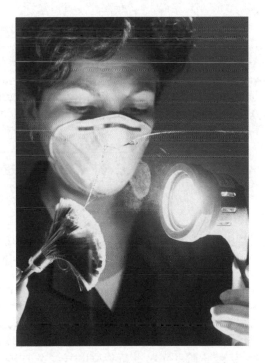

Figure 2.1 *A crime scene examiner developing a latent fingerprint with aluminium powder*
(Courtesy of the Director of the National Training Centre for Scientific Support to Crime Investigation © NTCSSCI 1996)

Forces are developing their in-house technical and interpretational facilities as well as subjecting their staff to regular performance reviews, both to maintain and improve professional skills and competence.

2.2.5   Training & Information

Since 1990 and 1992 respectively, Crime Scene Examiners and Finger-print Experts have been trained to approved 'National' standards at the National Training Centre for Scientific Support to Crime Investigation, based in Durham (N.B. now a part of 'CENTREX' – the Central Police Training & Development Authority). The Metropolitan Police in London see a particular need for a Crime Scene Examiner with enhanced skills in fingerprint examination and currently provide training for their own Fingerprint Experts/Identification Officers.

In many Forces, the 'Crime Scene Examiner' or 'Scene of Crime Officer' is taking an increasingly advisory and investigatory role in crime investigation using technical knowledge, crime intelligence data-bases and also by exercising their professional judgement. As this wider role is recognised, they may be increasingly referred to as 'Scientific Support Officers'.

In addition to specialised training, a valuable support and information service is provided by *Scenesafe Evidence Recovery Systems*. They make available a regularly updated *Scenes of Crime Handbook* in a handy pocket-sized format, and also other awareness material. In addition, they provide a comprehensive range of evidence recovery materials and containers which meet the requirements for subsequent scientific examination.

2.3   A BURGLARY: AN EXAMPLE OF A VOLUME CRIME SCENE

2.3.1   Case Circumstances

Cornelius Joseph Elliot, dressed in a track suit top and trousers, trainers and a woollen 'bobble' hat, had a small rucksack on his back containing the tools of his trade. He jogged down the road to create the impression that he was exercising. Elliot specialised in breaking into houses on housing estates where the majority of occupants were likely to be out during the day. On this day he selected one house where the rear was not overlooked, had high hedges and where there was a garden gate leading to a lane, thus providing an alternative way out of the back garden.

At the rear of the house, there was a kitchen window big enough to climb through. He pulled a dustbin into position, stood on the lid, and attempted to force open the window with a chisel so that he could then slip a screwdriver through the gap, and spring open the window catch. A concealed security lock prevented this. He removed his track suit top and, using this to muffle the sound, broke the window with a small hammer. Elliot listened carefully. Satisfied that there was no one in the house and that he had not been overheard by neighbours, he carefully removed enough pieces of glass to enable him to climb into the house. He stepped onto the work surface and jumped down to the floor. He unlocked the rear door using a key hanging on a keyboard, thus making sure of his line of escape. He quickly made a search of the areas which experience had taught him were likely to contain cash and valuables, and was lucky in finding cash in a kitchen cupboard and a box containing jewellery at the back of a drawer in a bedroom. He tipped the box out on to the bed and, hurriedly emptying the ring boxes and protective wrappers, stuffed the contents into the pocket of his rucksack.

The sound of a car pulling into the driveway alarmed him. Elliot quickly ran down the stairs and out through the rear door. In his hurry, he did not notice stepping on a child's comic lying on the kitchen floor, just as he did not notice that his woollen hat had fallen off as he ran down the stairs. He took the shortest route to the back gate, across the lawn and through the vegetable garden. He ran through the gate, brushing the hedge aside to do so, and then on down the lane. Elliot had no time to notice that the overgrown hawthorn bushes were in bloom.

2.3.2 What Happens at the Scene?

2.3.2.1 The Householder. As Mrs Jones pulled up in front of the garage of her home, she was startled when she saw in the gap, between the garage and the rear of the house, a dark haired figure who ran across the back garden. She got out of the car and started down the path but, as soon as she saw the rear of the house and realised that it had been broken into, she quickly returned to the car, got in and secured the doors. Then she called the police on her mobile phone.

Whilst she waited for them to arrive, she wrote down on the back of her shopping list everything about the figure that she could remember: white male; not more than medium height – he'd been able to run under the washing line; blue track suit – with a white stripe down the leg; some sort of bag over his shoulder – possibly canvas?; dark hair; not wearing any head covering or gloves that she could see.

Mrs Jones was greatly relieved when eight minutes later a police car pulled into the drive behind her and she could tell the officer what had occurred.

2.3.2.2   The Police Officer.   The first police officer to arrive in response to such a phone call would be logged at the police Control Room as the 'First Officer Attending' or FOA. In this instance the FOA was PC 1153 Brown and an essential part of his job was to reassure the victim, Jane Jones, to check the house and to outline to her what was to happen next. At the same time, he was starting the process of the investigation by thinking through the burglar's likely actions, asking himself such questions as 'What did he do?', 'What evidence might he have left behind?' and 'Where might he be now?'.

The FOA informed the Control Room that the burglar might still be in the area and that he would require scientific support. PC Brown asked Mrs Jones not to touch anything or to walk around whilst he assessed whether any evidence needed protecting. Clearly the intruder had stood on the dustbin lid which was outside and exposed to the weather; this PC Brown removed to the security of the garage. Then PC Brown and Mrs Jones entered the house through the front door, thus ensuring that they did not disturb any evidence in the kitchen. Leaving Jane Jones in the doorway, PC Brown quickly looked round the house to check that no intruder remained in the building and to make an initial assessment.

He then walked round it again, this time more slowly and accompanied by Mrs Jones so that she could confirm what had happened whilst she was out and also what appeared to have been taken. They were both careful of where they walked and of what they touched.

He gave clear instructions to Mrs Jones so that no potential evidence would be destroyed or compromised by her actions. He was aware of her wish to tidy her home and to make it secure and so offered practical advice about which parts of her home she could enter. Before he left, PC Brown also told her exactly what was going to happen in the next stage of the investigation. He obtained, *via* his personal radio, an estimate of the time a Scientific Support Officer might arrive and left for him a written note summarising his own thoughts and actions. PC Brown had now discharged his basic duties at a crime scene of assessing and controlling the situation, protecting the evidence from disturbance and contamination and communicating with those who would now take the enquiry forward. He then completed his required personal and Force records of the incident.

2.3.2.3   Arrival of the Crime Scene Examiner.   Later the same day, scientific support in the form of a specialist Crime Scene Examiner, Ian Williams, arrived and made an independent assessment of the crime scene. He had done this type of work for more than ten years. Although some crime information was available in the form of PC Brown's message, the scene examiner should never accept anything he sees, or reads, or is told, wholly at face value. Every scene should be assessed with an open and enquiring mind. The Crime Scene Examiner makes his own enquiries and assessments, decides on the points of the scene to be examined, and the type of examination to be made. At the end of his examination he will have:

1. Recorded all the features of the scene by diagram, notes and photographs.
2. Searched the scene and recovered and recorded all the physical evidence.
3. Packed all the physical evidence in secure and labelled containers.
4. Made sure that he has a detailed record of where each item was found, which also indicates its potential significance.
5. Looked for and recovered all traces left at the scene by the offender.
6. Taken samples, described as 'control samples', of all the materials the offender could have taken from the scene.
7. Formed a clear impression of how the crime has been committed.

2.3.2.4   The Crime Scene Examiner at Work.   As Ian Williams drove his Crime Van with all his equipment, and his sealed sample containers, back to the Police Station, he reflected on the words of that famous detective of literature, Sherlock Holmes – 'It has long been an axiom of mine that the little things are definitely the most important' (*A Case of Identity*, 1891). This had been a good scene for little things:

1. The dustbin lid showed traces of a footmark and he was taking it back to the Force crime laboratory where, using specialist lighting techniques, he could produce an actual size photograph to be compared with the burglar's trainers. Alternatively, the whole lid could be sent to the Forensic Science Laboratory.
2. The chisel, used by the burglar to lever the window, had left clear indentations in the edge of the kitchen window frame, both as a compressed mark and as scratches in the paint from the leading edge. Ian Williams had opened the broken window with Mrs Jones' security key. He had photographed the scratch marks against a scale measure to enable him to produce accurate, actual size photographs back in his studio. He had also taken a cast of the chisel impression using a dental casting preparation. He had decided against taking away the whole window.

3. Williams had looked for clothing fibres on the sharp edge of the glass still remaining in the window using a hand lens, but had seen none. He decided to use clear adhesive lifting tape on these edges, just to play safe, and then to stick this tape to a clear plastic sheet. He initialled the edges of the tape using a permanent marker pen, added the case details and then packaged, sealed and labelled it. The Forensic Science Laboratory might be able to find some fibres with their powerful microscopes. He had looked for, but not seen, any bloodstains on the glass.

4. Because it looked as though it might rain, he decided to complete the examination of the outside of the house first. Elliot, when he had removed the glass from the window, had carefully placed it on top of a coal bunker instead of knocking it out, to minimise the noise. Ian Williams carefully picked up each piece, at the edges, with gloved hands and placed it, secured in position, in a rigid container to be examined later at the police laboratory where its evidential value would be assessed.

5. Then he had carefully 'dusted' the area round the point of entry (the window), using a grey flake powder and a fine, fibre glass brush, and had been pleased to see some distinct fingermarks appear on the inside of the frame. These were in exactly the position where the burglar could have pulled himself in through the window. He used a clear adhesive tape to 'lift' the developed fingermarks from the frame and, using a small roller, secured this to a sheet of clear plastic. This he labelled with the details of the time, date, address and position of the fingermarks. He added his own name and then added a serial number together with his own initials and a crime reference number. Next he drew a small diagram on the plastic sheet and, by using a small arrow, indicated the position of the ground in relation to the finger-marks and then accurate measurements to show the exact position of the fingermarks. In addition, he noted all these details in his handwritten record of his visit to the crime scene. He had trimmed the edges of the adhesive tape, used to lift the fingermark, in his own individual manner and signed his initials over all the edges with an indelible marker. There was now no possibility that anyone, including a lawyer in court, could suggest that the tape had been interfered with. It would not be possible to move the tape and then replace it without it showing. Later, at the conclusion of his examination, he would place all the recovered tape 'lifts' in a secure package to which he would append all the case details and a signed exhibit label.

6. Again, with one eye on the weather, Ian Williams had a look round the rear of the garden and recognised the significance of a series of footmarks, embedded in the crumbly soil of a well dug vegetable plot and heading towards the garden gate. Carefully selecting two of the footmarks which showed the greatest detail, he placed measurement scales at right angles around the foot-mark, checked the vertical alignment of his camera and recorded the image. Again he was confident that actual size images could be produced but he decided to back these up by taking a 'cast', using a dental casting material, which would accurately repro-duce the detail of the footmark. Finally, he took 'control' amples of the soil in case the burglar had taken away soil on his shoes.

7. As he straightened up, Ian noticed the hawthorn blossom over-hanging the garden gate. It occurred to him that anyone over about 1.75 m tall must come into contact with the blossom as they passed through the gate. It was a long shot, but if he didn't take samples as controls now, it would be too late.

8. When Ian Williams returned to the house, he continued to follow the trail of the intruder. He quickly realised that it would be impossible for anyone to step directly through the window and down onto the floor of the kitchen, without first stepping onto the work surface. This was a smooth finish, beige colour For-mica and as such, quite likely to retain a footmark. By shading the outside light with a large piece of card and shinning his powerful scene of crime lamp horizontally across the surface, he could see a faint footmark, possibly caused by a damp shoe sole. He decided it was worth trying to enhance the mark by 'dusting' it carefully with a fine black powder applied on a soft, squirrel hair brush. Gradually, the clear imprint of the sole and trade-mark of a well known trainer shoe appeared. It was of a similar pattern to the footmark in the soil and, Ian thought, also the footmark on the dustbin lid. Again, he placed scale markers at right angles to the footmark and set his camera vertically above it to record the image. Finally, he decided to recover the devel-oped footmark on a gelatin lifting film, which he then secured in a rigid container. Having sealed the container, he attached an exhibit label on which he wrote the details of the incident.

9. He then turned his attention to the child's comic which lay on the floor. Crouching down, Ian shone his crime scene lamp across the floor in the direction of the comic and was able to see some impressions in its surface. He was aware that two kinds of shoe

impression could be present. One could be an indented impression caused by pressure from the person walking on the paper and the second could be a surface imprint of the pattern of the sole of the shoe, in the dust which had been carried on the shoe. Ian fetched from his van a portable electrostatic lifting apparatus, known as 'ESLA', and used it to transfer the dust footmark onto a black, foil backed, plastic sheet. The pattern of the sole was clearly visible. He photographed this and then he fastened the plastic sheet into a specially designed holder, which he sealed and labelled. The indented footprint, Ian thought, could best be recovered by the Forensic Science Laboratory using an electrostatic detection apparatus known as 'ESDA'. He therefore secured the comic in a large flat cardboard box which he then sealed and labelled.

10. On the kitchen work surface was a stone-ware jar, lying on its side and surrounded by a number of envelopes. These were labelled 'Housekeeping', 'Ballet' and 'Scouts' and the householder confirmed that these had contained money for the family's regular activities. In order to get the money out, the burglar must have handled the envelopes. With this in mind, Ian carefully packaged them, making sure that he sealed and labelled them correctly. These he would submit to the in-Force laboratory for chemical treatment to recover fingermarks. The grey stone-ware jar had a rough surface unsuitable for fingerprint detection, however, in the powerful light of his crime scene lamp, Ian Williams could see a distinct but faint, reddish-brown stain. He rubbed a corner of the stain with absorbent filter paper then added two drops from his blood testing kit and saw a strong colour reaction develop. This suggested to him that the stain was almost certainly blood. The Forensic Science Laboratory was the best place to deal with it, so he packed the whole item carefully in a strong brown paper bag, sealing and labelling it – including a 'Bio-hazard' label – before placing it in a box for protection in transit.

11. The next area which drew Ian's attention was the staircase. As he approached it, he saw a woollen hat lying on the carpet of the bottom step. He picked it up and placed it carefully into a small, clean brown paper sack. He called to Mrs Jones and asked her if the hat belonged to any member of her family. She said no, and so Ian sealed and labelled the paper sack. The paper sack had a 'see through' clear plastic panel so that the hat could be shown to witnesses without opening the seal, thus avoiding possible

contamination. Looking at the steepness of the staircase and at the white painted banister rail, Ian had felt that there was a strong possibility that the intruder would have held onto the rail either when going upstairs, or as he ran down. An application of the grey flake powder with the fibre glass brush did indeed reveal a number of fingermarks, however, there was an equally strong possibility that they might belong to members of the household.

12. Ian Williams' last task was to examine the one bedroom which showed signs of having been disturbed, and even here the signs suggested that the burglar had only had time to search one chest of polished wood drawers before being disturbed. Three drawers were pulled open and gave the impression that someone had rummaged through them. The other two drawers had been pulled out and their contents had been tipped onto the bed. Lying on the bed was a plastic box which Mrs Jones said had contained items of jewellery as had the other various ring boxes and plastic bags, all of which were empty now. Ian resisted the temptation to 'dust' any of these items with the grey fingerprint powder, deciding that these would be better dealt with by the in-Force laboratory, using its greater range of equipment, light sources and facilities for dedicated chemical treatments. Using a range of different and sequential processes, the laboratory should be able to develop far better 'latent' fingermarks for identification on the items than Ian using powder alone.

13. Having packaged and labelled the ring boxes and plastic bags, Ian Williams then turned his attention to the polished surface of the drawer fronts, here he found what seemed to be another probable blood smear. Using the same method as before, he tested it, and the test showed a positive reaction; this time he decided to recover the bloodsmear. There were several ways in which he could have done this: by cutting out from the drawer the piece of wood on which the smear appeared; by removing the whole drawer and taking it away with him for examination and treatment; or by using a medical swab. Experience had taught him that the first two options were never popular with householders. This being the case, he carefully moistened a swab with sterile water and delicately rotated the tip of the swab in contact with the stain until he recovered all of it and the tip of the swab was coloured red-brown. This he sealed in its own container, which he labelled – again including a 'Bio-hazard' warning label – and then took a second swab and swabbed the surface next to the stain area, this was to provide a 'control'

sample for the Forensic Science Laboratory of the water and any surface deposits which might influence their test. This too was sealed in its own container. Throughout this examination, Ian was mindful of the fact that the smear might have been left by a member of the household, which was another reason for using the swabbing technique.

14. This then left him only the front of the drawers to fingerprint and for this he returned to 'dusting' with powder. He was disappointed, but not surprised, to find that all the fingermarks revealed were pointing downwards. This would be consistent with those left by someone closing the drawer rather than pulling it open. These were likely to belong to a member of the family but, having developed them, he was compelled to recover and submit them to the Fingerprint Department for examination and checking with the 'elimination' fingerprints of the household. Under the rules of disclosure of evidence, even though fingermarks such as these are not likely to be called as evidence by the 'Prosecution' or by the 'Defence', or even be considered in any way contentious, they must be retained and declared on any list of exhibits relating to the case, and likewise, must be made available for viewing by any authorised person.

15. Mr Jones, having been called from his office by his wife, had arrived home and accompanied Ian on a final check of the house, outbuildings and garden to ensure that nothing had been overlooked. All thorough and experienced Crime Scene Examiners will make this final tour of inspection even without the presence of the householder. His final acts before leaving were, firstly, to reassure the Jones' that they could tidy their home and to check that they were able to make it secure. Secondly, conscious that some fingermarks he had 'lifted' may belong to members of the household, Ian took sets of inked 'elimination' fingerprints from Mr and Mrs Jones and then showed them how to take fingerprints of others using the specially prepared 'self-take', pre-inked elimination fingerprint forms. He gave them sufficient of these forms for all other members of the family. Once taken they could either be delivered by hand to the Police Station or posted to it in the post-paid, addressed envelope. Knowing that some people were wary of giving even elimination fingerprints, Ian took a moment to reassure the Jones' that all their fingerprints would be destroyed as soon as they had fulfilled the purpose for which they had been taken and that, if they so chose, the family could witness the destruction. Crime Scene Examiner Ian Williams then checked

that all his notes and records were complete and that the exhibits were securely packed and placed in his van before returning to the Police Station.

2.3.2.5 What Happened to Elliot, the Suspect? During the early evening of the same day, Police Constables 2022 Roberts and 1244 Green detained Elliot whilst he was walking his dog in the street. He was taken to the police station. Elliot was still wearing his track suit and carrying his rucksack which the officers had recognised from Mrs Jones' description. At the police station, the Custody Sergeant received him from Constables Roberts and Green. He took the personal details of Elliot and confirmed with the Constables the details of the offence for which they had detained him. In Elliot's presence the Sergeant took possession of the rucksack and also of its contents, which he listed. The same process was applied to the contents of Elliot's pockets. His dog, meanwhile, was placed in police kennels. The sergeant then discussed with Detective Constable Griffiths, the 'Investigating Officer' in the case, which procedures were to be followed. When asked about the burglary, Elliot had denied any knowledge of the crime, nor were any of the reported stolen items in his possession. It was decided that before he was formally questioned, Elliot should be treated as a 'crime scene'.

The following procedures were then followed:

1. John Handley, another Crime Scene Examiner, joined DC Griffiths and together they took Elliot to an examination room.
2. Elliot's hair was combed onto a clean sheet of paper and the resulting contents sealed in a packet and labelled by Handley.
3. Elliot undressed, item by item. Each item of clothing and his shoes were separately wrapped and sealed in brown paper sacks, which were also labelled.
4. Elliot was provided with alternative clothing.
5. Because traces of what appeared to be blood had been found at the scene of the burglary, DC Griffiths wanted a blood sample from Elliot. For this he needed the services of a trained physician.
6. Dr Oliver O'Neil, the Forensic Medical Examiner, arrived and with Elliot's consent, examined him. He made a general medical examination during which Elliot was found to have a small, apparently recent, cut between the first two fingers of his right hand. This could have been caused by a small shard of glass from a broken window. It would, in the opinion of the doctor, have bled briefly although Elliot might not have been aware of it so doing. Dr O'Neil, again with Elliot's consent, took control

samples of hair and blood, then made a record of all he had done, and also of the fact that it was his view that Elliot was fit to be detained and questioned. The Custody Sergeant likewise made a note of the doctor's opinion.

7. Buccal swabs were taken. These are swabs or scrapings of the cellular tissue of the lining of the mouth inside the cheek. The Police & Criminal Evidence Act procedures require buccal swabs to be taken from anyone arrested for a recordable offence and the results are stored on the DNA Database irrespective of whether the information is used in evidence in the case for which it was taken.

8. John Handley received all the items taken by Dr O'Neil and checked that they were securely sealed and labelled. It would be his responsibility to see that they were correctly and securely stored. In the case of perishable samples such as blood, buccal swabs and pulled hair *etc.*, this would be in a freezer or refrigerator.

9. Elliot was asked if he would be prepared to give a set of inked finger and palm impressions and this he agreed to do. These were taken by Handley. Elliot was then taken to an interview room for questioning by DC Griffiths and another officer. The interview was recorded on audio tape. John Handley later visited the kennels and took a hair sample from the dog, which he packaged and labelled, before returning to the Scene of Crime department.

2.3.3   Comment

Before moving on, it is worth identifying a number of points of 'good practice':

1. The First Officer Attending the scene, PC 1153 Brown, deals only with the scene and uses his own police vehicle.

2. Crime Scene Examiner Ian Williams deals only with the scene and uses his own scene of crime vehicle.

3. Police Constables 2022 Roberts and 1244 Green bring Elliot to the Police Station in a different police vehicle to that used by PC 1153 Brown.

4. Crime Scene Examiner John Handley deals only with Elliot and his dog, and he uses a different scene of crime vehicle to that used by Ian Williams.

5. DC Griffiths has been brought into the case as the Investigating Officer, and he will deal with all aspects of the case from now on. Because he has been in the presence of Elliot, should he wish to

visit the scene of the crime, he will not do so until after all the evidence has been gathered from it.

6. All of these procedures are necessary to avoid cross-contamination. For example, if the FOA, PC 1153 Brown, had later detained Elliot or taken him in his police vehicle to the police station, it could be argued in Court that any trace evidence found on Elliot was derived not from the scene but from his contact with PC Brown or his vehicle.

7. DC Griffiths has decided that Ian Williams, the Crime Scene Examiner who visited the scene, will be the "Exhibits Officer" for this case, and as such, will take charge of all the physical evidence. Therefore, when John Handley returns to the Scene of Crime Office, he will formally hand all the items he has taken possession of to Ian Williams, together with the appropriate supporting documents.

8. Cross-contamination is then no longer a problem since all the items recovered are now in securely sealed bags and containers. For example, the glass samples would be placed in a rigid container such as a cardboard box composed of a bottom container and a secure fitting lid. If such a box containing glass were to be shaken, it would be possible for minute shards of glass to escape through the gap between the lid and the body of the box. To prevent this, the lid is securely sealed to the box with adhesive tape and a further strip of adhesive tape would be placed all the way around the lid to completely seal this gap.

9. The next issue to be considered is continuity. The Courts require that the progress of an exhibit from the crime scene until it arrives in court must be totally accounted for by provable documentary control. This continuity or chain of evidence is controlled in three ways. Using John Handley as an example:

   i) When an item of evidence (Exhibit) is first taken, it is sealed in a secure container and an exhibit label attached. The exhibit label contains details of what the item is, where it was found and when, and a specific reference to the case or incident by name or number. It may also be given an exhibit number and reference based on the initials of the person taking it. For example, the hair combings taken from Elliot by John Handley might be referenced JH/1, the shoes as JH/2 and JH/3 respectively. On the reverse side of the exhibit label, every person who has responsibility at any stage for that particular exhibit must sign and date the label in sequence.

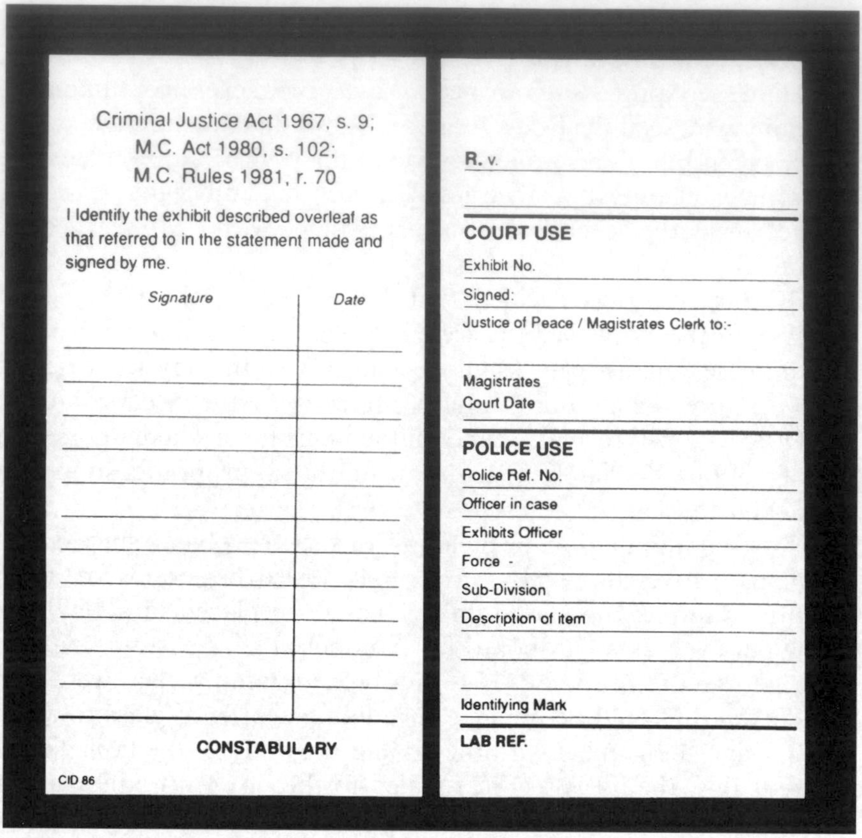

Criminal Justice Act 1967, s. 9;
M.C. Act 1980, s. 102;
M.C. Rules 1981, r. 70

I identify the exhibit described overleaf as
that referred to in the statement made and
signed by me.

Signature   Date

CONSTABULARY

CID 86

R. v.

COURT USE
Exhibit No.
Signed:
Justice of Peace / Magistrates Clerk to:-

Magistrates
Court Date

POLICE USE
Police Ref. No.
Officer in case
Exhibits Officer
Force -
Sub-Division
Description of item

Identifying Mark

LAB REF.

Figure 2.2   *A typical label used to identify an exhibit*

ii) John Handley will faithfully record in his detailed notes the examinations which he has made, the exhibits which he has taken, and what he has done with them.

iii) At the conclusion of his involvement in the case, John Handley will prepare a 'Statement of Evidence' for the Court. In it, he will describe his actions and account for the exhibits from the time at which he took them until they left his possession and were signed off to the next person in the chain.

2.3.3.1   *The Exhibits.*

1. Storage of Exhibits.
   On completion of his work, John Handley returns to the Scene of Crime Office, and personally hands his sealed exhibits to Ian Williams, who signs for them. They discuss the case. Ian Williams then places these exhibits into two containers, one holding the

general exhibits from Elliot and the other the perishable exhibits (blood samples *etc.*). Williams himself will also have two containers holding the general and the perishable exhibits taken from the scene. The Scene of Crime Office will have its own exhibits store. Ideally, Ian will have his own, lockable section complete with 'fridge/freezer. One of the senior officers will hold duplicate keys for emergency use only. In this way, the exhibits are kept under Ian Williams' personal control.

2. Examining the Exhibits.

   After DC Griffiths has interviewed Elliot, he will visit the Scene of Crime Office to discuss the evidential requirements of the case with the Exhibits Officer, Ian Williams. DC Griffiths knows what evidence he is required to put before the Crown Prosecution Service. Ian Williams has the experience to advise him on the best way to achieve this using the exhibits recovered.

   The papers, Fingerprint Exhibits, boxes and plastic bags from the crime scene, which may bear Elliot's fingerprints, are taken to the in-Force laboratory for chemical treatment so that any 'latent' fingerprints may be developed and made visible. Later they will be returned to Ian in a secure container together with photographs of any fingermarks found. These photographs, together with all the fingerprints 'lifted' at the scene, the inked sets of fingerprints taken from Elliot and those from Mr and Mrs Jones will be placed in a container sealed with an individually numbered security tag and sent to the Fingerprint Bureau for comparison. The expectation is that some of the fingermarks found inside the house, to which Elliot has no legitimate access, will be attributable to Elliot.

3. Forensic Science Laboratory Exhibits.

   All the other recovered material could be sent to the Forensic Science Laboratory, by courier, in sealed containers. The laboratory may be able to link the crime in the following ways:

   i) Traces from Elliot which could be linked to the scene:

   - Blood on the stone-wear pot and the drawer could be linked, possibly uniquely, to Elliot's blood sample.
   - The woollen hat could be linked to Elliot *via* his own head hair and also by DNA, and possibly by the hair of his dog.
   - Striations in the window paint could be matched to the cutting edge of Elliot's chisel.
   - The cast from the damaged window frame could be matched to leverage impressions from his chisel.

- Fibres from the tapings of the broken window edges could match Elliot's clothing.

ii) Traces from the scene which could be linked to Elliot:

- The trainer shoe mark on the dustbin lid, in the garden, on the kitchen work surface and on the comic on the floor, recovered by different techniques, could be linked, possibly uniquely, to Elliot's trainers.
- Glass shards from the broken window could be present on his clothing, particularly the track suit top.
- Paint from the window frame could be present on his chisel or screwdriver.
- Soil from the garden could be present on his trainers.
- Hair combings from Elliot could contain shards of glass from the broken window.
- Hair combings from Elliot could also contain blossom petals from the hawthorn hedge.
- Hair combings from Elliot could also contain wool fibres from the hat.

This case example demonstrates the work of the Crime Scene Examiners working for the police, and shows the operation in practice of Locard's 'Principle of Exchange' in evidence recovery. The following chapters of this book will describe how these typical evidential materials are examined in the laboratory.

2.3.3.2 Progression & Developments in Scientific Support. The manner in which scientific support is used in practice is now very much guided by the assessed needs of the investigation. Although the purpose of this book is to give an understanding of the breadth of scientific support, now that the reader has a grasp of evidence recovery, it is useful to consider some of the factors which control and influence its subsequent use as evidence.

Whilst it is important that the scene examination in evidence recovery terms is completed in full, decisions can subsequently be made to establish a priority which may limit further, and/or laboratory, examinations based on a number of factors:

1. Selection of 'best evidence' types to fit the needs of the case, for example, a fingermark inside the house places Elliot firmly where he has no right of access.

2. Careful selection of evidence and samples can limit the workload placed on the Forensic Science Laboratory or the Fingerprint Bureau allowing quicker examination to aid the investigative process. 'Elimination' fingerprints from the householder would fall into this category.

3. Developments in science occur rapidly and continually. For example, in the case of more sensitive DNA analysis, increasingly significant results are being obtained from both smaller samples and from different substrates. For instance, a partial or smudged palmprint may indicate a point of contact with the offender; after lifting the impression in the conventional way, wet and dry swabs applied to the underlying area may recover identifiable DNA.

4. The decision to charge for a forensic science provision on an itemised basis potentially brings into conflict budgetary restraints and the possible requirements of the investigation. This may tend to restrict the use of the Forensic Laboratory Service in favour of increased in-Force laboratory provision in areas where technical skills, rather than expensive scientific input, is required – and where no itemised charge is levied.

5. Due to concerns regarding balancing cost and 'best evidence' considerations, all Police Forces now have a Central Submissions Unit staffed by experienced personnel. After assessing the needs of an individual case, they deal with all submissions of exhibits for laboratory examination, handle the documentation and record keeping, and monitor the progress of cases. They are advised by a Forensic Liason Officer who, in addition to keeping in close contact with laboratories over scientific issues, also acts as a budget manager.

Increasingly, all aspects of command, control, investigation and information in the investigative process are becoming computerised by various 'Case Tracking Systems'. It is the current intention to link these systems together in a sophisticated manner such that, for example, an investigator who is about to interview a witness can be presented with, or select from, all the relevant evidence, intelligence and scientific information necessary for his task.

2.4   A MURDER: AN EXAMPLE OF A MAJOR CRIME SCENE

The previous parts of this chapter have established many of the general principles involved in the actions of scientific support officers at the crime scene. When police are faced with a major crime, or series of crimes, the basic principles of investigation remain the same but the sheer size of the task and the numbers involved introduce complex management and communication issues. It is only necessary, for our purposes, to be aware that a Senior Investigating Officer controls the enquiry with a management team to which all information, including that from scientific support officers, is directed. The information is controlled and stored through an 'incident room', using a computer system known as 'HOLMES' (Home Office Large Major Enquiry System). Two-way communications briefings take place twice a day and the results of enquiries are continuously monitored.

The Scientific Support Department also has to put its own management structure in place to manage its resources to meet the demands of the enquiry. A 'Crime Scene Manager' will exercise professional judgement over all the key activities whether they are concerned with techniques, personnel, contamination, continuity or, indeed, anything concerned with the examinations required, together with regular briefings and information flow. An 'Exhibits Officer' will be responsible for control, continuity and security of all exhibits. If the enquiry is more complex, then a 'Scientific Support Co-ordinator' will form part of the management team, as well as overseeing a number of 'Crime Scene Managers' who are involved with different parts of the enquiry. The Co-ordinator will be responsible for the overall development of scientific support, personnel and the communication of information between the Scene Examiners, the Crime Scene Managers and the Senior Investigating Officer and their teams. They will also maintain contact with forensic scientists or medical personnel such as the pathologist. Where a number of forensic scientists and possibly specialist laboratories become involved with the case, the Investigating Officer will ask for a Senior Forensic Scientist to join their team as an advisor and to oversee and co-ordinate the laboratory work. It is against this background that the crime scene examination may be carried out.

The level of management, documentation and control of all aspects of major crime investigation is such that it can be subjected to immediate 'in depth' review by an independent investigation team.

2.4.1 Case Circumstances

The victim of this crime had lived in a semi-detached house on a small estate in the town. When the victim returned to her house, she was surprised by an intruder. The intruder subsequently put her in his vehicle and drove to a secluded lay-by on a main road out of the town. He removed her from the vehicle, opened a gate to a field, carried her across to the far side and left her under a hedge. Her handbag and some items of female clothing were thrown over the hedge into the field beyond. The following morning, a farmer out shooting rabbits found the partially clothed body under the hedge and called the Police.

When the victim did not keep an appointment with a male friend that evening, he telephoned the house. There was no reply. The following morning, when he was still getting no reply to his calls, he became anxious and went round to the house. He saw that a window had been broken and was open. He called the police.

This simple sequence of events is typical of the start of many murder enquiries. At this point, it presents only questions and no answers:

1. Are the missing woman and the body in the field one and the same person?
2. Where did the woman die? – was it in the house, in a vehicle or in the field?
3. When did she die?
4. How did she die?
5. How was she attacked?
6. Was she attacked in more than one location?
7. What does the evidence in the field tell us?
8. Was the attacker known to her or was he a stranger?
9. Was the house really broken into or was it made to look as if it had been broken into?
10. Did an attack occur in the house?

It is also inevitable that the first person(s) at any scene must cause some disturbance. In this case we have the farmer who finds the body in the field, the male friend and the police officer who has checked the house for the occupier or an intruder. Only after these events can control of the scene be established.

Whilst the police make enquires to put together as much information as they can, the Scientific Support Department together with doctors and forensic scientists must build up a picture based on scene examination, trace evidence and scientific analysis. In doing so they must keep a totally open mind and 'read' the scene based on their knowledge and experience as the facts fall into place, testing each hypothesis as they go.

2.4.2 The Crime Scene

The police appear to be faced with three immediate problems in terms of the crime scenes:

1. *The Victim's House.* This appears to have been broken into for reason unknown and may contain evidence of an assault possibly leading to murder and traces from the intruder(s).
2. *The Field.* The field and the area surrounding the body may contain evidence from the murderer(s), from the vehicle used and also items from the victim.
3. *The Body of the Victim.* This will bear evidence of the cause of death, the nature of the assault and also traces of contact between the victim and her attacker(s). The victim will be a source of 'control' samples.

Subsequently, the police may expect at least three further, related, crime scenes:

4. *The Murderer's House.* The murderer may have returned after the crime taking evidence from the crime and traces relating to the victim and her house with him.
5. *The Murderer's Vehicle.* This may contain evidence from the victim who has been carried in the vehicle and possibly traces from the victim's house. There may also be evidence that an assault occurred in the vehicle.
6. *The Murderer(s).* When arrested, the murderer(s) may show medical evidence of the crime or injuries from a struggle. The person will be a source of control samples and clothing or possessions may be a source of evidence.

Each of these scenes will present its own special problems. In addition, there will be pressure to deal with the victim's house, the field and the body quickly since these may provide evidence of immediate value in locating or identifying the murderer(s).

2.4.3   Serious Crime Procedure

On receiving information of a 'suspicious death' or possible 'murder' one of the advantages for an investigator is that police forces have an established 'Serious Crime Procedure'. This enables most of the elements for the initial investigation to be put in place quickly. Each of the crime scenes in this particular incident has its own demands.

2.4.3.1   The Victim's House.   The 'First Officer Attending' (FOA) has the same responsibilities as before, to 'assess', 'protect' and 'communicate'. The main task here (having ascertained the victim is not in the house) is to preserve the scene. A tape barrier, or cordon, must be set up to restrict access to any area which might contain evidence and protect any evidence. The FOA must consider where any vehicle involved in the crime might have been parked and where the police vehicles should be positioned as they arrive. The FOA must also prevent contamination of any part of the scene by others, make a written record of all persons attending the scene, and maintain a communication link.

The Scientific Support Officer (SSO) on arrival will check the cordon which is in place, review the possibility of evidence external to the scene, assess the personnel and equipment needed to examine this scene, and then take the initial photographs. Wearing protective clothing to avoid contamination of the scene, the SSOs will search and mark an access pathway to the house, making notes and taking photographs as they progress. The FOA will ensure that no one else enters the scene until this is done. The Crime Scene Manager will assume responsibility and a decision will be made on how the house is to be entered to cause the least disturbance; the house will need to be searched thoroughly for evidence. The major difference between this scene and the burglary scene discussed earlier will be the slow, careful approach, the detailed recording of the scene with camera and video and the use of 'stepping plates' which allow the Scene Examiners to enter the building with minimal contact with the floor. This preserves footmarks whilst allowing reasonably quick entry into the house to assess the situation. In addition to the search for all types of trace evidence, the Crime Scene Examiner will be particularly interested in signs of a disturbance or struggle. Since at this stage there is little information from the body and other aspects of the case, a decision may be made to suspend this scene examination.

2.4.3.2 The Field. External locations present a major challenge, the control and preservation of evidence may require fast and decisive action because of existing or impending weather conditions. The FOA, who attends the outdoor scene to join the farmer in the lay-by, will be aware of the potentially overriding consideration to save life or to determine whether life is extinct. They will question the farmer accordingly knowing that if the FOA has to approach the body this must be done in a strictly controlled way. On this occasion, the farmer who has already approached the body is able to satisfy the FOA that this is not necessary. The FOA must now assess what damage they or others may do to this scene, decide how the scene must be protected, communicate to their supervisors by radio any special needs and confirm that a Senior Investigating Officer has been informed. In this case for example, although his vehicle and that of the farmer are already in the lay-by, there is a risk of evidence from the perpetrator's vehicle, such as tyre marks or footmarks, being further disturbed. The FOA may take the decision to close the lay-by with 'crime scene' markers and request Police Control to close the carriageway to allow access and parking only to police vehicles. It would then be the responsibility of the FOA to mark a route into the lay-by which all following personnel will use and to protect any significant evidence and access points to the field.

The FOA will commence a log of all persons attending taking note of all their actions and will be joined rapidly by other officers including the 'on call' Scientific Support Supervisor and a Detective Inspector. They will review the crime scene, decide on a course of action, and seek the approval of the Senior Investigating Officer who will appoint a Crime Scene Manager and confirm that a Home Office appointed pathologist will attend. It will also be confirmed that a 'Major Incident' vehicle with all the non-routine equipment, such as tents and floodlighting, is on its way.

The Crime Scene Manager, with Scientific Support Officers, will confirm that all protective cordons are in place. The Police Constable now maintaining the 'scene log' at the designated entrance to the scene is requested to allow no one into the scene area without clear purpose and who is not wearing protective clothing. At this stage it is not yet established that a murder has taken place. The priority is to examine and recover the body in a controlled manner with the minimum of evidence disturbance. Other issues such as searching the field and the lay-by can wait.

The first stage is to develop an access pathway to the body, known as a 'common approach path' (CAP). The most likely route to enter the field is through the gate where footmarks, tyre tracks and other evidence is likely to be. A hole cut through the hedge some distance from the gate opens up a clear pathway which would be marked across the field by a crime tape and cleared of all potential evidence material. At all stages a photographic and video record is made as is a record of the general area and conditions. When the body is in clear view, a decision would be made on the best line of approach as well as a review of equipment needs, for example a tent to cover the body and its surrounding area in order to protect the scene from wind and weather and public/press view. At a distance of say 6–7 m from the body, a further review with binoculars and photographs taken with telephoto lenses should make a clear record of obvious evidence, for example, the position of the body, perhaps indications of the cause of death, or signs of a struggle indicating that the victim was alive at this point. Such information both advise and constrain the final approach. Following Locard's Principle, evidence from the attacker is most likely to be on the body itself; unplanned and hasty approaches will disturb this evidence. The final stage of the CAP will be prepared so as to give access to the body itself and to allow for close-up recording of the scene. At all stages a video and photographic record continues to be made.

At this stage, the Crime Scene Manager, the Senior Investigating Officer present, the pathologist and forensic scientist (if called), will approach the body and review the situation to decide what action may be required before it is moved. The pathologist will be able to make an initial assessment and indicate his/her priorities. It is likely that the space around the body will be searched. The surface of the body will be searched and the clothing and, more particularly, the skin, will be taped with clear adhesive tape to recover any contact traces. Visible traces may be picked off and consideration will be given to sampling any individual stains seen on the body. When the body is ready to be moved, plastic bags will be placed over the head and hands and, if appropriate, over the feet, and taped into position to secure evidence. The body will then be lifted into a new, unused, plastic 'body bag' and the zip fastening sealed prior to transportation to the mortuary. An officer from the scene will accompany the body to the mortuary to maintain continuity and security of transfer, up to and including the post-mortem examination.

A decision will be taken on the extent to which the search will continue. It is likely that this will be concentrated on expanding the search area around the body and the area around the gate, leaving a systematic search of the field until later. Another Senior Scientific Support Officer will be responsible for this since the Crime Scene Manager will have accompanied the pathologist and Investigating Officer to the mortuary.

2.4.3.3 The Victim's Body. The victim is regarded as a scene and a post-mortem will have to be performed. The team for a post-mortem consists of the pathologist, mortuary technicians and the Scientific Support Officers who will record the post-mortem by photographs and video and receive, package and label exhibits ranging from clothing to body samples. The Investigating Officer, Scientific Support Manager and forensic scientist, if present, will view the post-mortem from a position in which they can observe yet be able to keep up a continuous flow of dialogue with the pathologist and the Scientific Support Officers. In accordance with health and safety requirements, all persons present at the post-mortem will wear full coverall protective clothing and face masks.

A post-mortem is a totally systematic examination which proceeds from the examination of the clothed body, the systematic removal of the individual items of clothing, and preliminary examination of any injuries which are recorded by note, diagram, photograph and video. Each stage of the examination will involve discussion and opinions on the significance of any findings. The pathologist will take all relevant samples from the body including blood, hair (head and body) and swabs from all the body orifices which may assist the enquiry. The pathologist will then make a full exploration of any injuries to the body concluding with a detailed internal examination. A Scientific Support Officer will take finger, palm and, if appropriate, foot impressions (particularly so if the body has been found barefoot).

At the conclusion of the post-mortem, the Senior Investigating Officer will ask the pathologist for a 'cause of death' and to classify it as 'accident', 'suicide', 'murder' or 'natural causes'. The nature of the enquiry is entirely dependent on this decision. In this case, the cause of death is opined to be 'murder' – death by stabbing probably following a sexual attack. The latter will be confirmed by examination of samples in the Forensic Science Laboratory. The consensus of opinion is that the attack did not take place in the field.

Note. To satisfy the legal requirements in England and Wales, a coroner must be informed of the death and assumes responsibility for the body until it is released for burial. The coroner may wish to be represented at the post-mortem either in person or through one of their officers. They will at some juncture, hold a 'Coroner's Enquiry' into the cause of death.

The body will, at some early stage, need to be formally identified to the pathologist and to the Coroner's Officer, most usually by a relative.

2.4.4   The Next Stages

The information from the post-mortem has made it clear that the police are dealing with a murder and they will proceed with their enquiries on that basis. The items from the post-mortem, including the clothing from the deceased, evidence taken from the body including medical samples, and the tapings recovered from the body at the scene, will be passed quickly, *via* the Exhibits Officer, to the Forensic Science Laboratory for urgent examination. The Forensic Liaison Officer and Central Submissions Unit need to be sufficiently flexible in their procedures to cope with such requirements.

Rapid examination of swabs taken from the body can confirm a sexual attack and provide information about the DNA 'profile' of the attacker(s). A similar examination of the tapings from the body may indicate the colour and type of clothing worn by the attacker(s) which the police must look for.

The immediate task facing the Scientific Support Officers is now two-fold: to recover everything of value from the field and its surrounding area and to search the house of the deceased to find out what happened there.

2.4.4.1 Searching the Field and the Lay-by. Whilst the post-mortem is taking place the Scene Examiners have extended the search area around where the body was found, recovering everything which may be of evidential value, including footmarks which may be from either the deceased or her attacker(s). At the same time, other Scene Examiners have recovered all the significant evidence from the gateway and lay-by, this time including tyre marks in addition to any footmarks.

The Crime Scene Manager now reviews the next stage. Items which might assist the enquiry will be dispatched urgently through the Exhibits Officer and submissions system to the Forensic Science Laboratory. Information on a type of footwear, a type and make of tyre, or, in the case of a vehicle track width, the make and type of vehicle may directly assist the progress of the investigation. Other recovered items will be passed to the Exhibits Officer to hold, pending an assessment of their value to the enquiry. As before, all items will be securely packaged, labelled and subject to strict documentary control.

The Crime Scene Manager is now faced with a controlled search of a large area and so therefore will call in a 'Police Search Advisor' (POLSA). This is an officer trained in structured searching techniques for large areas who, in turn, will call in teams of uniformed officers trained for this task. As this controlled and organised search takes place, Scientific Support Officers will be available to record, photograph and take possession of any item recovered. In this case, the search will recover the handbag and clothing items thrown into the adjoining field. These will be dispatched through the Exhibits Officer for urgent laboratory examinations which, in the case of the handbag, will involve a fingerprint examination of the bag and its contents.

2.4.4.2 Searching the House and Garden of the Deceased. The information from the body in the field and the post-mortem has indicated that there is a strong possibility that the death occurred in the house or that it was the scene of an abduction. It must also be determined how the intruder(s) gained entry to the house – by force or did the victim allow entry and has the intruder(s) left any trace of their presence? It will now be late in the day. The outside of the house having been searched already, the Crime Scene Manager, with the Senior Investigating Officer's agreement, may wish to suspend the examination of the house until daylight the following day rather than search in artificial light. This will also give the opportunity to bring together all the specialist support required. They may wish to have present, or on 'stand-by', forensic scientists with specialist skills and a 'Serious Crime Unit'. There are other, common sense, issues to be considered, such as obtaining spare keys to the property to give freedom of entry and access. On this occasion the house will be searched thoroughly from top to bottom in a controlled and sequential manner, initially concentrating on areas of disturbance, whilst continually reviewing the chronicle of events being revealed as questions are answered. At all stages, photographs, video recordings and detailed accurate notes are taken.

Questions to be answered and points to be considered will include the following important factors:

1. Was there a forced entry or could the intruder(s) have been let in by the victim? Was the forced entry genuine or made to look like it? The Scientific Support Officers will make this decision based on the findings of their examinations.
2. Are there signs of a struggle? Is there blood splashed about? Has the furniture been moved or overturned? A forensic scientist with experience of crimes of violence may be asked to assist in the interpretation of any blood splatter patterns.
3. The initial examinations will have been made by walking on a series of 'stepping plates'. Subsequently, using light, photography, electrostatic lifting and a variety of other techniques, footmarks may be identified on the floor or under the 'stepping plates'.
4. Footmarks found in blood require specialist recovery techniques. A Serious Crime Unit or a forensic scientist may be called in to assist with the detection and recovery of these marks.
5. A thorough examination for fingerprints will take place.
6. The house must be comprehensively searched from dustbin to attic.

This whole procedure may take several days during which time all the items recovered, documented, securely packed and labelled, will be passed to the Exhibits Officer. Selected items will be sent to the laboratory and to the Fingerprint Bureau for examination and comparison through a controlled and documented system.

At the conclusion of this part of the investigation, the Crime Scene Manager and the Senior Investigating Officer will 'walk through' the scene to review it before the scene examination is terminated.

2.4.4.3 The Suspect. The next phase of scene examination must wait until the police have a suspect or suspects. If found, a comprehensive examination of the suspect's house, car and of the suspect themselves will then take place.

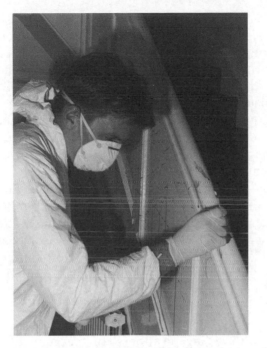

Figure 2.3 *A crime scene examiner recovering a blood stain by a swabbing technique* (Courtesy of the Director of the National Training Centre for Scientific Support to Crime Investigation © NTCSSCI 1996)

2.5 SCENE ATTENDANCE BY FORENSIC SCIENTISTS OR OTHER SPECIALISTS

As we have seen, the guiding principle requiring the attendance of trained Scientific Support Officers at the scene is the potential presence of contact trace material. The decision to call a forensic scientist to the scene arises from practical needs because of the type of physical evidence, the need for immediate advice and the complexity of the case. This may occur in cases where evidence is present as patterns of blood splashes, bloodstained footmarks or marks on bodies. Where bodies have been moved, there may be trails of blood and the body may be found in a different position or place after an attack elsewhere. It can be particularly helpful, where attempts have been made to destroy evidence, if a forensic scientist with relevant experience visits the scene. The application of new technology together with specialist training and experience will lead to a selection of the most informative evidence and to an interpretation of the scene which is of immediate assistance to the Senior Investigating Officer and the Crime Scene Manager. This can influence the direction and priorities of the enquiry.

A good example of this is the specialist firearms investigator in shooting incidents, especially if death has resulted. This expert's assistance at the scene is both valuable and often indispensable; it involves working side by side with the investigating team at the scene and with the pathologist at the post-mortem. The considerable activity which commonly takes place at a shooting scene and the nature of the operation of the weapon requires the accurate recording of the scene by document, photograph and plan. Close-up photography of wounds, bloodstains and the relative position of wounds, bullet holes, cartridge cases or shot damage is essential.

The examination of the scene of a shooting incident provides information about the type of weapon used and the distance over which it was fired, together with the positions of the firer and the victim. It will indicate whether the incident which occurred was an accident, a suicide or a murder. Information from residues on clothing and skin, together with scale photographs of marks and injuries, can be linked by test firing the weapon concerned to estimate the range and direction of a shot. The accurate reconstruction of the scene using all available evidence provides valuable information about lines of fire, the position of the victim(s), the firing position(s) of the gunman or gunmen and their various movements.

An excellent example of a multi-disciplinary team approach to the crime scene involves the forensic scientist who specialises in fire investigation. The investigation of arson and fatal fires, and their differentiation from accidental causes, requires expertise based on extensive practical experience in this field together with a sound knowledge of fire science as well as knowledge of the laboratory analyses which may be required. The investigation team may consist of fire brigade officers, police officers, Scientific Support Officers, forensic scientists and, where fatalities are involved, pathologists. All of these are trained to have a close appreciation of the knowledge and skills which the others can bring to the investigation.

The fire officer will initially seek to identify the cause and seat of the fire. His suspicions are not proof but he must be aware of the correct steps to preserve evidence. The police officer will initiate the investigation but will require the technical input of a Scientific Support Officer with the necessary background and experience. These three will make the decision whether the cause of the fire is self evident, whether the cause may become evident by analysis of possible incendiary materials or whether a forensic scientist attending the scene will help to determine the cause. Although all members of this team will 'pool' their knowledge and experience in the investigation, each will give evidence only in their own area of expertise when the case comes to court.

The investigation of the crime scene after an explosion is totally different from other types of scene because of the nature of the incident. Whilst trace evidence is still present, it has been dispersed by force from the source or centre of the explosion and the severity of the explosion can cause damage ranging from localised destruction, with a window being blown out, to the complete destruction of substantial buildings.

In investigation terms these may be considered as 'non-terrorist' or 'terrorist' offences. Non-terrorist offences arise from an accidental or deliberate act where a blast is produced by the ignition of gas, flammable vapours and combustible dusts or by the sudden release of pressure where a sealed container explodes. The cause can range from leaking gas in a domestic situation to an incident in an industrial plant. In these cases, which may be caused deliberately, by accident or through negligence, the investigator follows the 'team' approach of fire investigation with the addition of specialist investigators from the Government Health and Safety Executive or public utilities such as gas or electricity as appropriate. In such cases, the disturbance of the scene to recover the dead or injured, or to make the scene safe, does not interfere with the aim of the investigation which is to determine the cause of the explosion. If the cause of the explosion is a deliberate act, then any suspect will be treated as a crime scene in order to recover trace evidence from their person or clothes.

Where terrorist activity is suspected, possibly following the operation of an explosive device, the initial action taken is to contain and secure the scene. Safety takes precedence over all other issues and specialist advice is sought, although the scene may be disturbed by emergency services in their effort to save lives.

Initially, a Bomb Scene Manager will take control of the scene advised by specially trained military personnel, Scientific Support Officers, Anti-terrorist Officers and commonly a forensic scientist specialising in explosives. When the scene is considered safe, a search will be made particularly for traces of explosive material and the components of an explosive device. The latter may well be shattered into fragments over a large area. After clearing a common approach path to provide access, the scene will be divided into zones and searched systematically. The force of the explosion will cause evidence not only to be dispersed but to be embedded in walls, ceilings, furniture or people and thrown onto roof tops. Although some hand searching may be carried out, it is necessary to remove the debris, often in large containers, for the contents to be sieved and carefully searched by specialists at the appropriate forensic science laboratory for fragments which, when analysed, will indicate both the type of device and composition and source of the explosive used.

The handling of clothing and property of someone suspected of having been in contact with explosives requires an exceptionally high level of cleanliness and control. Because of the potential high risk of cross-contamination, the Scientific Support Officers concerned wear disposable protective clothing and keep themselves and suspects separate from areas in police stations, vehicles or indeed from other officers where these may have been in contact with firearms, explosives, ammunition *etc*.

There are many other areas where forensic specialists can assist at the scene which, in this text, can only be mentioned in passing, The forensic archaeologist can provide assistance in the controlled recovery of buried objects or bodies. The use of ground searching radar can indicate areas of disturbance. The use of computerised survey and plan drawing systems, electronic theodolites and photogrammetry techniques which can draw up scale plans from pairs of stereoscopic photographs, increasingly aid the recording of the crime scene. The forensic entomologist, who has knowledge of the type and life cycles of insects associated with dead bodies, can shed light on estimations of the time of death, the place of death, the subsequent movement of the body and, sometimes, the manner of death. The soil scientist, with the aid of the scanning electron microscope, can put new interpretations on traces of soil recovered from the scene and suspect.

One area of concern at crime scenes has only been touched on in this section in various parts and that is the issue of 'Health and Safety'. Where body fluids are present, post-mortems are carried out or buildings are unsafe through fire or explosion, these are areas of obvious risk and of which the Scientific Support Officer must be aware and must be mindful to alert others. It is, however, particularly important that a forensic scientist should be consulted and attend all aspects of the investigation of premises involved in the illicit manufacture of drugs where the hazards might range from inhalation, or contact with noxious chemicals, to the risk of fire and explosion.

2.6   CONCLUSIONS

Clearly, any person involved in crime scene examination has a duty not just to record the obvious, but to look beyond that and examine, observe and take notice of what the evidence tells them. Increasingly they must be prepared to combine their personal knowledge and experience with the skills of others and in so doing, adopt an holistic approach to crime scene interpretation.

Apart from the incidents described there are many more areas where specialist advice can assist the scene investigation and these will become apparent in the following chapters of this book.

2.7 BIBLIOGRAPHY

W.G. Eckert and S.H. James, *Interpretation of Bloodstain Evidence at Crime Scenes*, Elsevier, N.Y., 1989.

B.A.J. Fisher, *Techniques of Crime Scene Investigation*, 5th edn, 1993.

N. Hendall, (ed) *Gross's Criminal Investigation*, 3rd edn, Sweet and Maxwell, London, 1934.

H.C. Lee, T. Palmbeach and M.T. Miller, Henry Lee's Crime Scene Handbook, Academic Press, New York, 2001.

P. Margot and C. Lennard, *Fingerprint Detection Techniques*, 6th edn, Universite de Lausanne, 1994.

D.R. Redsiker, *The Practical Methodology of Forensic Photography*, Elsevier, N.Y., 1991.

Scene of Crime Handbook of Fingerprint Development Techniques, 2nd edn, Home Office, Police Scientific Development Branch, St Albans, 1993.

The Scene of Crime Handbook, 4th Edition, Forensic Science Service, London, 2003.

The Scenesafe Evidence Recovery Guide, The Scenesafe Unit, The Forensic Science Service, Washington Hall, Euxton, Chorley, PR7 6HJ, England.

CHAPTER 3

Trace and Contact Evidence

ANGELA GALLOP and RUSSELL STOCKDALE

3.1 INTRODUCTION

'Trace evidence' in the generally accepted forensic science sense is the term applied to normally very small amounts of material such as textile fibres, glass, paint, *etc.*, which can serve to link an item on which material is found with an otherwise unconnected source of it elsewhere. The finding of such a trace often implies that there has been direct physical contact between the item and the source, with a consequent transfer of material between them. The principle of interchange at work here is as indicated previously, attributed to Edmund Locard. He was the first person to advance the theory that when someone commits a crime they always leave behind something at the scene which was not there before and takes away with them something which was not on them when they arrived.

An everyday example of Locard's principle that every contact leaves a trace, and the logic which flows from it, might be the discovery by any of us who had just finished painting their house of fresh, sticky paint on the clothes of one of the children. No one would have any difficulty with the conclusion that the child in question had leaned up against our newly completed work and, having found the paint on their jumper, we would search for the point at which this had occurred in order to remove the jumper fluff and repair the otherwise pristine finish.

Textile fibres and paint (normally as dry flakes) are commonly encountered forms of trace evidence in a large portion of criminal cases examined in forensic science laboratories. However, almost anything has the potential to provide this sort of associative evidence, so long as it is capable of being transferred from one place to another, can be recognised for what it is, and is recoverable and amenable to meaningful analysis and critical comparison.

One of the functions of the forensic scientist is to search meticulously and painstakingly, often through a myriad of microscopic particles of debris collected from the crime scene, the victim, the suspect and anywhere else of potential significance, and to recognise and distinguish between what is seemingly 'in place', and what is 'out of place'.

There are two fundamentally different approaches to this. The first, and by far the most common, involves 'reactive' searching whereby potentially useful sources of trace material are identified and these materials are then searched for on other items in the case as a means of linking them together. In this way the forensic scientist is able to focus attention on lines of enquiry which appear most likely to yield evidentially significant findings.

In the alternative 'inceptive' approach, items, usually from a scene of crime, are searched for any trace materials which are distinctive and apparently unrelated to them and to their environment. These traces might point to a particular source and, therefore, eventually to the perpetrator.

3.2 TARGETING POTENTIAL TRACES

So far as reactive searching is concerned, potential sources of transferred material are identified in one set of items and targeted in the search of another set with which they are suspected of having been in contact. Examples of this include looking for textile fibres like those constituting the victim's clothes on the surface of those from the suspect or glass fragments from a broken window at the scene of a breaking and entering on the clothes of the suspected intruder.

Essentially the scientist has to assess, in evidential terms, whether and to what extent it might be worth looking for each of the various potential types of trace material which could have been transferred in a particular case and to select and pursue those which promise to be the most profitable. To look for each and every type of trace in any one case would turn it into a life's work.

To be selective the scientist needs to consider:

1. How much material is likely to have been transferred in the first place.
2. How well this might have persisted.
3. The ease with which the searcher is likely to find it.
4. The evidential value of any trace material found.

3.2.1 Amount of Material Transferred

The amount of material transferred will depend upon:

1. The amount of material available for transfer. This will be a function of the nature of the material itself, how readily traces of it become detached from its surface and the size of the exposed surfaces themselves. For example, peeling paintwork at a point of entry is more likely to yield tiny fragments of paint which can be transferred to and remain on the clothes of an intruder, rather than a smooth, undamaged finish (unless, of course, the paint is newly applied and still wet or soft).
   Obviously the larger the painted area the more opportunity there is for transfer to occur.
2. The nature of the recipient surface. So far as particulate material is concerned, traces are more likely to lodge on rough surfaces such as chunky knit jumpers than on smooth ones like shell-suits.
3. The nature and duration of the contact. Essentially, the more forceful the contact and the longer it endures, the greater the amount of material likely to be transferred.

3.2.2 Persistence of Material

Questions relating to the transfer and persistence of particulate material often feature clothing as the recipient surface. Nonetheless, just as almost anything can give rise to traces of evidential importance, so the surfaces on which to search for them can be of endless variety. According to the circumstances of the case at hand, interest might therefore focus on the persistence of glass fragments or fibres in hair, hair inside stocking masks or balaclavas, fibres on car seats and/or on broken glass and window ledges at points of entry, *etc.*

In any event, the length of time particulate trace material can be recovered from surfaces to which it has been transferred depends on several factors. These include:

1. The physical size of the material itself, with small particles tending to persist longer than large ones.
2. The nature of the surface of the material. Particles with irregular surfaces, *e.g.* glass fragments, or rough surfaces, *e.g.* wool fibres (by virtue of their overlapping cuticular scales), tend to remain longer than smooth-surfaced ones, *e.g.* nylon fibres.

3. The nature of the recipient surface. Just as transferred particles lodge more readily on rough surfaces than on smooth ones, so they tend to stay longer on rough ones. The weave of coarse wool cloth, for instance, is likely to be significantly more retentive of debris than the surface of a leather jacket or the bodywork of a car.
4. The time that elapses after the transfer occurred and before evidence of it is searched for. In particular, the longer the time between the transfer and when the recipient item is seized and safely packaged pending scientific examination, the greater the opportunity for transferred fragments to become dislodged and lost from its surface. This is especially true in respect of clothing if the garments in question were worn in the interim. The level of activity of the wearer may also play an important part in the loss of recently acquired particles.

Material which can produce smears as a result of contact, *e.g.* wet paint, blood *etc.*, forms a physical bond with the receiving surface. This means that, unless the item concerned has been washed or the trace removed in some other way, the material may persist for far longer than particulate material which tends merely to be temporarily lodged on the surface, *e.g.* between the fibres in woven cloth.

3.2.3 Finding the Material

Many forms of trace evidence are not readily visible to the naked eye, especially those on highly textured backgrounds such as the surfaces of some types of clothing and other fabric. This has led to the development of a range of techniques designed to recover tiny particles from a variety of surfaces, quickly, efficiently and in a suitable form for subsequent microscopic searching and individual recovery. The techniques most commonly used are described briefly in this chapter.

In any event, trace material which is likely to be eye-catching against the background of the recipient item itself, or amongst the superficial debris recovered from it, is clearly going to be easier to spot than something which is not. For example, more deeply dyed and unusually coloured textile fibres are easier to look for than pale coloured ones amongst a background of numerous assorted fibres, as tend to be collected from the surface of garments. By the same token, glass fragments are readily spotted on the surface of a black leather jacket, but difficult to search for in a bowl of sugar. Thus, much depends upon the specific items and materials in question and the contrasts between them.

3.2.4   Evidential Value of Trace Material

The evidential value of trace material in a case is related to its likely commonness. This is because the more commonly occurring it is likely to be, perhaps in a particular setting, the less strong the link it is capable of providing with one nominated potential source, to the point where it is sometimes simply not worth looking for the material at all. In the case of blue denim (cotton) fibres, for example, these are used so widely in the manufacture of popular clothing that their value as associative evidence tends to be very limited indeed. In consequence, blue denim jeans, jackets and the like are not normally considered as potentially useful sources of fibres in a fibre transfer examination and, because of these special circumstances, there would be little to be gained in looking for glass fragments on the clothing of a glazier. At the other end of the scale, where a relatively small batch of an unusual type of yarn was produced, the finding of individual fibre fragments which could have come from it might potentially be highly significant.

The alternative (inceptive) approach to targeting potential traces from known sources of them is to examine the material recovered from the surfaces of an item, selecting that which is eye-catching and appears foreign and then testing and screening possible potential sources of it emerging from a wider investigation. Intrinsically, this is much more difficult than reactive searching, because there is often no way of knowing whether what is conspicuous and captures the interest is necessarily related to the crime, as opposed to any one of a plethora of other, innocent recent contacts.

Nonetheless, in one notable case, the forensic scientist's eye was drawn to some unusual short, cut lengths of man-made textile fibres which were widely distributed over the body of a murder victim. These were eventually traced to the manufacturer of a particular sort of carpeting, who revealed that some off-cuts from this main consignment had been used to fit out certain cars. This led the police to focus their attention on one particular man who, amongst a large number of other potential suspects, had owned such a car at the material time. He was later convicted of the murder.

On the other hand, much time and effort was wasted in a case in which some tufts of unusual blue-black fibres were observed on the body of another murder victim. It was eventually discovered that these had originated from the decaying spine of the pathologist's own notepad, falling unnoticed onto the body as he recorded his observations at the scene! Some distinctive silver and brown paint flakes on the clothing of yet another murder victim turned from being significant inceptive evidence into a similar source of embarrassment. They had come from the Scene of Crime Officer's step ladder over which he had hung the garments to dry before sending them to the laboratory.

3.3 RECOVERY OF TRACE MATERIALS

There is no single technique for recovering trace material, rather there are various options from which one or more can be selected to suit a particular material and the surface(s) from which it is to be recovered. The procedures described here are among the most commonly employed on a routine basis. Nonetheless, the range of potential combinations of traces and the locations from which they have to be recovered is almost infinite and it is not unusual for the forensic scientist to have to devise a novel approach to suit a particular set of circumstances. Clear adhesive tape, *e.g.* Sellotape, has been around for a long time, but it was not until the early 1970s that someone used it to good effect to recover debris from garments. Similarly, it was discovered that the practice of combing a suspect's hair to remove fibrous material trapped in it could be enhanced by inserting cotton wool or lint-like material between the teeth of the comb.

3.3.1 Shaking

One of the simplest and most suitable methods for recovering loose particulate material such as glass fragments and paint flakes involves gently shaking a garment over, say, a sheet of paper or into a specially designed, inverted, metal cone and collecting the debris that falls off it. Such debris can then be placed in a convenient container, *e.g.* a clean piece of paper, folded 'Beecham Powder' fashion or a small covered glass dish where it can remain safely protected from contamination by other items with which a link might eventually be sought. The issue of contamination, which is of paramount importance in relation to trace evidence, is dealt with in greater detail later in this chapter.

3.3.2 Brushing

Shaking is not especially effective in removing particles from surfaces such as shoes and pocket linings. In these circumstances it is common for debris to be collected by brushing the surface with a (new/clean) tooth or paint brush, again collecting the debris on a piece of paper, or in a suitable container.

3.3.3   Taping

Tapings are used principally to recover clothing fibres and hairs. Essentially they are strips of clear sticky tape, *e.g.* Sellotape, applied sequentially to and then pulled off surfaces of garments, car seats, window ledges, the edges of broken glass at a point of entry – in short, almost any dry surface one cares to mention – in order to pick up superficial debris which may be resting there. The strips of tape are then stuck down onto clear plastic sheets and in this form the debris on them is amenable to searching at a later stage, with the aid of a microscope, for any fibres/hairs which seem to be relevant. After they have been taken, tapings are stored in sealed polythene bags as a precaution against contamination.

3.3.4   Vacuuming

In addition to tapings, vacuuming has also been used to recover minute particulate material in firearm discharge residues (FDR), for example, from the surface of clothing *etc.* (FDR are microscopic particles arising from the primer and propellant in a cartridge when a gun is fired). The vacuum appliances used have nozzles which focus the debris onto small filters which can then be removed and searched directly. Drug residues on bundles of cash acquired in the course of drug-dealing are also routinely recovered in this way.

3.3.5   Swabbing

Swabbing is an alternative technique for the collection of FDRs, although it is more commonly used to recover small amounts of smeared material – especially of blood and other body fluids, preparatory to traditional grouping or DNA profiling tests. Depending upon the material concerned, the forms of swabs encountered range from cotton wool buds on the end of a wooden or plastic stick, to single lengths of clean white cotton yarn.

3.3.6   Hand Picking

In cases where debris of interest is firmly lodged in place, where taping might be inappropriate because it could conflict with other examinations coming later, *e.g.* for fingerprints, and/or where it is important to observe and record its precise location on an item, material can be picked off by hand, usually with the aid of a low power microscope and precision forceps. Material recovered in this way is immediately transferred to a suitable and safe form of storage, sometimes directly onto glass microscope slides, until it can be examined in detail.

3.3.7 Extracting

For non-particulate and non-fibrous trace evidence such as oil and grease smears, cosmetics, shoe polishes, *etc.* and for blood and other fluids on fabric, the forensic scientist needs to dissolve out the trace by treating the stained material with a suitable solvent. The resultant extract, representing a solution of the material of interest, can then be analysed, although concentration of the extract may sometimes be required beforehand.

3.3.8 Liquids and Gases

Provided that there is sufficient material and it is on a suitable surface, liquids can be simply taken directly using a pipette or a swab or some other absorbent material. Because of their tendency to disappear rapidly into thin air, volatile liquids, *e.g.* petrol and various organic solvents, can present special problems. When dealing with these materials, as well as more persistent ones like paraffin, the item must be sealed in a suitable container and then the air inside the container can be sampled and analysed (head-space analysis). This approach is encountered in connection with fire debris analysis, when the scientist is likely to be searching for traces of volatile liquids suspected of having been used as an accelerant and, in quite another sort of case, in the analysis of blood for the presence of alcohol.

3.4 CHARACTERISATION AND COMPARISON

A wide range of techniques are in common use for the characterisation and comparison of trace materials, the ones selected depending on the nature of the material concerned and the amount of it available. Some techniques are used for more than one type of material, although they may be modified to accommodate individual and special characteristics.

In taking a closer look now at some of the more commonly relied upon types of trace material – the ones most likely to be encountered in forensic scientists' evidence, the ways in which they are normally analysed are also introduced.

3.4.1 Glass

Breaking offences, bank robbery, criminal damage and fights in which bottles and drinking glasses might have been broken can all result in those concerned being showered with glass fragments.

Figure 3.1 *A pin amid glass fragments of a size range typical of those examined in forensic science*

Where these fragments are sufficiently large, gross features such as colour, thickness, whether their original surfaces were optically flat (as in window glass) or curved (as in containers) and, sometimes, how the glass was manufactured can all be compared with potential sources. In addition, the sharpness and cleanness of their broken edges can provide important information as to how recently the fragments in question might have been broken and, if found on a smooth surface, how recently they are likely to have been acquired.

Further physical and chemical properties inherent in the glass itself can also be measured and compared. A starting point for such detailed analysis is normally refractive index, *i.e.* the extent to which the glass 'bends' light passing through it as compared to air. An everyday example of this property is the apparent bending of a stick when one end of it is dipped into water. Water has a higher refractive index than air, *i.e.* it has a greater tendency than air to bend a beam of light. If the stick were embedded in glass, which has a higher refractive index still, then the stick would appear even more bent. Typical refractive indices of glass are in the order of 1.5 and water 1.33, compared to that of air which is taken to be 1.0.

Some refractive indices occur more commonly in some types of glass than in others and this can be used as a means of assessing the rarity value and therefore evidential significance of the glass fragment at issue in any particular case.

Toughened glass, such as used to be found in car windscreens before laminated glass took its place, but which is still used for side and rear windows, has an 'artificially' high refractive index as a consequence of the way in which it is manufactured. This involves rapid cooling of the hot, molten sheet in order to set up internal stresses in the glass which impart to it the desired toughened property. The 'true' refractive index of toughened glass, and therefore another point of comparison, can be revealed by heating it and allowing it to cool slowly – the process is known as annealing.

In addition to physical properties, the chemical composition of glass can also be determined using the scanning electron microscope (SEM) coupled with energy dispersive X-ray analysis (EDX). In the SEM, a minute object can be magnified tens of thousands of times by using a beam of electrons instead of a beam of light as in a conventional microscope. At the same time, the object at the focus of the electron beam emits radiation which is characteristic of the chemical elements contained in it. By separating and monitoring these emissions, using EDX, the presence and relative amounts of the constituent elements can be established and compared with those in reference materials.

Essentially, glass is composed of silica (sand), soda and lime, with additions of other materials depending on the use to which the final glass object is to be put, and, of course, various impurities. It is the precise nature and amount of each of these components and impurities that in addition to refractive index, may allow discrimination between one glass and another.

A ready example of chemical differences would be the composition of glass intended for the manufacture of milk bottles as against that used in window glass. In the former, iron salts have been mostly eliminated because these impart the greenish tinge seen in a sheet of window glass when looked at end on and which would make the milk in a bottle appear distinctly unappealing. Another example is the addition of lead oxide to the melt that results in a glass with a high refractive index used in so-called 'cut glass' ornamental ware.

3.4.2 Textile Fibres

Textile fibre transfer has the potential to feature in all manner of investigations where the crime has involved clothing to clothing (or naked skin) contact, or contact between clothing and, say, bedding, car seats, points of entry to premises, *etc.*

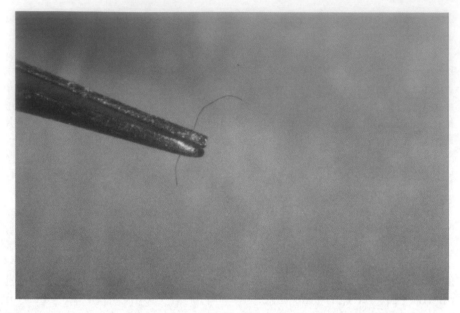

Figure 3.2   *An evidentially significant textile fibre held by precision forceps*

The textile fibres with which forensic scientists are concerned are not strands of yarn as might be commonly supposed but, rather, tiny broken fragments of the individual fibres, which tend to be extremely thin – finer than a human hair – and often no more than a millimetre or so in length. This means that, by and large, they are not readily visible to the naked eye.

The starting point for a fibre examination is normally the tapings referred to earlier. These are carefully searched under a low power microscope, *e.g.* at a magnification of about 20 × normal size, for any fibres like those in the potential sources which have been targeted.

Fibres that appear similar to these are individually removed, cleaned and mounted under thin slips of glass on microscope slides, in which form they can be examined and compared in greater detail with those from a suspected source.

This detailed analysis and comparison falls into three or four main sequential phases, namely: microscopy, microspectrophotometry (MSP) – sometimes referred to as visible spectrophotometry, thin layer chromatography (TLC) and, for man-made fibres, infrared spectroscopy (IR).

In the first phase (microscopy) the images of the fibres to be compared are brought together optically and examined side by side using a high powered comparison microscope. Magnifications commonly used are in the order of 400 × normal size, and features such as fibre type, *e.g.* cotton, wool, polyester, nylon, acrylic, and perceived colour are established and compared directly. The fibres are examined in (ordinary) white light and in ultra-violet, blue and sometimes green light, any or all of which may induce some fibres to fluoresce (glow). With fibres that do behave in this way, the colour of fluorescence itself serves as an additional comparison feature. Any recovered fibres that appear different from the possible source are rejected.

In the second phase (MSP) the perceived colour of a selection of microscopically matching fibres is measured by an instrument known as a microspectrophotometer. Essentially this records, in the form of a graph (a spectrum), the extent to which different wavelengths (colours) of light are absorbed as they are passed sequentially through the fibres. This is directly related to the colour of the fibres as perceived by the eye and serves as an objective means of measuring it. Recovered fibres producing different graphs (spectra) from those in the possible source, *i.e.* those with different colour characteristics, are also rejected.

In the third phase (TLC) the dyestuff in some of any remaining matching fibres must, if possible, be stripped out using an appropriate mixture of solvents. The resulting coloured extract is then concentrated into a spot on a thin layer of adsorbent powder bonded to a metal or plastic sheet. One end of this is then dipped into another solvent which slowly creeps up the plate, carrying the dyestuff with it and depositing its various components at different levels to form a line of coloured spots on a white background. In this way, the dyestuff mixture in one fibre can be separated out and compared directly with that in others. Of course, where individual fibre fragments are especially small or colourless, or they are dye-fast, TLC is not an option.

Recent advances in the design and performance of microspectrophotometers have served to cast doubt on the future of TLC as a routine procedure in fibre comparison. Some workers believe that fibres which match with MSP are unlikely to be distinguished by TLC, such is the precision and accuracy of the instruments now available. Nonetheless, opinion remains divided and, for the time being, many feel that there are insufficient grounds for confidently abandoning TLC and the scope for the independent discrimination that it can provide.

Finally, IR can be applied to man-made fibres, *e.g.* polyester, nylon, acrylic *etc.*, to confirm their chemical identity which will have already been largely established from their appearance under the microscope. With this technique, individual fibre fragments are squashed into a film through which infrared light is passed. The ways in which the chemical structure of the film reacts with the infrared light is again recorded in the form of a printed graph in which the various peaks can be used to determine the chemical identity of the material. In the case of acrylic fibres, for example, several chemically different forms can be identified and the differences between them within the broad group of acrylic fibres serve as additional comparison features.

IR is not applicable to natural fibres such as cotton, linen and other fibres derived from plants, or to wool, silk and other fibres derived from animals.

3.4.3 Paint

Paint can provide valuable associative evidence in a wide variety of cases. Crushed paint transferred to the business end of a jemmy, smears of vehicle paint transferred to a hit-and-run victim's clothes, or to another vehicle in a traffic accident, and paint flakes derived from a point of entry at a break-in are all common examples.

As with all forms of trace evidence, the observation and recording of gross physical properties is the first step in the chain of analysis of paint fragments. Colour is an especially important discriminator and it has been estimated that, in daylight, the unaided human eye can distinguish 10,000,000 different colour surfaces.

Figure 3.3 *Crushed paint on the business end of a jemmy. The paint is the light area on the right hand side of the bevel*

The type and surface characteristics/condition and/or contamination of the top coat will be noted, as will the sequence of any underlying layers, their colours, thicknesses, textures and the size and distribution of particles within them. Embedding of paint flakes in a resin and subsequent cutting through the resin and flake can be used to provide a section through the paint layers to improve examination of these parameters. In addition, some microscopical techniques, such as ultraviolet and blue light microscopy and MSP, can be bought to bear to distinguish between layers which are otherwise superficially similar or which can appear to be the same colour in certain lighting conditions.

While the microscope is the essential tool here, instead of relying on light that passes through the material (transmitted light), as in the case of fibres, paint flakes are compared using light reflected from their surfaces. Similarly, when used for paint examination, MSP is also employed in reflectance rather than transmittance mode.

Over and above all of this, there are different types of paint depending upon its manufacture and the purpose for which it was designed. For example, undercoats and primers in general have a higher proportion of inorganic components than top coat paints and these differences are reflected in their chemical composition.

A variety of techniques are used in the chemical analysis of paint including SEM EDX – essentially the same as that used for glass – which provides information about the chemical composition of the inorganic components of paint. These might be the pigments themselves, although inorganic pigments are not as widely used as they used to be, and any added so-called 'extenders' which serve several functions such as improving covering power and the strength of the dried film.

SEM EDX is sufficiently sensitive and precise to focus on and to analyse individual successive layers in a sequence, revealing the elements present in each one in turn. X-ray diffraction analysis (XRD) can also be applied here. Some of the compounds present (as opposed to merely the elements) in the paint are revealed by the effects their chemical structures have on X-rays being passed through them.

IR and pyrolysis gas chromatography (PGC) are both employed to learn about the composition of organic components of the paint, notably the resins responsible for the protective coating itself. The pigments in modern decorative top coats are also now mostly organic in nature.

IR was described earlier in connection with the chemical nature of man-made fibres. PGC involves burning some of the paint (pyrolysis) and sampling and analysing the resultant vapour. In the same way that burning rubber gives off a characteristic smell, the plastic-like materials in paint produce characteristic vapours which can be 'sniffed' by gas chromatography (GC) and electronically recorded in the form of a graph (chromatogram).

The information generated from all of this, in addition to the gross physical features mentioned earlier, can provide powerful associative evidence. Nonetheless, with perhaps few exceptions, none of this is capable of telling us, for example, that one layer of paint necessarily came from one specific manufacturer and another from another although, on occasions, the manufacturers themselves and the Paint Research Association are able to provide additional help with this.

Where the paint has been transferred in the form of a smear, which tends to involve the top coat only, for example on the clothing of a hit-and-run victim, there is no layer structure to examine. Similarly, a jemmy used to force an entry may have crushed paint deposited on it obliterating any original layer structure. In these circumstances, the scientist is thrown back on considering colour and appearance and principal chemical constituents.

3.4.4 Hair

Forensic scientists deal with both human and animal hairs. So far as the former is concerned, they tend to crop up mostly in, for example, balaclavas and stocking masks used by robbers, on clothing and bedding *etc.* in sexual assaults, and on blunt weapons in crimes of violence. Animal hair investigations tend to be related to allegations of bestiality, theft of furs, sheep worrying and as an additional source of associative evidence linking one set of items with another on which there are hairs from, for example, the domestic pets of those concerned.

The basic structure of hairs, whether human or animal, is essentially the same. They consist of an inner core, known as the medulla, surrounded by a cortex and enclosed in a thin outer layer, the cuticle – all rather like a pencil where the medulla is the lead, the cortex is the wood, and the cuticle the paint on the outside.

Considering each of these parts in turn, the medulla is particularly important in distinguishing human from animal hairs and different sorts of animals from each other. This tends to be achieved on the basis of the relative width of the medulla, and precise structure and arrangement of its component parts.

The cortex is usually more important when considering from whom a particular hair could or could not have come. It is in this substantial layer that the pigment granules reside, which give the hair its colour.

The overall colour of hairs can be compared and the size, distribution, and density of individual pigment granules within them provide additional features for comparison.

Clearly, if the hairs in question have been artificially altered in some way, *e.g.* by bleaching or dying, then points of similarity with others and, therefore, their evidential significance, will be increased. Not only can the colour of the hair be compared and, in the case of dyed hair, the artificial colour extracted and analysed by TLC just like the dyes in textile fibres, but other parameters such as the amount the hair has grown since the colour/bleach was applied can also be measured and used for comparison.

The cuticle (the outer, paint layer of our pencil) is not continuous, but is arranged as a series of separate, overlapping scales like the tiles on a roof. Although the margins of these scales can be elaborately sculptured, the sculpturing itself, like the different forms of medulla, is normally more useful in distinguishing between hairs from different sorts of animals than in associating human hairs with particular people.

Although as we have seen there are many and varied features of hair that can be compared between one sample and another, the value of human hair evidence is normally very limited. This is because, taken altogether, the hairs from any one person's head encompass a range of natural colour and microscopic features such that no two hairs, even from the same person will look exactly alike. There is often considerable overlap between the features exhibited by one person's hairs and those of another to the extent that, in general terms, an individual hair cannot be ascribed to one person to the exclusion of everyone else. Furthermore, it is sometimes difficult confidently to exclude someone as being the source of a particular hair, such is the uncertain nature of sampling hair for reference purposes.

As might be expected from all of this, there is no satisfactory statistical basis for measuring the commonness or otherwise of hairs of a particular sort. This stems partly from our inability thus far to define hair characteristics in an agreed, uniform and objective way and partly because of the extreme variability along the length of a single shaft and between hairs from the same head.

Only in the case of hairs with fresh roots may some traditional blood grouping and/or DNA profiling be possible. These are more objective approaches to determining from whom hairs could or could not have come. In addition, the main body of the hair contains a rather different sort of DNA (inherited only through the maternal line) and new technologies look set to be able to unlock the information contained in this as well. In the future hair evidence may well be transformed from being generally fairly weak to very strong indeed.

3.4.5 Oils, Greases And Waxes

Oils and greases can feature as evidence associating a hit-and-run victim with, for example, the underside of a car, as linking evidence in almost any sort of case involving machinery and tools of one sort or another and as lubricants in some cases of sexual assault. Usually they are present as light smears on, *e.g.* clothing, and apart from their obvious greasiness, perhaps their colour and any visible contaminants, there is little characterisation that can be achieved without chemical analysis.

This analysis usually begins with infrared (IR) spectroscopy which is essentially the same as that applied to man-made textile fibres. IR can distinguish natural, *i.e.* plant and animal products, from mineral oils and greases and wholly synthetic ones. Plant oils include those used to manufacture margarines and cooking oils and animal ones include butter, lanolin *etc.* Mineral oils are widely used as engine lubricants as are synthetic ones including silicone, also commonly used on condoms.

More detailed analysis tends to be provided by GC, which gives much more information about the components of the material and their relative proportions to one another and which can be very important in identification and comparison. It can reveal trace amounts of impurities/additives which may serve to define more closely the material recovered and to provide increasingly strong links with a potential source of it.

TLC and high-performance liquid chromatography (HPLC) are also used in the analysis of oils and greases in relation to additives such as dyes and preservatives. TLC was described earlier in the chapter as applied to the analysis of fibre dyes. While HPLC employs the same principles as TLC – taking up the material in a solvent, separating the dissolved components by virtue of the rate at which they pass through an adsorbent substance, and recording the results – HPLC is carried out under high pressure using a long thin tube into which the sample solution is injected at one end, the various components being detected and recorded electronically as they elute at different times from each other.

Depending on the amount of discriminating information that can be extracted from the sample in question, oils, greases and waxes may be precisely characterised and critically compared with other samples of a broadly similar nature. Some substances, for example microcrystalline waxes of the sort used in the manufacture of paper cups, and others used to provide the protective temporary wax coat to new cars in transit from the factory, are an especially rich source of information when analysed by GC. This proved to be the undoing of one criminal who specialised in stealing factory-coated cars on their way to the distribution centres. He would steam-hose off the dull wax prior to selling on the cars and was caught when precisely the same wax as used by the factory was discovered on his clothes.

3.4.6 Soil

The analysis of soil traces has perhaps not received the detailed attention that it might deserve. There is no doubt that soils have proved difficult and often unrewarding in many crime investigations, but much potentially useful information is locked up in even small amounts of soil, although technology to aid our understanding of it is being developed by agricultural and academic soil scientists rather than in the forensic science community itself. One of the main problems with soil as evidence has always been in the interpretation of the findings because of the patchy distribution of different types of soils throughout the country and local influences on them. However, these difficulties, once they are better understood, might cast soil evidence in a new light.

Soils tend often to be found as smears on clothing and deposits on shoes. As matters stand, perhaps less attention is paid to its chemical composition than to its nature and distribution and how this may relate to the circumstances of the case. For example, mud staining on the knees and elbows of the clothing of an alleged rapist, assuming that it is generally similar to that at the scene, can provide useful evidence in support of the prosecution's case.

So far as any analysis is concerned, and depending upon the amount of soil available, this may range from a simple comparison of colour and texture, to an analysis of, primarily, its mineral content (clays, chalks and sands) and, possibly, its organic content (derived from plant material) too. Where only a small amount of soil is present as light smears on clothing, then the scientist will make similar test smears with the reference sample on a clean area of the same garment and compare these directly with the original ones. This gets round any influence that the background colour and/or texture of the garment itself may have particularly on the colour of the soil smear.

Larger quantities of soil can be separated into different particle size fractions by sieving. The relative proportions of each fraction in the recovered material can then be compared with those in the reference sample and individual fractions in each pair compared in greater detail.

As a first step, the colours of the samples in their natural states might be compared, both moist and dry, and after being subjected to high temperatures (ashing) in a small furnace. Ashing burns off any plant material and has a dramatic effect on soil colour, usually resulting in bright oranges or reds.

The fractions can also be examined and compared microscopically when different sorts of grit particles can be identified and amounts and ranges in different samples established.

More precise information as to the chemical nature of the samples can be determined by chemical profiling of its inorganic content, either by atomic absorption spectroscopy (AA) or inductively coupled plasma mass spectrometry (ICP). Although very different instruments are used in these techniques, both rely on the characteristic emissions produced by inorganic substances when they are burnt. Soil scientists compare other components in soils such as humic acids and nitrogen isotopes, but these are currently not routine procedures in forensic science laboratories – again, probably because of uncertainties surrounding the likely significance of the results obtained.

3.4.7 Vegetation

Vegetable material can be important in a wide variety of cases as a means of linking people or property with outdoor sites where particular plants are growing, with damage to wooden doors and window frames *etc.* at points of entry and, historically, with blown safes. Safes used to contain sawdust as the packing or ballast between their inner and outer metal skins, but in modern ones this has been replaced entirely by mineral-based materials such as special types of concrete sometimes reinforced with metal. Safes now tend to be cut open by disc or thermal cutters rather than blown apart. The safe-cracker will still be showered with debris but this tends to be of concrete and tiny fragments of metal – sometimes melted into blobs – as opposed to wood.

Any type of plant material may feature as trace evidence: roots, leaves, pieces of stem/trunk, flowers and fruits and even pollen. The extent to which fragments of any of these can be identified depends upon their size, how much diagnostically useful information they contain and, last but not least, the experience of the scientist who is carrying out this particularly specialised work.

Probably the most informative fragments are those derived from wood (stem/trunk) encountered as splinters, fragments of chipboard and turnings from drilling *etc.* These often contain a host of microscopical features which, if present in sufficient variety, can enable the original type of timber to be identified, sometimes even down to the species.

Where the wood fragments are sufficiently large (but still very small in every-day terms, perhaps less than the cross-sectional area of a match-stick), wafer thin sections can be cut from them by what amounts to a very sharp knife mounted on a jig (a microtome). These sections, cut from different selected planes which correspond to the original orientation of the fragment in the parent tree, can then be stained to make some features stand out from others which all helps the scientist to recognise individual ones microscopically.

In one case of sexual assault, the victim alleged that her assailant had inserted a stick into her vagina. When she was medically examined shortly afterwards, a splinter of wood was found there and it was possible to show that this was a piece of hawthorn, as was the stick with which she alleged she had been assaulted. This proved very useful evidence in support of her complaint.

Most evidence based on vegetation is only corroborative but there are circumstances in which it has provided an absolutely certain link. One example concerned 'leek-slashing', which can be fairly prevalent, at least in the North East, at around the time of the annual leek shows. It was possible to match stained sections taken from the stump of the vandalised leek with a discarded top found elsewhere – a microscopical jigsaw fit. This left no doubt that the top belonged to the stump.

Of necessity this section has dealt with a limited range of types of trace evidence, albeit some of the more common ones. Nonetheless, it is important to remember that almost anything can provide evidentially significant traces, depending on the circumstances, but many of these will be recovered and analysed using the same techniques as those described.

3.5 ASSESSMENT OF SIGNIFICANCE

Once the scientist has identified, recovered, analysed and compared the trace material and is satisfied that it is indistinguishable from the source proposed, its significance in the case has to be assessed.

Unless there is a physical (jigsaw-type) fit between the two, it is not normally possible to ascribe a trace material to one particular source of it to the exclusion of all others. This means that in assessing the significance of his findings the scientist needs to consider:

1. The extent of the analysis and comparisons that have been performed.
2. The relative rarity of the material itself.
3. Whether the findings reflect what might reasonably be expected, given the circumstances of the case, or whether they are inconsistent with what has been alleged and, perhaps, point towards something else entirely.

4. Whether other scientific evidence contributes to the link that has been proposed.
5. Whether and to what extent alternative, innocent sources could account for it.
6. Whether there is any scope for accidental contamination to have played a part.

3.5.1 Extent of Comparison

The greater the number of discriminating parameters which can be measured and compared and which are found to agree, the closer the link between the trace and the proposed source, or ones like it, becomes. Clearly, the amount of trace material available may dictate the extent to which the chain of analysis and comparison can be completed. Where the paucity of material means that the chain has been foreshortened, there is a question mark over whether it would have continued to pass muster against the reference sample had further analysis and comparison been possible.

With textile fibres, for example, it may be possible only to compare the fibres at issue by microscopy and by MSP. This leaves open the question as to whether or not they would also have matched those in the reference sample in terms of their dyestuff composition had it been possible to proceed to TLC. This sort of limitation on the available information must be taken fully into account when assessing the likely significance of associative evidence and the weight to be given to it. All of this presupposes that the scientist conducting the work has sufficient knowledge, experience and technical resources to be able to carry out the necessary work in the first place.

3.5.2 Rarity of the Trace Material

In assessing the relative rarity of the material in question scientists rely firstly on their experience of just how often, in very general terms, they think they have seen its like before. To augment their own impressions, they may seek advice from experienced colleagues, who will inevitably have something to contribute simply because of their long experience with a different set of cases. Scientists may interrogate any relevant databases generated from similar casework material or manufacturers' details over the years and, in addition, may carry out their own researches including direct approaches to manufacturers themselves if that seems appropriate.

So far as databases are concerned, these have their limitations which should be recognised. In the case of glass, the FSS and police laboratories have for a long time gathered data about the refractive index of glass featuring in day-to-day casework. From this, some glasses clearly appear to occur far more commonly than others. But the database(s) cannot take into account changes in glass utilisation as a result of, for example, shifting patterns of importation and/or manufacture, all of which may contribute to local variations in building stock and produce unexpected peaks and troughs in apparent overall distribution. In other words, merely because, according to a database, glass of a certain refractive index is fairly rare country-wide or even in a particular laboratory area, this does not mean that there could not be a preponderance of it in a particular location.

In the past and in relation to textile fibres, police and Home Office laboratories all contributed to a data collection which recorded the number of times a particular fibre type and colour (measured objectively using MSP) had been encountered as a proportion of all textile items examined in general casework. Although figures derived from these data have been quoted in evidence, their validity has always been highly suspect, not least because the total number of items, mainly clothing, never exceeded about 7,500, a tiny fraction of the number in use everyday. Furthermore, experience showed that the collection was unable or too slow to respond to local variation and fashion trends. All in all, it was not possible to determine how the data related to what was being worn on the streets at any particular time and location. These data have not been added to for several years and the collection is now out of date, although there have been suggestions recently that it might be resurrected in some form or another.

One case which demonstrates how careful one has to be with these sorts of databases concerned the finding of a black acrylic fibre at a point of entry. The records showed that this fibre type had only been encountered in a very small number of items examined in the recent past and it seemed to provide reasonably good evidence of a link with the defendant. Nonetheless, it was possible to show that it was equally indistinguishable, in every particular, from quite another innocent and unconnected source. This coincidence served further to undermine the safety of unquestioning reliance on databases in general and this one in particular.

So far as manufacturers' records are concerned, these can be helpful in certain circumstances but, again, they need to be treated with caution. One example where they provided invaluable assistance concerned the case of the murder victim referred to earlier and whose body was covered in regular, short, cut lengths of nylon fibres.

On the other hand, manufacturers' records can be misleading as evidenced by one recent case in which a man had been linked to a shooting by means of some of the textile fibres found on his clothes and which could have come from a sleeve mask left behind at the scene. The (man-made) fibres in question were traced to their manufacturer and the precise dye-batch from which they were thought to have come was identified. From there, the yarn-spinner was named – he had apparently received the entire consignment of the dye-batch concerned – and the number of garments which had been made from it calculated. Nonetheless, further enquiries on behalf of the defence in the case revealed that the yarn-spinner had not spun the yarn used to knit the sleeve mask at the centre of the case. This meant that there must have been at least one other untraced source of the fibres at issue, a factor which cast a long shadow over the safety of this aspect of the prosecution's evidence.

3.5.3 Expectations

The extent to which forensic scientists are able to predict what would reasonably be expected in terms of the transfer of trace evidence in a particular case is governed partly by the information they have to hand about the case and partly by their experience. As to information, this may be relatively limited for the police scientist, who is often engaged in the work at an early stage and before all the relevant facts are available. As a matter of history, there was a period in the 1980s when the prosecuting authorities admitted their nervousness about supplying the scientist with 'too much' information for fear that his judgement might be prejudiced!

This reflects a common misconception that forensic scientists faced with a series of anonymous items can somehow plan and execute an appropriate programme of work, produce results and interpret them in a meaningful way – all without knowing the essential details of what is alleged (and disputed), where the items came from and what their history is likely to have been. This is like asking an architect to design a building but not telling him anything about its proposed purpose and location. Science without context and/or out of context is meaningless at best and dangerous at worst.

In contrast to the position of the police scientist, there have been and are no such restrictions on the information available to the scientist appointed by defence – who normally will be provided with all information available about the case, including the defendant's latest word on the matter. However, defence scientists are disadvantaged in another way because, in most cases, they will not have had an opportunity to examine items in their original state, or to visit the scene of crime or participate in the painstaking archaeological excavations at the scene of a fire. This means that by the time they see them, blood stains, for example, which may be crucial to the case in terms of their physical appearance, may have been cut away and extracted. Glass fragments will have been shaken, brushed or otherwise removed from their original resting places and loose textile fibres and other important debris will have all been taken up on tapings and vacuumings. The defence scientists must therefore rely instead on the detailed records made by the prosecution's scientists at the outset. Thus, the latter bear a heavy responsibility to the justice system in ensuring that their records are complete and available for informed scrutiny. But, importantly, and flowing from all of this, the scientist appointed by the defence must know intuitively what their colleague's examination strategy is likely to have been and how they will have set about achieving it. In short, the scientists must personally have had many years of exactly the same sort of experience as the prosecution scientist, *i.e.* total immersion in cases of the same general sort, day-in and day-out, year-in and year-out, if they are to do justice to the task which should be required of them. Merely spending the time checking the accounts of others who have gone before is just not enough to qualify the scientist who would be appointed by the defence.

3.5.4 Combination of Evidence

As a matter of common sense, as the number of different types of apparently transferred material, each of which could form a link between two sets of items, increases, the likelihood of finding them in combination merely by chance diminishes rapidly. Common examples would be the finding of multi-layered paint fragments or, say, several different types of fibres on one item, all of which could have come from another item in the case. These combinations provide especially powerful links where the transfers appear to have taken place in opposite directions (two-way transfer), *e.g.* glass like that in the broken window at the point of entry on the suspect's gloves, and fibres like those comprising the gloves on the broken glass found at the scene.

3.5.5 Alternative, Innocent Sources

The finding of apparently relevant transferred traces carries with it the assumption that there is no alternative, innocent explanation for it. The extent to which this assumption will be tested depends upon the range of other possibilities presented to the scientist. But this in turn now has direct cost implications for the police who, faced with tight financial budgets, may wish to draw the line increasingly sooner than later. The effect of this sort of selectivity may, without the knowledge of the scientist concerned, artificially skew their findings and their perceptions of them to the point where their evidence may become misleading.

Returning to the example of the sleeve mask, there were problems relating to manufacturers' records which could have given rise to the incorrect assumption that there was a single source of the fibres at issue. In addition to demonstrating that this was not the case, the further enquiries conducted on behalf of the defence revealed that the defendant himself owned two jumpers, entirely different from each other and from the fabric of the sleeve mask, which contained precisely the same sort of fibres. This meant that either of these jumpers could have been the source of the apparently incriminating fibres and, against this background, what began as powerful associative evidence crumbled away. At the end of the day, this part of the prosecution's case was withdrawn.

3.5.6   Contamination

Over and above considerations as to whether trace material found on one set of items did in fact come from the source proposed for them, it is always important to investigate if any of the material in question could have arisen as a result of accidental contamination between one item and another. Such contamination (secondary transfer) can occur if the same police officer, for example, seizes, handles or otherwise comes into contact with items from more than one location. In these circumstances traces may have been unwittingly transferred *via* the officer's hands and clothes. Accidental contamination can also occur if the items concerned are packaged or repackaged in the same room, since those which arrive later, assuming that they are not all brought in at one and the same time, may pick up traces left behind by ones which were dealt with earlier.

Properly trained and experienced forensic scientists are only too well aware of the dangers of contamination. Only routine procedures and protocols, designed to avoid contamination and rigorously adhered to, will ensure that there can be a high level of confidence about the integrity of items examined.

3.6   SAFETY OF TRACE EVIDENCE

All of this serves to refocus attention on some of the consequences of the entry of forensic science into the marketplace without there being any system of personal accreditation and regulation of the profession in place. Historically, defence lawyers have experienced difficulties in identifying properly qualified forensic scientists to check the prosecution's scientific findings from amongst all those who would claim to be competent. But now this has been extended to the investigating and prosecuting authorities themselves. Not only does it remain difficult always to ensure that the work is properly checked, it may now be difficult to ensure that it was properly done in the first place. In these circumstances, it seems likely that quality and standards may be sacrificed in the name of economy, and nobody should be surprised if trace evidence remains very much in focus as a possible contributor to future cases in which it is shown there has been a miscarriage of justice.

Against this background, lawyers should be cautious about forensic science evidence in general and trace evidence in particular. If prosecuting, they should make sure that:

1. The work was done by a reputable forensic science laboratory.
2. The scientist concerned is aware of all the relevant circumstances in the case.
3. The work has not been artificially shaped or curtailed through lack of funds.

If defending, lawyers should.

1. Never accept scientific evidence from whatever source at face value.
2. Always ensure that the scientist appointed is properly qualified and experienced and has investigated in a primary sense *many* cases of the same general type and not merely looked over the shoulders of those who have.
3. Take a very robust line with legal aid authorities over funding for the work and not be fobbed off with the requirement to take the lowest estimate unless it can be established that the scientist concerned knows what they are doing.

3.7 BIBLIOGRAPHY

C. Catling and J. Grayson, *Identification of Vegetable Fibres*, Chapman and Hall, London, 1982.
M. Grieve, Fibres and Forensic Science – New Ideas, Developments and Techniques in *Forensic Science Review*, Central Police University Press, 1994.
R.C. Murray and C.F. Tedrow, *Forensic Biology*, Prentice Hall, New York, 1992.
J. Robertson, *Forensic Examination of Fibres*, 2nd Edition, Ellis Horwood, Chichester, 1999.
R. Saferstein, *Forensic Science Handbook*, Prentice Hall, New York, 1982.
B. Caddy, *Forensic Examination of Glass & Paint: Analysis and Interpretation*, Taylor and Francis, London & New York, 2001.

CHAPTER 4

Marks and Impressions

KEITH BARNETT

4.1 INTRODUCTION

Everything we do leaves a mark on the world. It could happen as we are walking, driving, working or sitting in the chair. While we carry out these tasks our shoes, clothes, hands and even the tools we use can leave unique tell-tale signs to the expert eye. A car will leave tyre tracks, the shoes on our feet will leave footwear impressions and even the screwdriver used to take the lid off the paint tin will leave its mark. The simple principle that Locard proposed was 'every contact leaves a trace' and in the field of marks and impressions put two items together and they are likely to leave a mark on one another.

Sherlock Holmes said in *A Study in Scarlet* that 'there is no branch of detective science which is so important and so much neglected as the art of tracing footsteps'. The same is also true for the examination of tyre tracks and the comparison of plastic bags of instrument and glove marks. In all these cases there are features present that can be used to form a unique connection between suspect and scene or stolen property. The following chapter gives an insight into the methods used to recover and present evidence at court in some of these areas.

This evidence can be divided into two distinct groups: damage based evidence and non-damage based evidence. In the first group, which includes evidence from footwear and instruments, an item has to acquire damage in order to leave behind a unique impression. In the second group, as with fingerprints, a combination of inherent features provides the unique link. Any damage to the features of the fingerprint, such as a scar, is not used in its classification as it may fade with time. Just as it is true that no two fingerprints have been found to be the same, even in identical twins, so in the examination of damage features no two items, no matter how similar they are initially, are likely to acquire the same random damage features during use.

In the first part of this chapter the evidence provided by damage and wear will be addressed, specifically evidence that can be obtained from footwear, instruments and mass-produced items. In the second part, fingerprints and the evidence they provide will be discussed.

DAMAGE BASED EVIDENCE

4.2 FOOTWEAR IMPRESSIONS

4.2.1 Introduction

Every time a person takes a step, whatever the surface they are walking on, they will leave behind a footwear impression. An impression could be defined as the retention of the characteristics of an item by another object. Hence over soft ground shoes impress themselves into it and leave behind their characteristic impression. The impression left behind is not always obvious and can be difficult to find without the aid of a specialist technique. It is also vital that as many impressions as possible are recovered from the scene of a crime as this will increase the chance of finding an area corresponding to the area where the damage features are located on the undersole of a shoe. It is possible that the undersole has only one or two of these damage features present and they may all need to be found to provide a significant link between a suspect's shoe and the scene of a crime.

4.2.2 Recovery of Impressions from Scene of Crime

There are several mechanisms by which a shoe can leave an impression behind. On a two-dimensional (flat) surface, such as a tiled floor or a piece of paper, material from the undersole can be deposited and remain for a considerable length of time. Sometimes this is due to a static electrical charge produced on the undersole transferring particles to a surface or the wet deposits on an undersole being left behind on the surface. A three-dimensional impression is formed if the surface over which the shoe passed was soft and the undersole sank into it before moving on. The different types of impression require a variety of techniques to recover them as no single method can cover all eventualities. Recovery can be achieved simply, by casting with plaster of Paris or by applying fingerprinting powder, but often the impressions are fragile or transient in nature and these methods will not be successful. Recently, a number of new and novel techniques have been applied to the problem with startling results. Some of these techniques were developed to aid fingerprint enhancement and recovery, others specifically for the enhancement and recovery of footwear impressions.

4.2.3   Impressions in Two Dimensions

In all cases where an impression of this type has been left behind on a flat surface, it will be visible only when there is contrast between the background and the material of the impression itself. To use a simple example, if a footwear impression is made on plain white paper with talcum powder it would be almost invisible. Treat this with a black powder, say finely powdered charcoal, that adheres preferentially to the white talc and the impression will become visible because of the black and white contrast created. To enhance most impressions so that they are fully visible requires the production of this contrast. The impression found can be photographed successfully, with a scale along side it to provide a permanent record for court purposes. Often it is possible to seize the item on which the impression has been made and submit it to the laboratory for examination. On other occasions the enhancement has to take place *in situ* and a suitable method has to be used at the scene of crime to improve the visibility of the impression. The detail produced has to be of such a quality that it can be photographed.

Porous surfaces such as paper, cardboard and carpet cannot be treated in the same manner as non-porous surfaces such as glass, linoleum and tile. On most of the latter surfaces dyes or chemicals can be applied, but on the former these would be absorbed into the surface and obliterate the impression.

Impressions in dust need to be treated with special care because any contact with a brush or spray would destroy them. The recent development of methods involving electrostatic treatment has allowed the full recovery of even the most delicate impressions of this type on the most unlikely surfaces.

Impressions made in blood are likely to be the most important of all those found at a scene of crime because they can be relevant to a particular offence. Often a person has legitimate access to the premises where the crime was committed and therefore any dust or dirt impressions are of little relevance as they may have been there for some time. If, however, during the crime someone has bled, the suspect's presence at the relevant time can be proven if they have stepped in that blood and left a subsequent footwear impression. Heavy impressions left by an undersole that has made contact with a pool of blood contain little for the forensic scientist other than the possibility to determine the size and pattern of the shoe concerned. They contain none of the fine damage detail that is required to make a significant comparison. It is now possible to enhance the detail of even the faintest of impressions left in blood by using some of the new techniques available.

4.2.4 Methods for Enhancing Two-Dimensional Footwear Impressions

The search at the scene of crime for footwear impressions is a systematic business. Apply only one technique and the chances are that just a few impressions made by one particular mechanism will be found. The more techniques used the greater the chance of recovering a large range of different types of impression that may be present at the scene. It is usual to start the search with techniques which will cause little or no damage to the impressions present and then progress to those which are likely to cause irreversible change to them.

The simplest and in some cases the most effective starting point is to shine a light obliquely across the surface of interest. Oblique means a light source positioned close to the surface giving a low angle of incident light. This will cause shadows to be produced by the deposits adhering to the surface and in turn this shadow provides the vital contrast. The method will show an array of impressions formed by a number of different mechanisms without any disruption occurring to them.

It is possible to illuminate the surface with a number of different light sources. Some of these are certainly more practical for use in the laboratory than at the scene of crime. Now, however, it is possible to take a portable low powered argon ion laser to a scene and use it, with great care, to illuminate surfaces of interest. Light from a laser is intense, monochromatic, coherent and polarised. The intensity of the radiation and the fact that it is monochromatic give rise to fluorescence and phosphorescence of some materials either present in an impression or in the background on which the impression has been made. In addition to this, impressions can be stained with a suitable dye to make them fluoresce under the effects of laser light. The monochromatic properties of the radiation also enable considerable enhancement to be made of faint impressions. In effect an argon ion laser acts like an extremely powerful light source that can be tuned to give out light at a few specific wavelengths. The main drawback is the amount of energy it produces: place a piece of paper in front of the light beam and you could soon have a fire on your hands. Therefore a method is required to dissipate this energy and this can be achieved by using a fibre or a liquid optical light guide. This has the added advantage that the direction and spread of the light can be controlled.

If a difference in fluorescence is created between the deposit and the surface then the impression becomes visible and a photographic record can be produced. UV and IR light sources can be used in a similar way and again the results can be photographed.

4.2.5 Dust Impressions

The most fragile impressions present at a scene are likely to be those in dust, therefore it is important that the scene is examined for these next. In 1981 research was instigated to determine if it was possible to recover impressions electrostatically. This proved successful and a portable device, an electrostatic lifter, was produced that could be used at the scene of a crime.

The device contains only a small 12 volt battery, but is capable of producing a variable high voltage. The voltage, which could be anything up to about 15,000 volts, is then applied to a thin conductive film. The film comprises a sandwich that includes an upper layer of thin aluminium foil and a lower layer of black insulating plastic. When the high voltage source is turned on, an electrostatic charge is produced in the aluminium layer of the lifting film. This charge causes the dust or other particles of the footwear impression to jump onto the black underside of the lifting film. When carefully removed and turned over, the film can hold a complete dust impression as it was on the originally examined surface. Enhancement is achieved because most of the dust impressions contain large amounts of dead skin that drops from our bodies, so the impressions are light in colour and on the black background of the lifting film a colour contrast is obtained. The technique has been used successfully on a variety of surfaces including linoleum, carpet and even car body panels.

4.2.6 Other Deposits

Most of the impressions that would remain after this treatment are resilient and require more vigorous methods to recover them. These methods of enhancement include the application of aluminium powder with a brush, as in the treatment of fingerprints, or treatment with chemicals such as ammonium thiocyanate to detect any iron present in soil deposits or gentian violet which reacts with grease.

Figure 4.1(a) shows a footwear impression on an advertising box. It is present in dried mud and covers both a white area and a dark area. Therefore photography has shown little of the detail of the impression.

Figure 4.1 *Footwear impressions: (a) photograph of footwear impression left in dried mud on an advertising box at the scene of a crime; (b) test impression of the suspect's shoe made in the laboratory; (c) enlargement of the heel area of the test impression, the arrow indicates the unusually large damage feature*

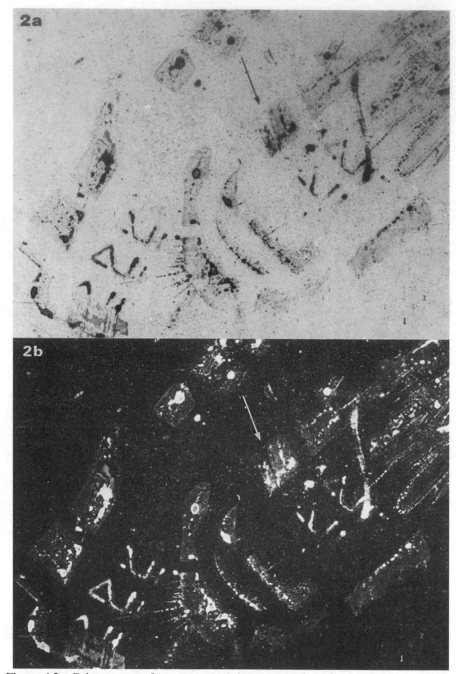

Figure 4.2   *Enhancement of impressions: (a) gentian violet enhanced footwear impressions lifted from the advertising box featured in Figure 1(a); Arrow indicates
damage features as seen in Figure 1(c); (b) Laser illumination of gentian
violet enhanced footwear impressions. Arrow indicates damage features as
seen in Figure 1(c)*

In Figure 4.2(a) however, the impression has been treated with gentian violet and lifted from the box. The detail of the impression in this dark area is now much clearer. When this impression is exposed to laser light as in Figure 4.2(b) the detail of the impression, including the spray away from the heel of one of the impressions, is clearly visible.

4.2.7 Impressions in Blood

One special material left at the scene of crime is worth considering in a little more detail. Blood shed at a scene can fall onto two types of surface mentioned earlier, porous and non-porous. When an undersole comes into contact with blood it will pick up some of it and then deposit it at various places until the blood becomes dried or is all used up. It is best thought of as a rubber stamp being inked and then being used to produce a number of similar stampings and then occasionally being re-inked.

4.2.7.1 Blood on Non-porous Surfaces. Blood marks on a non-porous surface can be enhanced in two ways, either with protein dyes such as amido black or fuchsin acid, or with a haemoglobin stain like diaminobenzene (DAB). The latter stain is difficult to use at the scene of crime on the large areas that require treatment. Blood contains a large amount of protein, so by choosing a protein dye, which will stain the blood impression a totally different colour in contrast to the surface on which it was deposited, enhancement is achieved. Protein dyes are the preferred enhancer at a crime scene as they are easy to work with and cause less health risks than treatments like DAB. The impressions found can defuse when they are treated in this manner but this can be avoided by including a blood fixative, 5-sulphosalicylic acid, with the dye when it is applied. Obviously substances other than blood such as milk, eggs and meat contain protein, therefore it is necessary to show that the impression which has been enhanced contains other blood constituents such as haemoglobin.

The presence of haemoglobin can be demonstrated by treating the small areas of impression with DAB. Research has recently shown that some of the latest DNA profiling techniques, like Low Copy Number DNA (LCN) can produce results even from the minutest amounts of DNA that are still present after fixation and staining have been carried out.

4.2.7.2 Blood on Porous Surfaces. To recover impressions in blood from a porous surface, the forensic scientist has either to attempt to separate some blood from the surface on which it was deposited or stain it so that it can be visualised without the impression leaking away. This can be achieved with a limited degree of success by lifting the blood with another impervious substrate. Gelatine sheets can be used, with the lifted impression then transferred to a piece of nylon protein bonding membrane or it has been found that it can be lifted directly onto protein bonding medium using 5-sulphosalicylic acid as both a fixative and a moistening agent. The lift is then treated with DAB to give a strong brown colour even though only minute quantities of blood have been lifted.

Recently research has been carried out with a number of staining agents. Some, such as luminol, need viewing in the dark but others like ninhydrin can be viewed in normal light. Ninhydrin reacts with the amino acids present in blood and the mark produced does not leach out from the porous material in which it was made. Therefore it can be used to good effect on sections of carpet or clothing. Stains such as luminol do not produce so fixed a mark and their health and safety implications are not fully documented. This along with their difficult viewing requirements does not make them so attractive to use at a scene of crime.

4.2.8   Other Impressions on Porous Surfaces

To treat impressions on porous surfaces such as paper and carpet the scientist has to resort to techniques such as electrostatic lifting and illumination with different light sources. There are, however, other lifting methods that will separate the impression from the surface.

The lifting of a particulate footwear impression involves the transfer of the two-dimensional impression from its original surface to one that will provide a better contrast. There are a number of lifting media available but the commonest are adhesive tape, gelatine (either plain or coloured) and the conductive film used in electrostatic lifting. Apart from the last the others rely on the sticky nature of the material applied and therefore it would be unwise to try to use an adhesive tape lift on an impression made on paper because the tape is too sticky. A gelatine lifter would work well in this case because of its less sticky surface. One danger with lifting only the deposit without fixation is that the deposit may migrate in the adhesive or gelatine causing it to become invisible. This does not occur when impressions are dusted with aluminium powder.

4.2.9   Three-Dimensional Impressions

The only recourse for the satisfactory recovery of impressions of this type is by casting. This was traditionally carried out using plaster of

Paris but had the problem of being fairly soft even when set and therefore it was always difficult to clean the cast afterwards. Now, however, plasters are available that become that extremely hard when set so that it is possible to clean them off under running water.

If a mark at the scene is impressed into soft soil it can be cast easily with one of these new plasters. The plaster is weighed out into a heavy duty plastic bag and mixed with the correct amount of tap water. The corner of the bag is then cut off and the plaster squeezed gently out into the mark in the same way that you would ice a cake. Once set, the cast can be pulled up along with any adhering soil and returned to the laboratory for examination.

4.2.10   Conclusions

The methods given here are just a limited number of many that are available to the forensic scientist in their search for footwear impressions both at the scene of crime and on items submitted to the laboraory.

Once a footwear impression has been treated to obtain the maximum amount of fine detail, the forensic scientist has to consider the other significant exhibit, the shoes from the suspect!

4.3   INFORMATION AVAILABLE FROM A SHOE

What information can we obtain from a suspect's shoe that will be of significance when we come to compare it to a footwear impression recovered from the scene of a crime?

There are four main aspects of the undersole of a shoe that the forensic scientist will consider when making their final opinion:

1. The pattern.
2. The size.
3. The degree of wear present.
4. The random damage present.

4.3.1   The Pattern

A pattern match between the undersole of the shoe and the impression at the scene is imperative because without it the comparison needs to go no further. Today, there are a myriad of different undersole patterns present on shoes because they are not just something to keep your feet warm and dry but are a major fashion accessory. Only a small proportion of shoes today have plain leather undersoles. By far the largest section of the footwear market is dominated by training shoes that have increasingly complex undersole patterns. Each of the major manufacturers tries to make the pattern of the undersole specific to them and each new upper style manufactured may have a new undersole to go with it.

Collections of the different undersole patterns encountered are kept as a database by laboratories. These have been established in an effort to answer two vital questions:

1. What make of shoe could have made the impression at the scene of a crime?
2. How common are these undersoles and how do their pattern components vary?

The collection allows the forensic scientist to give vital early evidence to the police about the style of shoe a suspect may be wearing. It also provides assistance to the scientist in gauging the strength of their evidence by showing how common a particular undersole pattern is and how this pattern may vary with size. With the large number of impressions in a collection the scientist may even be able to determine small differences in the design of the items used to produce different undersoles within a particular size.

The methods employed to manufacture an undersole can also impart some degree of individuality to the pattern present. There are two methods commonly used to make undersoles. One is to cut a number of units from a large piece of pre-moulded rubber, the other is to inject molten material into a pre-formed metal mould.

With the 'cut outs' variation can occur as each unit is individually cut from a large piece of rubber using an undersole shaped cutter. The operator has some latitude in the positioning of the cutter on the sheet and therefore variations can occur in the pattern present particularly around the edge of the undersole.

With the injection moulding process every undersole unit that is produced by a single mould should be identical. However, because of the large quantities of shoes that will be produced in a particular size, the manufacturer can often have a number of moulds to make any one given size of shoe. This allows the manufacturer to produce several undersoles at the same time in an automated process. There can be considerable variation in the fine pattern detail from one mould to another, even within the same size, as the moulds are often handmade.

4.3.2 The Size

Size is felt to be an important feature of a shoe. However, to the scientist determining this, it is a two edged sword. The investigating officer always wants to know 'What size shoe am I looking for?' And the scientist finds this a very difficult question to answer. Why? Well it is not the size of the undersole that determines the size of the shoe printed on the label but the space inside the upper that accommodates the foot. There can be a large variation in undersole size amongst the shoes worn even by one person. Look at the variation in your own shoes. The other problem is knowing how the shoe undersole varies with size if only a partial impression of the complete undersole pattern has been recovered.

The length of an undersole is however one of the factors to be taken into account when judging if a significant connection is present. Some sizes are much more common than others; in adult males in the United Kingdom the average shoe size is eight or nine but only a few percent of that population wear size six or size twelve shoes.

4.3.3 The Degree of Wear

As a shoe wears, the pattern on its undersole changes. These changes are only small, but if an unworn undersole is compared with one that has been worn for a few months a distinct difference will be noted in the pattern features. Some bars or blocks may have become wider and some detail may even have disappeared altogether. This is often seen with the fine surface stippling present on the undersoles of some new shoes such as Dr Martens boots.

People walk in different ways, placing their weight on different parts of the undersole. Some have deformities of the foot, or maybe a limp, and all these peculiarities will show up in the wear pattern on the undersoles of their shoes. More rapid wear usually occurs in the areas where the weight upon or scuffing of the undersole has been greatest and this causes a more rapid change in the pattern. Some of these wear features can be specific to a person's shoe undersole but more often they can provide a further significant connection between a shoe and a scene of crime.

4.3.4 The Damage Detail

Continual contact between a shoe and the ground causes the undersole to acquire cuts, scratches and other damage features that happen completely at random. As a consequence two shoes that start with identical undersole patterns will, over a period of time, start to gain damage features in different places and so obtain a degree of individuality. Each impression they leave can now be unique to that particular shoe. As time passes this damage record will continue to change with some marks being removed and others added. By the examination of this unique set of fine damage features the forensic scientist can, even from some small fragment of an impression, tell if a particular shoe did or did not make a given footwear impression.

Using the knowledge gained we can now look at how the forensic scientist carries out their comparison between a shoe's undersole and the impression recovered from the scene of a crime.

4.4   COMPARING AN IMPRESSION WITH A SHOE

To compare the undersole of a shoe with an impression made at the scene of a crime the scientist has to produce a new impression from that shoe's undersole in the laboratory. Without these test impressions the contact points of the shoe may not be clear and the same can be true of any damage or wear features.

4.4.1   Making a Test Impression

There are several methods available for the scientist to make these impressions. One of the simplest, quickest and most commonly used ways is to apply a film of light oil uniformly to the undersole and stepping onto a sheet of oil impregnated foam rubber. The undersole is then pressed onto plain white paper and the oily mark produced dusted with a powder mix of fine black powder and magnetic iron filings, using a magnetic brush. Because the brush is magnetic it does not have to come into contact with paper. The impression produced will show all the wear and damage features present on the undersole and, if required, other test impressions can easily be made to try to accurately reproduce the manner in which the impression was made at the scene of crime.

If a three-dimensional impression has to be compared with a suspect shoe then it may be necessary to reproduce the test impression in a similar way. Perhaps, for example, the impression at the scene was made in soil. This effect can easily be achieved using a tray filled with fine damp sand. The resulting impression of the shoe in the sand can then be cast and used to carry out the comparison.

As a general rule the scientist should try to reproduce as accurately as possible the mechanism used to make the 'scene impression' when making their own test impression. This can sometimes be difficult as there are numerous variables that may cause slight variations in the impressions that are left at the scene.

4.4.2 Comparing Impressions

Figure 4.1(b) shows a test impression made from the undersole of a Fila casual boot. The word Fila is present three times in the heel of the boot. In the edge block adjacent to the letter A in the middle word Fila, see Figure 4.1(c), there is a large damage feature that travels at an angle across this block. This is a highly significant feature. The size, pattern and degree of wear exhibited by the undersole in the test impression agrees with that shown in the impression that has been enhanced with gentian violet and photographed using a laser light source as in Figure 4.2(b). Looking at this impression one will see the corresponding damage feature accurately reproduced in the same block in the impression. Given these agreements between the undersole and the impression there is no doubt that this particular Fila boot made the impression enhanced from the scene of the crime.

4.5 INSTRUMENT MARKS

As with footwear, the presence of damage on any instrument used in the commission of a crime can be utilised to form a connection between that instrument and the mark it has left behind. The instruments normally encountered fall into two main categories: those that cut and those that act as levers. The first category includes bolt croppers, drills and knives, the latter, jemmies, screwdrivers, and chisels. One or other of these groups are often used to gain access to a scene of crime, be it to carry out a burglary or a murder. The following sections will deal with each group of instruments and the special requirements for their examination and comparison to marks left behind at the scene.

4.5.1 Cutting Instruments

The commonest cutting instruments examined by the forensic scientist are bolt croppers and drills. The latter are usually auger bits used to drill doors or window frames. However, with the recent production of battery operated power drills the use of high speed drills is not uncommon. No matter what the cutting instrument encountered the method of comparison is normally the same. A test cutting is produced with the item in question and any damage features it contains are compared with those on the mark recovered from the scene. The final examination is usually carried out with a comparison microscope.

4.5.1.1   Construction.   It is important to know how the particular instrument has been constructed as this knowledge can be invaluable when carrying out a comparison. It is not always necessary to know how the whole item is constructed but it is important to know how the final cutting edges are formed during manufacture.

A pair of bolt croppers has two long metal handles that are joined at a fulcrum, and these then connect to the cutting blades. This form of construction allows the blades to be adjusted to give the best action and for the blades to be replaced as they become damaged or worn (see Figure 4.3(a)). Bolt croppers can vary in length. The smallest that can easily be hidden in an inside pocket are about 12" long, the largest up to four feet long. The length of the handles determines the amount of force that can be generated at the blades, but the size of the item that can be severed is limited by the hardness of the blades. As a criminal is not bothered about the size of the item they are cutting, the blades of the bolt croppers often suffer severe damage when they are used.

Even though most bolt cropper blades start with a forged and ground cutting edge, within a short period of time they acquire damage features that are unique to them. Even the grinding of the final cutting edge when the item is manufactured may impart some areas of individuality.

Auger bits also have this final finish and the same 'manufactured' areas of individuality may be present. Surface damage to the bit during use may also produce some areas of individuality.

4.5.1.2   Marks at the Scene.   The items severed by cutting instruments at scenes of crime are most likely to be shackles of padlocks, chain link fencing and telephone wires. Drill bits are normally used to cut holes in window frames to allow the release of internal catches or to drill holes around door locks to weaken the door so that it becomes easier to force.

In all these instances the damaged items should be seized if possible and submitted to the laboratory for examination. As the blades of the bolt croppers shear through the smooth metal of a shackle or the auger bit cuts through the painted wood of a window frame the damage features on their edges will have caused a series of parallel grooves which produce scratches on the item. Some of these scratches may only be a fraction of a millimetre in width and they are known collectively as 'striation marks'. This pattern will be produced again and again by the cutting item until it acquires further damage and the pattern changes. The damage features can be used to connect the severed edge of a padlock shackle, as seen in Figure 4.3(a) inset right hand side, with the test mark produced by the suspect's bolt croppers (Figure 4.3(a) inset left hand side).

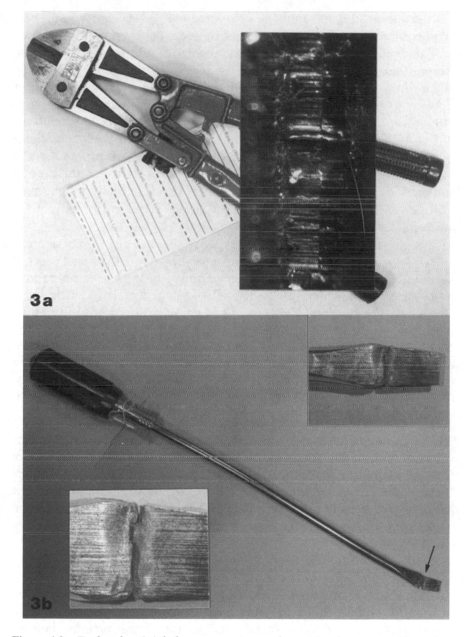

Figure 4.3 *Toolmarks: (a) bolt croppers; inset enlargement shows comparison photo-graphs of the severed edge of a padlock shackle on the right hand side and on the left hand side the test mark made by the suspect's bolt clippers; (b) screwdriver broken at the scene of crime; inset enlargements show the grinding detail as well as the physical fit between the two broken items*

If the damaged item cannot be recovered to the laboratory to make a direct comparison with the instrument in question, then taking a cast of the damaged portion can produce a record of the mark. It must be stressed that casting should only be used as a last resort because it may not produce a clear record of all damage features present. The object should be cast rather than photographed as a photograph cannot be printed from a negative with enough accuracy to reproduce the relationship of all the fine striation detail normally contained within the damaged surface. Illumination of the mark is also an important factor when carrying out a comparison and it is often difficult to produce good lighting conditions even in the laboratory. Therefore, the production of a properly illuminated image at the scene of crime would be almost impossible.

Casts are normally made with a type of silicon rubber. These compounds have similar properties to those used by dentists to take casts in the mouth. They consist of a hardener and a base that combine in a similar way to the fillers used to repair damaged motor vehicle body panels. For the casts to be effective the casting material has to be sufficiently runny to flow into the mark but robust enough to be handled when set. A point of note here is that the marks, however they are produced, are a negative (inversion) of the item that made them. Therefore when a comparison is carried out the scientist must make test marks of their own so that they are comparing like with like. It is not possible to compare a cast directly with a mark.

4.5.1.3 Comparison. Probably the most difficult part of the comparison examination for the scientist is the production of control marks from the submitted instrument. Often they will not have seen the mark *in situ* at the scene and will therefore have to produce a number of test marks to give the full range of the detail that could be produced by the damaged cutting edge of the instrument. The production of test marks with bolt croppers and auger drill bits does not present as much of a problem as those to be made with a levering instrument. The difficulty will be discussed later in the section on levering marks.

It is also important when carrying out a comparison that the forensic scientist illuminates both the questioned and test marks in the same manner. Oblique lighting can be used to make the various damage features of an instrument mark cast shadows in a similar way to the oblique illumination of a surface for footwear impressions. The light striking both questioned and test mark must originate from the same direction or the shadows produced will appear different. It would not be possible to obtain any match if the marks were not illuminated in the same manner.

Initial comparison of the test and scene marks is often carried out using a standard bench search microscope with the items obliquely illuminated. In the final examination of the test and scene marks a comparison microscope is normally used.

A comparison microscope consists of two identical sets of microscope lenses linked together by an optical bridge. The questioned sample is placed on an adjustable stage below one set of optics and the test mark is placed on a similar arrangement below the other. With the same magnification in use on each side the fine striation detail present in the marks can be set side by side in the viewing field produced by the optical bridge.

Figure 4.4 consists of two photographs taken using a comparison microscope. The split between the different sides of the microscope can

Figure 4.4 *Photographs of marks made by an auger bit as seen with a comparison microscope: (a) test drilling made in laboratory; (b) drilling marks taken from the scene*

be seen running at a slight angle from the top to the bottom of the picture. The object on the left (Figure 4.4(a)) is a test drilling made in the laboratory. The object on the right (Figure 4.4(b)) is a drilling taken from the scene of the crime. On the scene drilling the concentric detail is readily visible and it will be seen that some of this also occurs on the test drilling. Given the number of fine striation details that agree there is no doubt that the auger bit used to make the test drilling also made the drilling taken from the scene.

4.5.2   Levering Instruments

Usually the levering instruments encountered by forensic scientists in their work are jemmies and screwdrivers. Whatever the type of levering instrument the method of comparison used is normally the same as with a cutting instrument. A test mark is produced using the item in question and the detail present compared with the detail in the mark recovered from the scene. The initial comparison is carried out using a standard bench microscope and the final one using a comparison microscope. The use of a comparison microscope is not always necessary as photographs can often be produced that show the unique characteristics required to prove a match. An overlay or side by side montage could perform the job equally as well.

4.5.2.1   Construction.   Once more it is important to know how the particular instrument has been constructed as this knowledge can be invaluable when carrying out a comparison. It is not always necessary to know how the whole item is constructed but it is certainly important to know how the blade and edge of the lever are made.

Jemmies and screwdrivers are normally forged from a length of metal bar to produce a flat blade and the final finish to form the edge is usually produced by grinding (see close ups in Figure 4.3(b)). Some modern screwdrivers are cast and then ground to provide the finished edge. This difference is normally unimportant except where a mark has been caused by impression of an instrument into a soft surface because this may show the face of the blade as well as its edge. If this is the case the detail shown on the blade may not be random but, as with a moulded shoe, may be the same for all screwdrivers produced from a particular mould. If a comparison has to be carried out on these features then this limitation in the strength of evidence should be borne in mind and marks examined for damage rather than mould features.

Screwdrivers and jemmies are made in a wide variety of shapes and sizes, but once again the criminal is not concerned with this. Often the item will be used for a purpose for which it was not designed and become damaged and pieces may be broken from the tip or it may be dented. When this damage occurs the item has obtained its individuality and the mark it produces, as with a cutting instrument, will be unique and will not change until it acquires further damage.

4.5.2.2 Marks at the Scene. Items commonly attacked with a levering instrument at a scene of crime include window and door frames. These surfaces are relatively soft so the levering instrument tends to impress its detail into the surface rather than scratch it. However in some cases, where the levering item has slipped in use, a combination of both scratch and impressed marks can be produced. If possible the items that have been levered should be recovered from the scene and submitted to the laboratory for examination. In the following section we will concentrate on the comparison of impressed detail as we have touched on scratch mark comparison earlier in the section on cutting instruments.

The mark produced with a levering instrument takes the form of the edge or, in some cases, the full blade of the object used to carry out the levering, therefore any damaged present on this part of the item can be impressed into the surface. The unusual shape of a damaged tip or blade can be also recovered by casting as with cutting instruments. With an impressed mark the cast produced is likely to show good detail for comparison whereas this is not always true of casts of cutting marks.

4.5.2.3 Comparison. The comparison can be carried out in a similar way to that employed for cutting instruments. Test marks are produced in a way that tries to mimic the mechanism by which the mark was left at the scene. The tip of a blade can be quite thick and any alteration to the angle in which it is presented to the surface can cause a significant variation in the detail produced, therefore, a number of test marks will be required. If striation detail is also produced then test marks have to be produced with the blade at various angles to the surface. This is necessary as the combination of striation detail may only have been made with one small area of the tip.

Again the test and questioned mark can be placed on the stages of a comparison microscope and the details compared side by side. Some microscopes allow the two images produced to be superimposed one on the other and with impressed detail this is often the better method for carrying out the comparison.

Because the detail in the marks may be quite large, and in some cases the marks may be almost flat, it should be possible to produce good, accurate photographs of the damage detail present. A photographic transparent overlay can then be produced of one of the marks and this can be placed over the photograph of the other mark to accurately show the corresponding damage features. Sometimes it may be sufficient to put the two photographs side by side to be able to highlight the similarities in the test and questioned marks.

Casts of test marks for a jemmy taken from a suspect and the window frame at the scene of a crime are shown in Figure 4.5. If the edge of both marks is followed then it will be seen that they both agree with each other, with large areas of damage being present. There is no doubt that the jemmy from the suspect made the impression found at the scene.

4.5.3   Conclusions

The value and strength of the evidence gained from the examination of instruments and instrument marks can be enormous. Given the presence of matching damage detail it can, like in the examination of footwear and footwear impressions, provide a unique connection between an instrument and the scene of the crime.

4.6   BRUISING

One special type of mark not already mentioned is the bruise. It is quite common to find bruising on the body of someone who has been assaulted. Sometimes the clothes they are wearing being forced into the

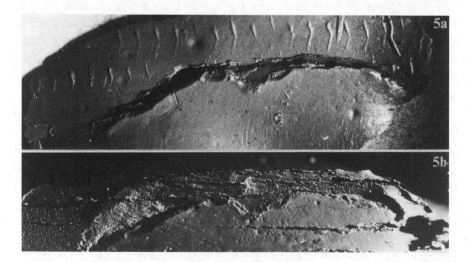

Figure 4.5   *Marks made by a levering instrument: photographs of silicon rubber casts of (a) test mark made in the laboratory and (b) those from a damaged window*

skin by a blow cause these bruises, but often they are caused by the impact from the undersole of a shoe or an instrument used by their attacker. Car and lorry tyres can also leave behind their tell-tale traces.

A mark made on the skin by the undersole of a shoe or an instrument causes a temporary deformation of the skin to occur. However, a record of that contact may still be found in the form of a bruise that will exist for some time. This bruise can show characteristics from the pattern of a shoe's undersole if the person has been stamped on. A bruise may also show the shape of the instrument used to strike the victim. These marks can be photographed with a scale alongside them and used for comparison with the shoe or instrument thought to be responsible. Colour photographs are the best choice because it may not be possible to distinguish a bruise from blood smears in black and white photographs.

The degree of detail present in the bruise will depend on the force with which the blow was struck and the shape or pattern of the object used. A light impact causes some of the blood vessels at the point of impact to be broken and show a reddening of the area. Usually the mark on the skin appears as the negative of the object that struck it. This happens because the blood released from the damaged blood vessels is forced away from the contact point into the non contact areas. A light blow from a shoe that has a pattern of fine grooves on its undersole will cause each of the groove edges to produce a slight bruise along the edge (see Figure 4.6(a)). With a heavier impact (see Figure 4.6(b)), it is likely that these two bruises will be squeezed together because of the pressure involved, to form a single heavier line. This usually sits in the non-contact void between the two grooves and gives the image of a negative mark. When a very heavy impact occurs (see Figure 4.6(c)), so many blood vessels are broken that all the blood produced fills the contact area and leaves little detail, other than a large area of reddening.

A similar mechanism applies when an object is used to strike the body and produce a bruise. The visibility and information present in the mark can improve with time and a second visit to the mortuary to look at the bruising on a body, a few days after the initial post mortem, can prove very revealing.

4.7 PHYSICAL EVIDENCE

Another aspect of forensic science that can provide significant evidence is a physical fit. This occurs when it is possible to fit two pieces of a broken item back together like a jigsaw. For example, during a burglary

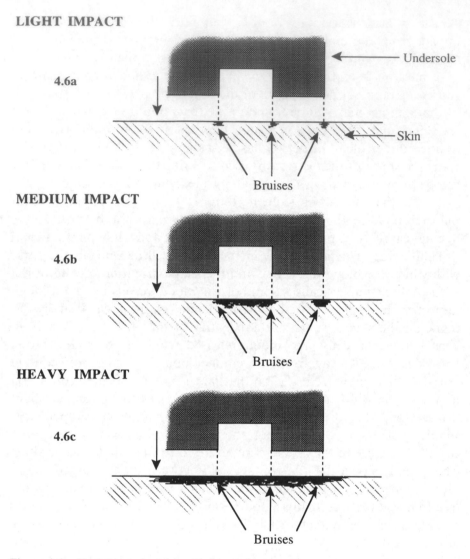

LIGHT IMPACT

4.6a

Undersole

Skin

Bruises

MEDIUM IMPACT

4.6b

Bruises

HEAVY IMPACT

4.6c

Bruises

Figure 4.6 *Variation in bruising with force of impact*

if a screwdriver breaks and one piece is left at the scene and the other part, usually the handle, is taken away it can be recovered from the accused (see Figure 4.3(b)). The break that occurs has an uneven surface and by showing that the detail of each piece can be matched together there will be no doubt that they were originally one complete item. Any attempt to fit items directly together should not be carried out until all other records and comparisons have been completed as this may damage the delicate surface at the break.

4.7.1 An Impressed Fit

Sometimes it is not the fact that the item has been broken that is important but that the two items have pressed against one another for a period of time. This may produce dirt or discolouring patterns that can be used for comparison purposes.

For example, a pickaxe handle was used to strike a man during an aggravated burglary, and was left by the offender as he fled from the scene. A short while later the police, acting on information they had received, searched the house and garage of a suspect. In the garage they found the head of a pickaxe and the question was asked, 'Did this at one time fit onto the handle left at the scene?'. Examination of the inside of the hole into which the handle fitted revealed forging and hammering marks caused as the item was hand-produced. The inside of the head was cast and seen to have detail present which was a negative image of that on the pickaxe handle. There was no doubt that the two items had at one time been fitted together.

4.7.2 Mass-produced Items

The products from almost all mechanical processes, even those in which items are mass-produced, can be examined to show a degree of evidential similarity between them. Cases as diverse as those involving the comparison of stolen metal waste with the stamping machine that produced it, to the comparison of stolen silver ingots and the corresponding mould used to cast them, have been undertaken in the laboratory. Even small items such as a pierced ear sleeper manufactured by a jeweller can provide telling evidence. One such item was found in the boot of a motor vehicle of a suspect smash-and-grab raider. The jeweller had made a shaped die to his own individual design and the shape and pattern of a sleeper it produced was compared with the one found in the vehicle. The fine detail agreed and therefore gave a conclusive link between the suspect's vehicle and the scene of the crime.

4.7.3 Plastic Bags and Film

The presence of manufacturing extrusion detail on plastic bags and metal pipe can also be used to link items together. With the increase in drug trafficking the need to be able to connect items that have been used to package drugs has become increasingly important. The packages are often made up from lengths of cling film or plastic bags and sealed with adhesive tape.

Recently, methods have been developed to compare some of the physical properties of cling film and plastic bags such as the die lines present on them when they are made. Cling film is produced by a method known as blown film extrusion. The process involves molten plastic, usually polyethylene or polyvinylchloride (PVC), and this is extruded through a circular die in the form of a continuous tube. The tube is then slit to form a single sheet. In the manufacture of plastic bags the extruded tube is heat sealed and perforated at regular intervals. The cling film or bags are then wound onto a cardboard former and cut into individual rolls.

During the course of a production run, solidified fragments of plastic build up around the edges of the die and cause faint lines or striations to be present in the finished product. These striations vary in position, number and intensity during a production run and will differ between production runs.

Comparison of the striation patterns may then be used to assess the likelihood that products have arisen from the same production batch when evidential links between items are being sought. These striation lines can be best seen by producing shadow graphs of them to give a photographic record of the results.

Heat seals that form the ends of plastic bags can be treated in a very similar way to other impressed detail. Any damage or faults in the seals of the seized packages can be compared with the bags taken from the suspected source in order to prove a connection.

4.7.4 Conclusions

The examples here are just a sample of the many different types of unusual comparisons that can be undertaken. It is worth emphasising again that any items that come into contact with one another can leave a mark behind. This almost insignificant occurrence can be used to form a unique connection between the items and this may be all that is required to solve the most serious of cases.

4.8 ERASED NUMBERS

The identification of stolen property and its recovery is an important aspect of the police investigation of crime. When personal identification by the owner is not possible it may be necessary to resort to the restoration of an object's obliterated serial number to prove its identity. This happens most frequently with stolen cars, when engine numbers are erased by the thieves and substituted with new ones to give them a new identity. On most items the manufacturer's serial number is stamped into the metal but now serial numbers can also be stamped into the plastic of such items as video recorders and mobile phones.

We shall discuss briefly the aims of the investigating scientist and current areas of interest. We will also look at how punches and stamps can be linked to the restored numbers.

4.8.1 The Erasure

When marks punched into metal have been erased their restoration can usually be achieved by etching the surface with acid. The strength of acid used depends on the chemical reactivity of the metal involved.

All metals are crystalline and when an indentation is punched into the metal's surface a large localised stress is produced that alters the crystal structure immediately around and below the mark. There is a plastically deformed area around the mark and an elastically deformed area surrounding that. Therefore, even if the surface has been removed by filing or grinding, in order to obliterate the mark, it is likely that the plastically deformed area will still exist. With the application of a suitable acid etching reagent, this area, because it is more electrochemically active, will be more readily attacked than the surrounding one.

With cast-iron engine blocks from motor vehicles heat can be used to restore the erased numbers. Again the heat acts preferentially on the area where these numbers had been stamped. An oxy-acetylene torch that is moved over the surface normally provides the heat. After the heating is completed and the surface has cooled, it can be rubbed down to reveal the erased numbers.

Other surfaces can also be treated in an effort to restore stamped serial numbers. The identification numbers in polymers are usually produced with a heat stamp that when applied causes the polymer to shrink. Experiments have revealed that the heated regions possess a higher swelling capacity under the action of solvents than the unheated areas. As more and more electrical items, such as mobile phones, are being stolen this method of number restoration will continue to be of great significance.

A similar method has allowed the restoration of numbers on wooden items where steam is used as the swelling agent.

4.8.2 Connecting Punches to Marks

We have already discussed the methods of connecting two items together by using their damage features. During use a punch will often become damaged or the chromium plating may deteriorate and flake off, producing areas that are unique. These unique damage features will be stamped into the object and may be used to connect the punch and object together. Comparison can be carried out in a similar way to that used to compare cutting and levering marks left by bolt cutters and jemmies.

NON-DAMAGE BASED EVIDENCE

4.9　FINGERPRINTS

The unique nature of fingerprints was noted around the turn of the century and first used in a trial in this country in 1902. However, long before that the Chinese were aware of the qualities of fingerprints. Sir Edward Henry is credited with producing the modern fingerprint classification system and setting up the first Fingerprint Bureau at Scotland Yard. Since the first fingerprints were used as a unique method of identification several million have been taken and catalogued and in that time no two prints taken from different fingers, thumbs and palms have been found to be the same.

The ridges on the hands, fingers and thumb and the patterns that they form are unique to each individual; even identical twins have different fingerprints. The patterns are present at birth and are formed in the early stages of pregnancy. They remain unchanged for life unless some deep damage occurs to the skin such as a burn or severe scaring. They are one of the last features to be lost from the skin as it decomposes and therefore can be used to identify dead bodies, months, even years after death.

Fingerprints can be left at the scene of crime when touching surfaces or handling objects. They are produced from a mixture of natural secretions that come from the various glands in the skin and these are then set down by the ridge detail on the surface of the skin. Other areas of the body like the soles of the feet and the palms of the hands also have ridge detail and therefore they can also leave behind a characteristic pattern. These patterns are also believed to be unique to an individual but occur much less often at a scene or on a recovered item. So the next time you kiss yourself on the bathroom mirror just look to see the lip ridge detail present!

Fingerprints will also be caused when the finger has been contaminated with another material such as blood and others can be produced when the finger has been pushed into a soft material like putty.

Methods for enhancing latent fingerprints will be discussed in more detail later.

4.9.1 Why Are They Unique?

The individuality of fingerprints is still the subject of some controversy, especially in some of the countries that do not require a minimum number of 'points' to show proof of a connection. A 'point' is a classifiable characteristic and we will find out more about these later. Some countries require only a few of these 'points' while the British system is based on the sixteen points of comparison rule which was adopted in 1953. It is said that if such a number is found then there is no likelihood of another finger having made that print.

Fingerprints can be split into three groups by their general pattern. These patterns are aches, loops and whorls. The words quite accurately explain the shapes you would be looking at. These large groups can be sub-divided further by smaller differences in this basic pattern. For arches, these are plain or tented, and small elliptical, twin, composites, lateral pockets or accidental for whorls. Again the terminology explains fairly clearly the different patterns seen.

In all patterns, other than arches, there are points where the ridges break or bifurcate known as deltas: loops have only one delta each and whorls have two or even more. Deltas give rise to further characteristics based on the number of ridges between a delta and the core of the pattern. By taking these properties of the fingerprint into account they can be classified, catalogued and held on a fingerprint card or database.

To be able to take a comparison further, a series of classifiable characteristics of 'points' are required. These are described as identifiable peculiarities that occur in certain positions in the print. There are four basic characteristic bifurcations: where a ridge divides, a ridge ending, a lake and an independent ridge. These are the points referred to earlier which, when taken in combination with each other, can provide the unique fingerprint every police officer wants to have recovered from their crime scene.

4.9.2 Current Developments

A new fingerprint system, to replace AFR (Automatic Fingerprint Recognition) has now been taken up by most of the police forces in England and Wales. The forces along with a number of others worldwide now subscribe to the computer based system called NAFIS (National Automated Fingerprint Intelligence System). The computer

LOOP RIGHT

Information for classification   Identification
                                               (ridge characteristics)

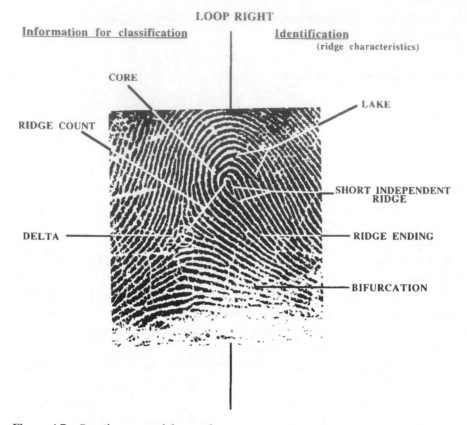

CORE

LAKE

RIDGE COUNT

SHORT INDEPENDENT
RIDGE

DELTA

RIDGE ENDING

BIFURCATION

Figure 4.7 *Details extracted from a fingerprint*

system allows details of fingerprints to be checked quickly and simply by and between the various fingerprint bureaux in the country. The computer uses a special algorithm to select a number of characteristics that can determine the closest matches between a suspect's fingerprint, held in the database, and a scene print. A fingerprint expert then has to decide which of these possible matches is a genuine one. Until systems like this and AFR were developed all of a bureau's fingerprint records were kept in a massive series of card indices. They may have been divided into smaller categories by age of offender or offence type but they had to be manually examined for each fingerprint comparison.

A fingerprint expert is likely to carry out a comparison using a photograph of the scene of crime mark and a photograph of the suspect's fingerprints viewed side by side on a piece of equipment that is a variation of the comparison microscope. A report of the result will be issued and if necessary the expert will be called to give their evidence in court.

4.9.3 Enhancement of Fingerprints

Like footwear impressions, latent fingerprints can also be enhanced to make them visible. In the early years they were often directly photographed or powdered with a variety of materials to make them visible and then photographed and presented as evidence. This meant that only a limited number of prints were available for comparison purposes. Now it is possible to use a wide range of chemical treatments or different types of light source in order to visualise a greater number of prints.

With fingerprints, unlike footwear impressions, it is possible to adopt a sequential approach to the enhancement of any print present. When a fingerprint is laid down it normally contains a number of different compounds that are excreted through pores in the skin; therefore, a search for one set of compounds such as lipids will not necessarily be to the detriment of other materials present such as amino acids. The proportion of the chemicals produced may vary from one person to the next.

4.9.3.1 Enhancement Techniques. Latent fingerprints left at a scene of crime or on an item that has been handled, such as a cheque, contain a mixture of natural secretions, which come from the various glands in the skin. Prints may also have been caused by contamination of the hand by another material or because the hand has been pressed into a soft material.

There are three main types of sweat gland in the human body; these are the eccine, sebaceous and apocrine glands. Each of these excretes different compounds along with a large amount of water. Most natural fingerprints contain a mixture of substances from the eccine and sebaceous glands. Some of the prints made from these materials will persist on the touched item for a considerable time; others may be transient or even decompose. Water is the first component to be lost from a fingerprint so there is little to be gained in trying to look for this in fingerprints that are a few days old, but fatty material is likely to be found for several days or possibly longer.

A number of circumstances can alter the quality of a print left behind during a crime. These can include the elapse of time, the nature of the surface on which the print was made along with factors such as storage of the item, its exposure to light and humidity after it was recovered

Some of the techniques employed by the scientist are specific for individual compounds while others detect any oily or fatty products. Since the chemical and physical nature of the print is not known when it is examined a systematic approach must be adopted for the full recovery of any prints present. A number of different complementary routes have now been determined that should facilitate the maximum recovery of the latent prints.

The following are just a few examples of the techniques that are currently available. Apart from the application of aluminium powder, which is frequently seen in use by the TV detective, the other commonly used technique is to treat the object with ninhydrin.

Ninhydrin is a compound that will react primarily with amino acids. It produces a purple coloured print, the development of which may be speeded up with the use of heat and humidity. This heat treatment can also cause further fingerprints that are present to become visible. Some prints, however, can require a longer period of time to appear. The fingerprints found are then individually photographed and recorded for comparison. The ninhydrin technique is very effective on paper and other porous surfaces but it is not successful on wetted items or silk finish painted surfaces. Ninhydrin can interfere, like most enhancements, with any further forensic examinations that are intended including those for indented impressions, body fluids, hair and other particulate material. The person carrying out the treatments must be aware of the implications for the detection of other scientific evidence that may be present and decide on the order of examination of the particular object in each case.

Fluorescence techniques such as exposure to laser or UV light can also prove very beneficial. The resulting print can be enhanced further by the application of a dye such as gentian or crystal violet. The dye will stain the fatty constituents of sebaceous sweat to produce an intense purple coloured print that can then be illuminated with laser light which will cause an increase in contrast (see Figures 4.2(a) and 4.2(b) in Section 4.2.6).

Gentian violet is a very useful dye even when used without special illumination. It can be effective on adhesive and pressure sensitive tapes as well as crockery and light coloured metals such as aluminium, particularly if there is any oily or greasy contamination of the metal surface.

The vapour given off by superglue, either ethyl or methyl cyanoacrylate, will produce a white deposit with some latent prints. The process is humidity sensitive and items to be treated in this way are often humidified as part of the treatment. Fluorescent dyes can be applied to an object after exposure to superglue has taken place to make visualisation easier as any print present is now effectively fixed in place.

Vacuum metal deposition can be used on smooth non-porous surfaces like polythene, leather, photographic negatives and prints. Thin films of gold and zinc are deposited on a fingerprint when only monolayers of fats are present. The use of this technique is limited as it requires extremely expensive equipment and there is also a limit to the size of the items that can be placed in the vacuum chamber. The treatment is effective because of the disturbance in the physical and chemical nature of the surface by the fingerprint. This is shown by different growth rates of the gold film on the surface of the object. The contrast is improved by exposing this gold film to vaporising zinc in the same chamber.

One surface from which most investigating police officers would love to be able to recover fingerprints is human skin. A number of methods have been reported in the literature as having potential but none appear to have any particular merit, so the search is still on to solve the problems encountered with this vital surface.

4.9.4 Future Developments

These sections on fingerprints and their enhancement can only touch on the value of prints to the forensic process. Fingerprints can now be recovered for examination from the most unlikely surfaces, increasing their value in linking a suspect to a crime. Since the development of these new techniques has led to a resurgence of work with fingerprints, a number of police forces are forming enhancement laboratories. The Forensic Science Service has a Specialist Fingerprint Unit that attempts to work with prints the police are unable to tackle and also those where more radical treatments, even with radioactive chemicals, can be of value. The research continues!

4.10 CONCLUSIONS

The comparison of any mark or impression, whether from an instrument or from the undersole of a shoe, is probably one of the most difficult and challenging aspects of forensic science in which to work. From the brief outline of the techniques and the examples given this may not seem to be the case. However, the results that are obtained from these examinations rely heavily on the knowledge and experience of the forensic scientist who carries out the work. In view of this, the Forensic Science Service, in England and Wales, has set up a quality management programme that covers most of the examinations it carries out including that of marks and impressions. It became the first such organisation to receive NAMAS/BSI accreditation for carrying out this work. Quality trials are undertaken by individual scientists to cover the range of marks and impressions they are likely to encounter. The scientist who produces the report will have undergone suitable training and have several years of experience in these types of comparisons before they are allowed to report their findings to a court.

I hope this chapter has given at least an insight into the type of work a marks and impression examiner can be asked to perform, both at the scene of a crime and also on items submitted to their laboratory. So, the next time you take the lid off that paint tin, just remember what you could be starting!

4.11 BIBLIOGRAPHY

R.O. Andahl, The Examination of Sawmarks, *J. Forensic Sci. Soc.*, 1978, **18**, 31–46.

W.J. Bodziac, *Footwear Impression Evidence*, Elsevier Science, New York, 1990.

J.F. Gower, *Fiction Ridge Skin*, Elsevier Science, New York, 1983.

P.S. Hamer, B. Gibbons and D.A. Castle, Physical Methods for Examining and Comparing Transparent Plastic Bags and Cling Films, *J. Forensic Sci. Soc.*, 1994, **34**, 61–68.

S.H. James & J.J. Nordby, *Forensic Science – An Introduction to Scientific and Investigative Techniques*, CRC Press, 2003.

H.C. Lee and R.E. Gaensslen, *Advances in Fingerprint Technology*, Elsevier Science, New York, 1991.

P. McDonald, *Tire Imprint Evidence*, Elsevier Science, New York, 1989.

P.D. Pugh and S.J. Butcher, A Study of the Marks made by Bolt Cutters, *J. Forensic Sci. Soc.*, 1975, **15**, 115–126.

L.W. Russell, C.J. Curry and D.A. Castle, A Survey of Case Openers, *Forensic Sci. International*, 1984, **24**, 285–294.

CHAPTER 5

Blood Pattern Analysis

ADRIAN EMES and CHRISTOPHER PRICE

5.1 INTRODUCTION

Bloodstain Pattern Analysis (BPA) is the term most commonly used to describe the examination, identification and interpretation of patterns of bloodstaining in relation to the actions that caused them. For many years BPA has played an important rôle at scenes of violent crime, where it forms an integral component of the reconstruction of events. DNA profiling now has extremely high discriminating power and this has had the effect of focusing the attention of police, scientists and the courts more frequently on questions concerning how bloodstaining was caused. A consequence of this is that BPA is now applied as often in the laboratory to the distribution of blood on items such as clothing and weapons as it is at scenes of crime.

The identification of bloodstain patterns is, by its nature, subjective, but it is underpinned by sound scientific principles. Each and every bloodstain pattern is unique, but by applying these principles, together with the application of experience based on experimental work, the identification and interpretation of patterns can be made sufficiently objective to be used with confidence in legal situations.

The following are examples of how BPA may assist the investigation at scenes where assaults are known to have occurred:

1. It may prove possible to establish the relative positions of assailants and victims and the possible sequence of events.
2. Disturbance to the scene subsequent to bloodshed is often apparent in the staining.
3. The presence of spattered, clotted blood gives a good indication of an interrupted or prolonged assault having occurred.

4. Recognising patterns of staining greatly increases the chances of locating blood from different sources and it is this understanding that frequently determines the samples that are taken for DNA profiling.
5. It may be possible to assess the degree to which the assailant would have become bloodstained during the assault.

There are, however, limitations. For example, it often proves impossible to determine the direction of a trail of dripped blood between two sites of attack from an examination of the resultant stains. Determining the number and sequence of blows to a victim is usually much better achieved by a medical examiner or pathologist than by use of BPA.

Looking at the overall usefulness of BPA to an investigation, at one end of the spectrum police may be confronted with, for example, just a bloodstained room and little or no information as to how the bloodstains got there. Analysis of the bloodstain patterns will usually be limited to addressing such issues as whether the staining is the result of an accident or a crime and, if the volume of blood can be estimated, whether the person who shed the blood is likely to be dead or alive. On the other hand, in situations where a defendant or suspect offers a version of the events that caused the bloodshed, there is the opportunity to test defined hypotheses that relate directly to the core concerns of the prosecution and of the defence.

This chapter gives a brief outline of the basics of BPA and illustrates how it may be used in the investigation of violent crime. For the reader who wishes to find more detail on the subject a number of textbooks have been published in recent years.

5.2 CLASSIFICATION OF BLOODSTAIN PATTERNS

Only by understanding the dynamics of the actions that cause bloodstain patterns can the scientist give reasoned answers to the questions asked by investigators and the courts. Over the years bloodstain pattern analysts have developed systems for classifying the various patterns of staining encountered, and the following is widely used amongst those involved in teaching BPA:

1. Single drops.
2. Impact spatter.
3. Cast-off.
4. Arterial damage stains.
5. Large volume stains.

6. Physiologically altered bloodstains.
7. Contact stains.
8. Composite stain patterns.

We shall now consider in turn these eight categories of bloodstain patterns.

5.2.1 Single Drops

This category concerns primarily the formation and properties of blood drops falling vertically under the effect of gravity. This includes the dripping of blood from wounds and weapons, trails of blood and the phenomenon of secondary spattering. Consideration is also given to the characteristics of stains formed by blood drops landing at an angle onto a surface. Many of the characteristics of falling drops of blood are applicable to understanding the behaviour of droplets of blood projected by forces other than gravity, such as by impact.

5.2.1.1 Drops of Blood. Blood can drip from any surface where the force of gravity can overcome the surface tension retaining the blood at the dropping site. In free fall, a drop of blood is spherical because this is its form of least energy. A single drop of falling blood will not break up in flight unless it is acted upon by another force.

The volume of a drop is determined predominantly by the shape and size of the dropping site. If there is a large surface area the surface tension holding the blood tends to be greater than if the surface area is small, and consequently the drops formed will also tend to be larger. If the available surface is extremely small (*e.g.* a single hair), then it can become impossible for gravity alone to overcome the surface tension, so that no drops can form.

Single drops of blood formed under the influence of gravity alone may vary in volume, from maybe as small as 15 microlitres (µl) up to 100 microlitres or more, depending on the nature of the dropping site and the amount of blood available.

5.2.1.2 Stains from Single Drops. When a drop of blood falls onto a smooth surface the resulting stain is circular, with no distortion at its edges. A rougher surface will result in an approximately circular stain with irregular edges. If the surface is sufficiently rough, the surface tension of the impacting drop will be overcome enough to allow the production of spines radiating from the stain, or secondary spatter in the form of discrete stains (sometimes referred to as satellite stains). Wet blood itself provides an example of a surface that promotes very pronounced secondary spatter when other blood drops impact on it and the effect of the underlying substrate is often obscured.

The size and shape of the stain produced by a drop of blood falling under the influence of gravity perpendicular onto a target surface will depend upon three variables:

1. The volume of the individual drop – the greater the volume the larger the stain for the same dropping height and the same target surface.
2. The dropping height – increasing height produces an increase in stain size until the terminal velocity is reached, which is approximately 7 m for a 50 µl drop.
3. The nature of the target surface – non-absorbent surfaces will lead to the formation of larger stains as the entire volume of the drop spreads over the surface, whereas absorbent surfaces will tend to produce smaller stains as a proportion of the blood is absorbed within the substrate.

In general terms, therefore, the size and shape of a stain can give no indication of the dropping height unless the other variable parameters of the drop's volume and the effect of the target surface are known.

5.2.1.3   Secondary Spatter.   Secondary spatter consists of small droplets of relatively uniform size, usually 1–2 mm in diameter. The drops are projected at low velocity and on non-absorbent surfaces their stains often retain a dense, globular appearance when dry. The distance travelled by the droplets depends on the volumes of blood involved, the dropping height and the nature of the target surface. Experiments have shown the limits of travel for spatter caused by the extreme case of blood into blood to be approximately 1 m horizontally and 50 cm vertically, with the density of spatter decreasing away from the origin of the spatter. Thus it is a common occurrence when an injured person drips blood onto the ground for secondary spatter to impact upon the footwear and lower clothing of a person nearby in the form of small blood stains.

If DNA tests show this to match the injured person then the presence of the secondary spatter is strong evidence to support the view that the two people were close together (within 1 m) when bleeding occurred. On the other hand, the secondary spatter itself says nothing significant about the events which caused the bleeding.

5.2.1.4   Case Example.   This case illustrates the importance of understanding the behaviour and significance of dripping blood. It concerned a murdered young boy, found lying on the floor of a garage having

Figure 5.1 *Blood that has dripped onto the ground, and secondary spatter staining the footwear of a nearby individual. The large stain on the ground is the result of several drops that have pooled together*
(The Forensic Science Service® ©Crown copyright 2003)

been repeatedly stabbed. The suspect had a deep cut to the palm of one hand which he claimed was received when he disarmed the boy who was attacking him with a knife. He also stated that he had not stabbed the victim and that the boy was still standing when he left the garage. Many bloodstains on and around the body were grouped, and this showed that while most of the blood was the victim's, some of the staining by the body and in a trail leading from the garage was the same group as the suspect. Of particular interest were stains on the upper back of the boy's shirt that were clearly the result of blood drops falling vertically under the effect of gravity. These matched the suspect and showed clearly that he had stood over the victim dripping his own blood.

In addition to the strong evidential value of the stain pattern, this case illustrates two important aspects of scene work and preservation of evidence. Firstly, the importance of regarding the body as an integral part of the scene, and secondly, the importance of removing the clothing before moving the body. If the shirt had not been removed before transporting the body to the mortuary, the vital pattern of staining would almost certainly have been obscured by additional blood leaking from the wounds.

5.2.1.5 Bloodstain Shapes from Drops at Angles Other than Perpendicular. The shape of a stain produced by a vertically falling drop of blood striking a target surface at an angle other than perpendicular will depend upon the factors outlined above, together with the following:

1. Landing at an angle, an impacting blood drop produces an elliptical, elongated stain with the tapered end of the stain pointing in the direction of travel. The tapered leading edge of the stain may show considerable distortion, whereas the back edge is more smoothly elliptical.
2. The angle of impact that the blood drop makes with the target surface will affect the shape of the stain. This angle is the internal angle between the plane of the target surface and the path of the impacting blood drop. The more acute the angle, the greater the elongation of the stain formed.
3. Drops landing at angles of less than 40° often produce secondary or satellite stains by a process termed wave cast-off. These secondary stains tend to be very elongated and can be at some considerable distance from the parent stain, though still in line with it.

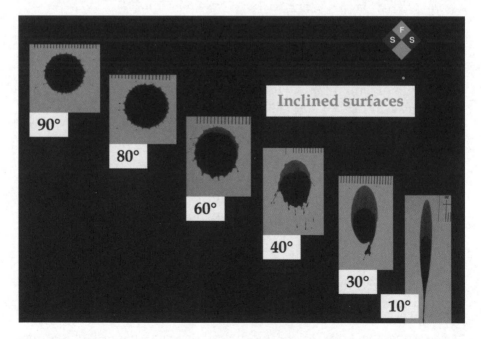

Figure 5.2 *Stains made by single drops of blood landing at different angles to the surface*
(The Forensic Science Service® ©Crown copyright 2003)

Study of the shape and measurement of the width-to-length ratio of a stain enables the angle of impact to be calculated. These relationships apply equally well to spattered droplets of blood striking a surface at an angle and are thus of great practical use in determining the area of origin of impact spatter (see next Section).

5.2.2 Impact Spatter

This is the most common type of pattern encountered in casework and can be caused by a wide range of actions, such as kicking, stamping, beating, punching and shooting. The cause of impact spatter is a force impacting directly into wet blood that breaks the liquid blood into small droplets of varying volume. The droplets are dispersed radially from the impact site along the paths of least resistance. They have various trajectories and velocities. Generally speaking, the greater the force applied the smaller the average size of the droplets formed. It is important to remember that the force referred to is that applied at the site of impact and this does not necessarily imply the application of violence or extreme energy in the overall causative action.

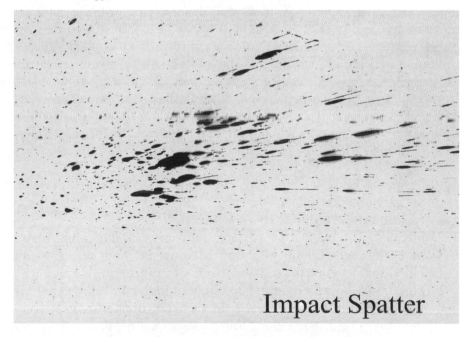

Figure 5.3 *An example of impact spatter on a horizontal surface, such as this page when laid flat. The spatter has originated from an area just above the left edge of the photograph. The individual droplets have radiated from the origin, and formed stains of various sizes randomly positioned within the pattern* (The Forensic Science Service® ©Crown copyright 2003)

5.2.2.1 Factors Influencing Impact Spatter. A number of variable factors will affect the distance, direction and quantity of blood spattered by an impact. These include the amount of wet blood at the impact site, the position of the impact site relative to the attacker, the shape and size of the weapon and the speed and angle of the weapon at the moment of impact.

Assuming the droplets have the same initial velocity and trajectory, smaller droplets will travel less distance than larger ones because of the effects of air resistance. There will be considerable variation in the sizes of the stains in the pattern formed on a target surface; most will be quite small, usually 3 mm or less, although occasionally stains as large as 10 mm may be found. As the distance from the impact site increases the density of the stains within the pattern decreases. The stains from a single impact will show directionality from a common area of origin, although the number and distribution of the stains within the pattern will be random. Estimates can be made of this area of origin, and the methods used to do this are described later.

In the 1970s, Professor Herbert MacDonell, working in the USA, classified impact spatter on the basis of the velocity of the impacting object. However, experience shows that in some circumstances this can be restrictive and possibly misleading. It can be demonstrated that the effects of the varied geometry that occur at the area of contact between a weapon and the bloodstained surface often have a greater effect on the range of drop size than does the velocity of the impacting object. Consequently many workers have adopted a classification proposed by Terry Laber which is based on the observed stain sizes within a pattern (not just impact) and does not link these to specific velocities or actions. The following forms the basis of this classification:

1. Large – a bloodstain pattern consisting of individual stains that are predominantly 6 mm or larger. Blood dripping from objects typically show stains of this size range.
2. Medium – a bloodstain pattern consisting of individual stains that are predominantly 2 mm to 6 mm in diameter. Cast-off bloodstaining is characteristic of this size range.
3. Fine – a bloodstain pattern consisting of individual stains that are predominantly 2 mm or smaller in diameter. Impact spatter resulting from a medium velocity impact such as a beating and spatter from a high velocity impact such as a gunshot are both capable of producing stains of this size.
4. Mist – a bloodstain pattern consisting of individual stains that are predominantly smaller than 0.1 mm in diameter. Due to the

pronounced action of air resistance on the minute drop size of this spatter, the droplets will travel only a very short distance from the area of origin. Spatter arising from a high velocity impact such as from gunshot is characteristic of this size range.

5.2.2.2 Information from Impact Spatter. Detailed examination of the patterns can provide significant information to the investigation. Recognising impact spatter allows the identification of sites of attack and the determination of the relative position of objects or people at the scene. The patterns may also offer information on the nature of the impact that caused the spatter. Most importantly, they can give an indication of the likelihood of bloodstaining being present on the assailant or others present at the scene, thereby helping to determine the significance of bloodstains found on a suspect's clothing.

Figure 5.4 shows the shirt of a man who bludgeoned his victim with a club hammer. This is an extreme example involving many blows with a heavy weapon and considerable spatter has been directed back towards the assailant. However, it is quite possible for the majority of spatter from a beating to be directed forwards or to the sides, away from the attacker. In situations where there have been a great many blows it is very likely that the assailant will be bloodstained to some

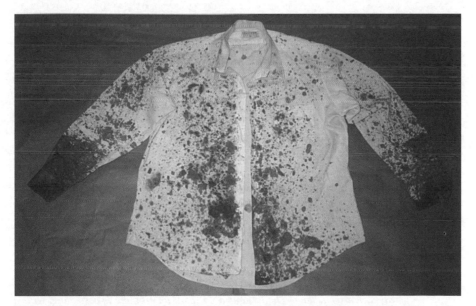

Figure 5.4 *The heavily spattered shirt of an assailant who bludgeoned his victim with a club hammer. Note the dense spatter on the cuffs, the parts of the shirt that would have been nearest the origin of the spatter*
(The Forensic Science Service® ©Crown copyright 2003)

extent, but with a single or just a few blows it may be that no blood at all is directed onto the assailant.

With kicking and stamping a wide range of patterns may be encountered, typically on footwear and trousers. Figure 5.5 shows a 'classic' kicking distribution on a shoe. There is evidence of forceful contact staining where the blood has been forced into the crevices around the seams and other recessed areas. This is associated with some degree of spattering. Stamping with the heel may well produce spatter that travels up inside the lower part of the trouser leg. This is a most significant finding, as is the presence of contact stains caused by hair.

With punching, the amount of blood on the assailant will depend on the nature of the injury and the amount of blood available to be spattered. Punching into wet blood will cause impact spatter that may be seen around the cuff and is particularly significant if it is on the inside. It may also commonly be found on the upper sleeve, across the chest area and other parts of the front. Note that in the example shown in Figure 5.6 the most significant evidence will be lost as soon as the hand is washed, leaving only a partial pattern on the cuff. It should also be remembered that a bleeding nose may produce a significant amount of blood but soon afterwards there may be no visible injury.

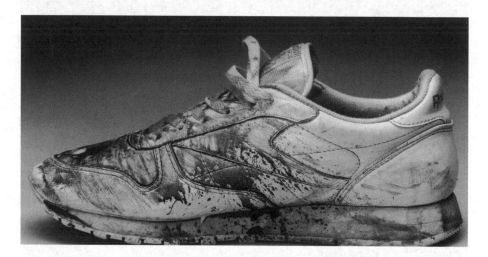

Figure 5.5 *Impact spatter on a shoe caused by kicking the head of a person who was already bleeding. The large area of contact staining is associated with a small amount of impact spatter on the side of the shoe. Most of the spatter produced at the moment of impact would have been directed away from the shoe*
(The Forensic Science Service® ©Crown copyright 2003)

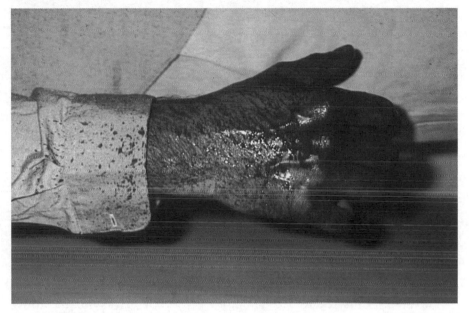

Figure 5.6 *Bloodstained fist and shirt cuff caused by punching into a heavily bloodstained surface*
(The Forensic Science Service® ©Crown copyright 2003)

With an assault by stabbing, a single stab to the body rarely causes blood to be transferred to the assailant, as most of the bleeding will be internal or absorbed by the victim's clothing. However, if there is subsequent contact with the victim or multiple stabbing occurs, there is likely to be a transfer of blood, often from impact spatter produced from the build up of blood on the surface of the skin.

When a person is shot in the head from close range the bloodstain patterns that may be produced are affected by a number of factors. These include the site of the injury, the type and calibre of the weapon and the distance between the muzzle and the skin.

A contact or near-contact injury from a shotgun will probably include extensive spattering of blood, tissue and bone fragments. Most of this will travel away from the victim in the general direction of the initial shot, away from the person firing the weapon. This is referred to as forward spatter and is usually dramatic in appearance. However, because it is directed away from the shooter, that person may receive little or no blood on their clothing.

A contact or near-contact injury to the head from a shotgun may also produce an aerosol of blood droplets that travel back towards the person firing the weapon, and this is called backspatter. The mist-like blood droplets comprising backspatter are unlikely to travel more than 1 m from the injury site and will not easily be seen on clothing.

5.2.2.3 Determining the Origin of Impact Spatter. Locating the position from which an impact spatter originated may be an important element in reconstructing a scene of crime. In many instances an experienced BPA examiner can estimate this position by eye. In some cases a more accurate measurement of the origin is required, for example to locate the position of a victim at the moment he was shot. For many years the technique known as 'stringing' has been used, with varying success. The principles of this technique are soundly based on trigonometry, but the method has practical difficulties, and cannot easily be checked by others.

The following illustrates the principles of 'stringing' for an impact spatter onto a horizontal surface. A straight line is drawn through the long axes of a number of selected stains, and where these lines intersect is termed the area of convergence, as shown in Figure 5.7.

The actual impact site will lie somewhere on a perpendicular above the area of convergence. To determine this, the width and length of the selected stains are measured. This allows the angle at which they landed on the surface to be calculated – the ratio of the width to the length closely approximates the sine of the angle of impact. Using a protractor and lengths of string, these angles are projected from the stains to the perpendicular to define the area of origin as shown in Figure 5.8.

This method assumes that the trajectories of the blood droplets are straight lines. In reality all the droplets are affected by gravity and their trajectories are curved downwards. Therefore the strings in this method will project back to an area higher than the actual impact site. It is a cumbersome technique and is no longer in routine use in the UK. However, the principles of stringing form the basis for the method of determining the area of origin that is now recommended.

A Canadian physicist, Professor Fred Carter, became interested in BPA in the 1980s and in collaboration with The Royal Canadian Mounted Police, developed computer programs called Backtrack™ and Images™ to determine the area of origin. These programs have been modified and improved over the years and Professor Carter continues to collaborate with scientists, including staff at the FSS, to develop systems that are quick and easy to use at scenes of crime. A number of selected stains are photographed using a digital camera, and the Images™ program calculates the angle at which the stains landed on the surface. These data, together with data on the position of the stains within the pattern is processed by Backtrack™ to produce a graphical

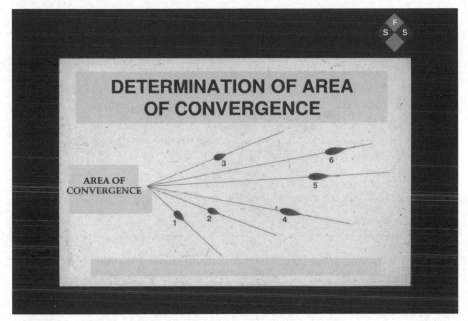

Figure 5.7 *Determining the area of convergence of an impact spatter – top view showing the convergence of lines through the long axes of selected stains* (The Forensic Science Service® ©Crown copyright 2003)

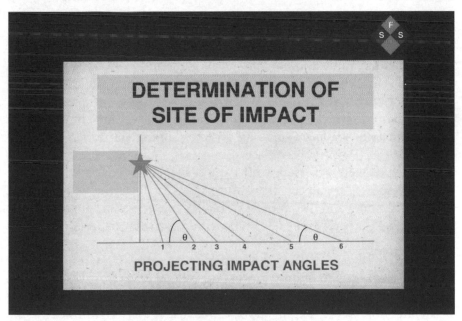

Figure 5.8 *Determining the area of origin (or site of impact) of an impact spatter – side view showing strings from the selected stains set at calculated angles to a perpendicular marker above the Area of Convergence* (The Forensic Science Service® ©Crown copyright 2003)

representation of the area of origin. In the very near future these programs will be in regular use in casework in the UK.

The ease of use of these programs means that the analysis of the data can be carried out quickly and, if necessary, remotely from the scene. Although these methods utilise a more sophisticated understanding of the physics involved than in the old 'stringing' method, it must be remembered that at the present time it is impossible for any method to calculate the exact trajectory of a blood droplet in the vertical plane. The methods do, however, work at a level of accuracy sufficient for practical purposes and allow the confident estimation of areas from which blood could and could not have originated.

5.2.3   Cast-Off

The term cast-off is applied to blood thrown from the surface of a moving object, either by the action of centrifugal force (swing cast-off) or by the object being brought to an abrupt halt (cessation cast-off). The patterns created can be varied and complex.

5.2.3.1   Swing Cast-Off.   This is most often associated with the swinging of a weapon, although blood on the hands is also a common source of cast-off. Centrifugal force will cause wet blood on the surface of the weapon to run towards its far end, allowing it to pool at one or more sites, dependent on the shape of the object under consideration. When the momentum applied to the mass of blood is sufficient to overcome its surface tension, the blood will be cast from the surface, rapidly forming a series of spherical droplets. When these land on a surface they will form a pattern which is generally linear in shape and comprised of stains which are similar in size – an 'in-line staining' pattern as shown in Figure 5.9.

The size of the droplets cast off from a swinging weapon will depend on the shape of the object, the nature of its surface, the velocity with which it is swung and the amount of blood that is present. Observations from many experiments have identified certain aspects that are relatively constant.

Firstly, it is true to say that pronounced cast-off staining at any distance from the site of attack, for example on ceilings, is usually only seen with long, relatively light weapons. Short, heavy weapons tend to be swung more slowly and in shorter arcs, using the weight to cause the damage, and are less likely to produce cast-off.

Secondly, as more force can usually be applied on the forward swing of a weapon than on its back swing, it is generally easier to generate

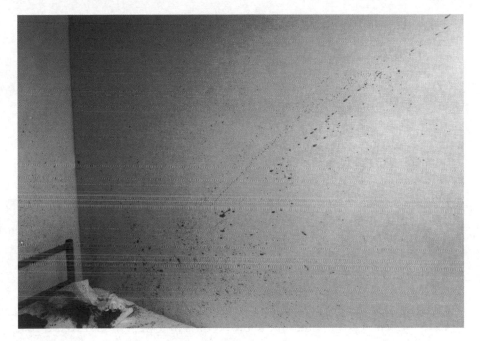

Figure 5.9 *Several lines of swing cast-off can be seen on the wall, cast-off from the swinging pickaxe handle used to attack the victim on the bed. Most of the staining on the lower part of the wall near the bed is impact spatter* (The Forensic Science Service® ©Crown copyright 2003)

cast-off when a weapon is swung in the forward direction. Also, as more force is being applied, the average drop size from the forward swing tends to be smaller than from the back swing. In practice one sees more cast-off due to back rather than forward swing and this is simply because there is more blood available. After blood is cast-off on the back swing little or none is left to be cast-off on the subsequent forward swing. Particularly large forces are generated when a weapon rapidly decelerates and accelerates, as at the end of its backswing, and this often leads to pronounced cast-off, sometimes of particularly large drops that can travel quite considerable distances from the site of attack.

The physics of the cast-off process is complex and the trajectory of the drops leaving a swinging object is determined by competing vector forces. Thus, depending on the velocity of the blow and the amount of blood present, drops can leave the surface at any angle between radial and tangential to the arc of the swing. Generally speaking, all of the cast-off blood is projected away from the arc of the weapon's travel, so it is not surprising that in most instances no blood lands on the wielder, who is inside the arc. A person kneeling to attack may be an exception, as the lower leg is then often positioned outside of the arc and thus can become stained. It is also possible for cast-off blood to be found on the back of the attacker.

Considering that there are so many variables affecting cast-off, it is not surprising that much variation is seen in the resulting patterns. 'In-line staining' is the only truly distinctive pattern resulting from swing cast-off and, in practice, a great deal of cast-off blood at scenes and on clothing does not form patterns that allow the mechanism of origin to be recognised. Even when dealing with 'in-line staining' patterns, the presence of so many variables should make for caution when interpreting the causative events.

5.2.3.2 Cessation Cast-Off. The term cessation cast-off is used for those commonly encountered situations where the swing of a weapon or other object is abruptly arrested, causing rapid deceleration and a large amount of cast-off. In this situation the arresting force does not act directly on the blood pool, but is transmitted to it, for example through the length of a weapon's handle. Although, by definition, this is a form of cast-off, the force acting on the blood acts in a way similar to a direct impacting force and the resulting stain patterns often show more characteristics of impact spatter than of cast-off.

The rapid deceleration of the weapon will often produce a character-istic pattern in the blood remaining on its surface. During the swing phase blood will tend to form into runs towards the distal end as centrifugal force acts upon it. The rapid cessation of the swing then tends to produce 'feathering' along the leading edge of these runs, and fine spines are projected forwards. This characteristic pattern is termed percussive staining and an example is illustrated in Figure 5.10.

5.2.4   Arterial Damage Stains

This category is unique in that the causative factor is internal rather than external, *i.e.* the pressure within the circulatory system forcing blood to exit from a damaged artery. A wide range of patterns can result, depending on the extent and site of the injury, the direction of spurting, and whether the victim was stationary or moving.

5.2.4.1 General Features. Most arteries in the body are generally afforded good protection against damage during normal day-to-day activities. However, in violent assaults and other incidents they may become damaged by sharp edged instruments or, where the vessels are near the surface, by blunt instrument trauma. Arteries commonly damaged in assaults include the following:

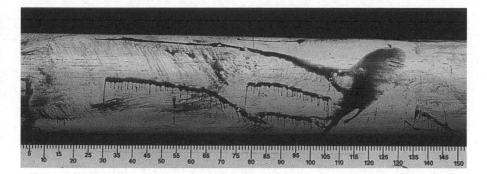

Figure 5.10 *Percussive staining on part of a white pole that had been used as a weapon. A number of larger stains were deposited on the pole, one of which is shown. The pole had then been swung causing the stains to run towards the distal end (from right to left). The downward feathering was produced when the pole hit its target and stopped suddenly and the blood continued to move in the direction of the swing*
(The Forensic Science Service® ©Crown copyright 2003)

1. Temporal – both sides of the forehead.
2. Carotid – both sides of the neck and commonly used for checking the pulse.
3. Radial – near the surface of the inner aspect of the wrist, near the base of the thumb.
4. Aorta – the main vessel leading from the heart, deep in the chest cavity.
5. Femoral – in the thigh running from the groin to the knee.
6. Brachial – on the inner aspect of the upper arm.

When an artery is damaged the blood is emitted in a column which soon breaks up into individual droplets of approximately equal size. Consequently when the droplets land on a surface they form stains which are also equally sized and parallel to each other. The droplets are readily affected by gravity and when they land on vertical surfaces they give patterns showing a markedly downward direction. Individual droplets frequently coalesce on the surface, causing pronounced runs of blood. The blood leaves the artery in a series of spurts that correspond to the beating of the heart and if the injured person is moving this can result in a 'V' or 'W' shaped pattern. An arterial gush results from a stream of blood hitting a surface, producing a large stain, maybe several centimetres in diameter. This may show secondary spattering, and is usually associated with a lack of movement by the victim. Arterial rain is the term used to describe blood that has spurted into the air and fallen to the ground under gravity.

Other factors that affect the range of patterns formed include the extent of the damage – small wounds produce small droplets, larger wounds produce larger ones. Small wounds also allow blood to be projected further than larger wounds. The appearance of the staining will be affected by the site of the injury and particularly by whether or not it is covered by clothing. Variation of the angle at which the arterial blood strikes the target surface will also significantly alter the appearance of the staining.

5.2.4.2   Simulated Arterial Bleeding.   In order to carry out realistic experiments and demonstrations of arterial bleeding an arterial pump that mimics the process was first constructed by Anita Wonder in Sacramento, California. This was further developed by us in London and has been used successfully in many training courses for forensic scientists. The operation of this pump is shown in Figure 5.11

Blood is drawn from a reservoir and circulated through silicone tubing by a peristaltic pump. The rhythmic action of the heart is simulated using a solenoid-driven arm that squeezes and releases the tubing, giving two pressures approximating a pulsing heart. The blood is pumped through a manifold and directed along one of several lengths

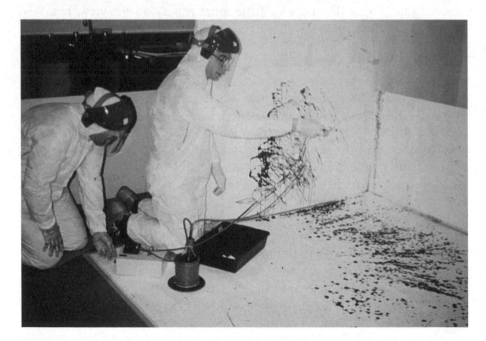

Figure 5.11   *The arterial pump being demonstrated by the authors during a training course*
(The Forensic Science Service® ©Crown copyright 2003)

of tubing, each of which has a pre-constructed hole or slit to mimic a wound of a particular size. As the blood spurts, the tubing can be directed at surfaces at varying distances and angles to demonstrate the characteristics and range of patterns produced by arterial bleeding. A typical pattern is shown in Figure 5.12.

5.2.4.3 Case Examples. Damage to a carotid artery will result in blood spurting from the wound and although there will be dramatic and probably fatal blood loss, this is highly directional so that it does not always follow that the perpetrator will become heavily bloodstained. In a case in London some years ago a victim, urinating in a public toilet, was attacked from behind and slashed across the throat with a craft knife, severing the carotid artery. There was extensive bloodstaining on the walls and floor at the scene and the victim died as a result of the injury. When arrested shortly afterwards the suspect was found to have numerous small spots of blood on his shoes that were thought to have resulted from secondary spatter from the blood spurting onto the floor. The remainder of his clothing was virtually unstained.

Figure 5.12 *A pattern caused by simulated arterial bleeding. The blood was spurting perpendicular to the surface and the 'artery' was moving from left to right resulting in five or six spurts indicated by the upper areas where the stains have run together*
(The Forensic Science Service® ©Crown copyright 2003)

Assaults by beating that cause massive head injuries may well result in bleeding from a temporal artery. Attempted suicide by slashing the ulnar or radial artery in the wrist is another cause of arterial bleeding. Some incidents involving arterial bleeding are particularly well documented, such as the death in London of Damilola Taylor, who was stabbed in the thigh just above the knee causing bleeding from the femoral artery.

An unusual situation occurred in London when on a busy main road two vehicles collided head on. The driver of one survived but the other was dead at the wheel and appeared to have sustained a major wound causing extensive blood loss. During the subsequent post-mortem examination he was found to have a penetrating wound to the arm, but there was no obvious cause of this wound from anything within the vehicle. The dead man's vehicle was black and blood was not obviously visible, but a detailed examination revealed that bloodstains caused by arterial spurting were present on the outside. This staining had arisen before the accident, and it was concluded that he had been the victim of a stabbing before escaping in his car. While driving down the main road he became unconscious and lost control of his vehicle.

5.2.5   Large Volume Stains

Arterial damage can cause large volume stains, but also included here are those stains caused by prolonged bleeding or by the sudden release of a large quantity of blood, typically from the mouth or directly from a wound. Stains in this category can also arise *post-mortem* when the accumulation of gases within a putrefying body eventually causes the emission of blood and other body fluids from wounds or orifices.

Sometimes, scenes with large volumes of blood are found with no victim present and little or no accompanying information. It may then be necessary to determine the volume of blood present in order to give an indication of whether or not the missing victim is alive or dead. Often there will be both dry and wet blood still present and several different surfaces of varying porosity will be involved. Consequently, in practice, calculation of the original blood volume by area and depth measurement or by weighing samples dried to constant weight, at best produce very rough estimated values.

5.2.6   Physiologically Altered Bloodstains (PABS)

This term is used to cover distinctive patterns that may arise when blood undergoes a physiological or biochemical change before or shortly after it is shed. Examples of this include clotting and admixture with other body fluids such as saliva or with vomit or excreta. The category has also come to include alterations to bloodstain patterns subsequent to spatter, such as that caused when blood lands on a wet surface or is disturbed by wiping, washing or other masking actions.

5.2.6.1 Clotting. The process of clotting involves a complex series of chemical reactions that occur in shed blood. This process occurs more quickly if the injury is accompanied by damage to the surrounding tissues. The actual process begins about 15 s after the injury is caused, but at this stage there will be no visible change to the blood. At some time, which maybe several minutes later, the blood thickens and the products of clotting are deposited around the injury. In extreme situations clot retraction will occur and a clear separation of clot and serum becomes apparent. Clotted blood has a distinctive appearance when it is spattered, and thus indicates that some time has elapsed since the original injury was caused. This time interval usually cannot be determined with certainty in any particular case as so many unknown factors will have affected the timing of clot formation.

When an impacting force acts upon clotting blood, the clot and serum components may be apportioned differently in the resultant droplets, producing some irregularly shaped stains with clumps of clot and other pale coloured stains of a watery appearance. These latter stains are easily confused with stains of diluted blood.

The detection of stains arising from clotted blood can be extremely important when reconstructing events. Care must be taken to distinguish clotting phenomena from those resulting from the chromatography of blood *i.e.* separation of the blood components seen on some porous surfaces, and those resulting from the presence in the stains of pieces of body tissue.

Over the years, situations have been encountered where blood has become admixed with all manner of other body fluids and subsequently been spattered to form patterns. In practice, however, only a few of these are regularly encountered.

5.2.6.2 Expired Blood. Blood will become mixed with saliva when there is a wound to the mouth or throat. The resulting stains have a diluted appearance (and the presence of saliva may be shown by other tests). There can also be limited frothing, leaving traces visible in the resulting stains. Penetrating chest injuries allow blood to become mixed with air which produces marked frothing. With injury to the lungs, surfactant from within the lungs may also become mixed with the blood and air mixture and this can cause pronounced frothing and bubbling visible in the stains. There may also be mucous present, producing a distinctive bead-like appearance to the stains. The act of coughing, sneezing or snorting when blood is present in the mouth or the airways will often produce a discrete pattern that may well resemble impact spatter. An example of expirated blood caused by coughing is illustrated in Figure 5.13.

5.2.6.3 Flyspeck.   Another type of biological action that can produce stains of distinctive appearance is that caused by flies, which are frequently found associated with a corpse a few days after death. The fly feeds on the blood and tissues of the corpse, regurgitating some of this from its proboscis, leaving small round stains. These stains are called flyspeck and may have a small tail like a comma or they may have a diluted appearance. Occasionally they may bear the impression of the end of the fly's proboscis. These stains may be mistaken for

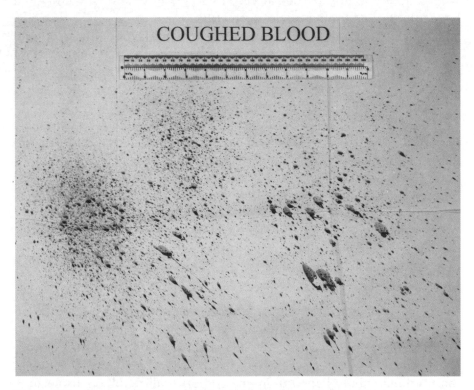

Figure 5.13   *An example of coughed blood. Detailed examination of the individual stains shows the presence of bubble rings, some degree of dilution with saliva and some irregular stains*
(The Forensic Science Service® °Crown copyright 2003)

blood spatter, although they are frequently found in warm places that do not appear to fit with the rest of the scene, such as around a light fitting or on a window frame.

5.2.6.4   Vomited Blood and Products of Decay.   Injury to the abdomen or ingestion of blood may well lead to vomit being mixed with blood. The acid environment in the stomach causes the red blood cells to form small clumps so that blood in vomit is characterised by its brown granular appearance resembling coffee grounds. The biochemical products of decay may also lead to stains with a distinctive appearance. Even though they contain blood, on close inspection they can usually be distinguished from normal blood by their dirty brown colouration.

5.2.6.5   Non-Physiological Actions.   It is not uncommon for someone to mix or apply a substance to bloodstains with the intention of disguising the presence of blood. Cleaning fluids of all kinds have been used in attempts to remove bloodstains from surfaces, and occasionally, in desperation, a covering coat of paint may also be applied. Washing or wiping blood may disguise its presence, but in nearly all instances will leave traces detectable by the use of specialised light sources and chemical reagents.

When blood is spattered onto a wet surface it will diffuse and form stains with an indistinct perimeter. This is in contrast to where the bloodstain has been allowed to dry, even momentarily, and then partially removed by wiping. A distinct halo or ghost stain will result, consisting only of the blood at the periphery.

5.2.7   Contact Stains

This category includes stains whose appearance is due to direct contact between an object wet with blood and another surface. During the examination of a bloody scene or of bloodstained clothing it is probable that any number of stains will be found that have no clearly defined shape and nothing of significance can be said about how these stains were caused. Contact stains are of particular interest if there is information about the bloody object that left the pattern, as, for example, with a fingerprint, shoemark, fabric mark or weapon impression. An example of a contact bloodstain is illustrated in Figure 5.14.

Finger and footwear marks are types of evidence that tend to be included as part of their own forensic disciplines rather than a BPA matter. Suffice it to say that a great deal of technological expertise is nowadays available to detect, enhance and retrieve such evidence and,

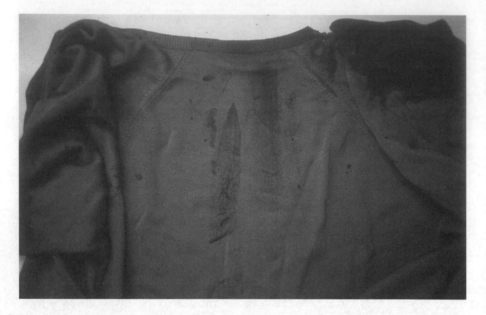

Figure 5.14 *A contact stain made by the murder weapon being wiped on the back of the victim's jumper. A number of knives were recovered during the murder enquiry and test wipes were made, but none was found that matched* (The Forensic Science Service® ©Crown copyright 2003)

in serious crime where the effort is justified, this is carefully integrated within complex strategies that aim to maximise the marks, DNA, and other forms of evidence retrieved.

5.2.8 Composite Stain Patterns

This category includes both superimposed stain patterns on a target surface caused by different actions, together with patterns caused by a complex action. An example of the latter is a person with arterial damage to their wrist who waves their arm around – arterial bleeding and cast-off would then be combined in a single simultaneous action and the resultant stain patterns may have features of both.

Composite patterns can be difficult to interpret, even for the most experienced analysts. Anita Wonder has published the results of her systematic study of many hundreds of bloodstain patterns caused by impact, cast-off and arterial damage, and has produced a list of objective criteria for the identification of the main pattern types. These criteria take account of the shape of the overall pattern and the size, alignment, and density of stains within it. This provides a useful key to refer to when faced with any pattern whose mechanism of formation is not readily apparent.

5.3   THE EVALUATION OF BLOODSTAIN PATTERN EVIDENCE

Police, forensic scientists and the courts have always been interested in the evidence provided by bloodstains. The ready transfer in violent crimes of blood from one person to another or from a person to a scene has, over the years, been a major stimulus for the development of techniques to individualise bloodstains. During the 1950s and 1960s this was limited to antigenic blood groups such as the well-known ABO system. The late 1960s and the 1970s saw the development of many biochemical blood grouping methods and in the late 1980s DNA came to the fore. Developments in DNA technology have since been rapid and spectacular and details of these techniques are documented elsewhere in this book. These have given us the ability to discriminate between individuals to the extent that if a DNA profile from a bloodstain is found to match a particular individual, the chance of a random match is quoted as being of the order of one in a billion (a thousand million).

This strength of evidence for the transfer of blood from, say, a victim to a suspect has meant that increasingly suspects and defendants are obliged to explain how the blood came to be on their clothing. For example, where an individual is suspected of assault by kicking it is not uncommon to find the victim's blood on their shoes. Faced with this evidence the suspect may offer an explanation, such as admitting to being present, but providing first-aid to the victim. How do the courts assess the validity of this explanation when the prosecution says that the blood got on to the defendant's shoes because it was he who did the kicking? In this hypothetical case it is no longer necessary just to show that the blood on the defendant's shoes matched the victim. The forensic scientist needs to evaluate the appearance and distribution of this staining.

Before carrying out this evaluation the relevant items need to be examined to locate any blood present and stains selected for DNA analysis. It is extremely unwise for forensic scientists to discuss the significance of bloodstains without first establishing that they are indeed blood and from whom that blood may have originated.

In recent years the question of how forensic science evidence should be evaluated has been considered at length and in the Forensic Science Service an Interpretation Group has been at the forefront of this work. The three fundamental principles of the evaluation process, which can be applied to all areas of forensic science, are as follows:

1. Evidence is evaluated in the light of other relevant information in the case, often referred to as the framework of circumstances.
2. To evaluate evidence it is necessary to consider at least two propositions which are mutually exclusive – these are usually the prosecution and defence propositions.
3. It is necessary to consider questions such as, 'what is the probability of the evidence if the prosecution proposition is true, compared with the probability of the evidence if the defence proposition is true?' The ratio of these probabilities is known as the Likelihood Ratio.

To apply this approach to the evaluation of bloodstain patterns it is necessary to have an expectation of what might be found in given circumstances. Returning to the case of alleged kicking, let us assume that blood matching the victim has been found on the defendant's shoes. The prosecution proposition is that the defendant kicked the victim and the defence proposition is that the defendant did not kick the victim but offered first-aid. What would we expect to find if the prosecution is correct, and similarly, if the defence is correct?

During many training courses we have carried out experiments whereby people wearing different types of footwear have simulated violent kicking attacks by kicking a bloody surface. Various characteristic features of the bloodstaining are regularly seen, indicative of the forceful contact. These include the presence of contact staining in the crevices of the shoe and directional, spattered blood associated with the contact stains. Therefore, if the suspect has this type of staining on his shoes, the evidence will support the prosecution.

But what might be expected if, as he says, he was just giving first-aid. Other experiments show that such incidental contact will produce contact bloodstains, but not of the type indicating that any force was used and with no associated spatter. So this pattern of staining on the shoes will support the defence proposition.

This is a simplified example of what is known as a Bayesian approach to evaluating evidence and for BPA it is still in its early days of use. For this to become more effective databases are required showing what bloodstain patterns might be expected in different sets of circumstances and giving a reliable estimate of the frequency with which characteristic features are likely to be present in these different scenarios. There are practical difficulties in establishing such a database and the work done to date has, perhaps, only established the experimental protocols that will need to be followed to make the data collected fully usable.

The ability of forensic scientists to assess and interpret evidence has become central to the development of many areas of forensic science, not just that of DNA where frequency data is readily available. BPA is proving to be a discipline where these Bayesian principles can be applied effectively and we expect this approach to be central to the development of BPA in future years.

5.4 BIBLIOGRAPHY

T. Bevel and R.M. Gardner, *Bloodstain Pattern Analysis With an Introduction to Crime Scene Reconstruction*, 2nd edn, CRC Press LLC, 2002.

A.L. Carter, The Directional Analysis of Bloodstain Patterns Theory and Experimental Validation, *Can. Soc. Forens. Sci. J.*, 2001, **34**(4), 173–189.

A. Emes, Expirated Blood – A Review, *Can. Soc. Forens. Sci. J.*, 2001, **34**(4), 197–203.

S.II. James (ed), *Scientific and Legal Applications of Bloodstain Pattern Interpretation*, CRC Press LLC, 1999.

T.L. Laber, Bloodspatter Classification, *IABPA News*, 1985 **2**(4).

H.L. MacDonell, *Bloodstain Patterns*, Laboratory of Forensic Science, Corning, NY, 1993.

A.Y. Wonder, *Blood Dynamics*, Academic Press, San Diego, California, 2001.

CHAPTER 6

The Forensic Examination of Documents

AUDREY GILES

6.1 INTRODUCTION

The identification of forgeries is one of the oldest of the forensic sciences with references made to it in Roman Law in the 3rd Century AD, and forgery being a statutory offence in Britain in the 13th Century. Notwithstanding this, the examination of documents in modern forensic science laboratories is an up-to-date science exploiting modern technology to cope with the ever increasingly sophisticated documentation of modern living.

The forensic scientist who specialises in the examination of documents will have a number of areas of expertise. These areas include the identification of handwriting and signatures, a knowledge of typewriting and modern office printers, the composition of inks, papers and the materials from which documents are produced and techniques such as electrostatic detection of impressions (ESDA) and infrared imaging techniques which allow the origin and history of documents to be studied.

Forensic document examiners are frequently misnamed 'handwriting experts' even though this is only part of their expertise. In recent years this term has been highjacked by graphologists. However, graphology has no relevance in forensic document examination. Graphology is a pseudo-science purporting to determine personality from handwriting. Its methods of interpreting handwriting are very different from the objective approach made by the forensic scientist and are demonstrably unreliable.

6.1.1 Qualifications and Training

There is no doubt that it is extremely difficult to become a fully trained forensic document examiner. The majority of forensic scientists specialising in document examination in the United Kingdom are trained and employed in the government laboratories. Graduate scientists join these laboratories and train for up to two years alongside highly experienced document examiners before handling their own casework. There are a handful of properly equipped private sector laboratories run by scientists who have completed training in the government laboratories.

There are as yet few formal qualifications upon which the legal profession or the public can rely when wishing to employ a forensic document examiner in the private sector. The Forensic Science Society offers a diploma in the field but problems in setting standards have not gained it recognition amongst the majority of practitioners. However, forensic document examiners can be registered with the Council for the Registration of Forensic Practitioners (CRFP) where peer review of current casework ensures that the registered practitioner is competent. In addition to registration with the CRFP a number of laboratories in both the private and public sectors have looked to external accreditation with standards such as NAMAS and the ISO 9000 series which have successfully provided benchmarks for assessing the quality of work in these laboratories.

It is still very much up to the individual lawyer or member of the public to take the trouble to ascertain that the expert is competent and credible. This is a very difficult job for the non-scientist and it is not assisted by the various so-called 'Registers of Experts'. Inclusion in such registers generally requires payment of a fee and a reference from a single satisfied customer.

The properly trained and qualified forensic document examiner will however be able to demonstrate a solid scientific background at appropriate degree level, a period of training in an established forensic science laboratory, possession of a fully equipped laboratory, a record of continued active participation in research and development in the field and will have CRFP registration.

6.1.2 Equipment

The majority of techniques employed in forensic document examination have been developed to extract as much information as possible from the document without damaging or altering it in any way. This is usually a prerequisite of any proposed examination since a disputed document is of considerable value – indeed its very existence may be evidence itself.

The forensic document examination laboratory will have the following basic equipment:

1. Good lighting sources including daylight which is essential.
2. Low power stereo-microscopy allowing magnification of ×5 to ×40.
3. Infrared, UV, high intensity and transmitted light sources for studying inks, alterations and latent marks.
4. Accurate measuring grids and graticules.
5. Electrostatic Detection Apparatus (ESDA) or equivalent for the study of impressions.
6. Oblique light source.
7. Methods of recording visual results by either photographic, thermal imaging or computer image capture techniques, as well as facilities for preparing demonstration materials for presentation in court.

More sophisticated laboratories will have available advanced techniques for ink analysis and electronic imaging, desk-top publishing facilities for producing illustrated reports, as well as databases of background literature, handwriting, typewriting, Transmitting Terminal Identifiers of facsimile machines and ink data. Forensic document examiners working within a general forensic science laboratory will also have access to yet further sophisticated equipment such as β-radiography, high performance liquid chromatography systems, lasers *etc.*, which are occasionally used in the examination of documents.

6.2   EXAMINATIONS

Forensic document examiners are asked to carry out examinations to provide information in a number of areas:

1. The identification of individuals from their handwriting.
2. Identification of signatures as genuine or forgeries.
3. The determination of the origin and history of documents – where and how they were produced and what has happened to them in the course of their existence.
4. Dating of documents.
5. The identification and interpretation of alterations, deletions and additions to documents.
6. The identification of counterfeit documents.

It may well be that the forensic document examiner will be asked to study a particular aspect of a document such as a disputed signature. This will not deter the experienced examiner from checking that other aspects of the document under examination are consistent. A disputed signature on the final page of a multi-page document may be identified as genuine. However, if there are unobtrusive differences in paper and typestyle of a signature page and previous pages, this may give a considerably different view of the origin of the signature.

The full and proper examination of documents therefore takes place in a properly equipped forensic document laboratory. There is a tendency for lawyers to misunderstand the level of sophistication available today in such laboratories. There is a commonly held belief that the 'handwriting expert' can merely look at the document in a lawyer's office or in the public area outside court and, possibly with the aid of a magnifying glass, give a full and positive opinion regarding the authenticity of the document. Examinations carried out without the benefit of proper laboratory facilities will always be to some extent inconclusive and may be entirely misleading. There is no substitute for proper examination in a laboratory.

The second most popularly held erroneous belief is that adequate examinations can be made using photocopies. Photocopies do not show all the details of the original documents and in most cases the conclusions which can be drawn from the examination of such material will nearly always be restricted.

6.3 THE IDENTIFICATION OF HANDWRITING

The identification of handwriting is one of the few forensic sciences which actually identifies the individual. With the exception of finger-printing and DNA profiling, the majority of forensic science sets out to establish links between individuals and places or objects.

It is an established fact that handwriting can be recognised. Most adults can recognise the handwriting of their immediate family and close friends. Every character of the alphabet, both block capital and cursive, and the numerals can be constructed in a number of different ways. Each person's handwriting therefore, displays a particular combination of character forms which gives that handwriting much of its individuality.

The basic shapes and construction of handwritings are taught in school. In some countries handwriting systems are adhered to rigidly, but in the UK, children tend to learn their teacher's version of handwriting. It is difficult to determine the nationality of handwriting with certainty. However, certain character constructions are more likely to be found in some nationalities than others.

Individual handwriting features begin to be introduced during adolescence when the young person begins to experiment with the appearance of handwriting and mimics attractive features from sources other than the standard taught systems. The handwriting of any individual tends to attain maturity in early adulthood and remains consistent in shape, structure and proportion over the years until changes are introduced as a result of old age.

6.3.1   Construction of Character Forms

Handwritten characters can differ in their construction, shape and relative proportions; of these, construction is probably the most important feature used to distinguish between handwritings. Some of the block capital forms such as the 'E', 'G', 'H' and 'K' can be written in several distinctly different forms as shown in Figure 6.1. It is unusual to find one person using more than one of these forms.

Different constructions of other characters are more subtle. For instance the point of entry and exit of the character 'O' or the numeral

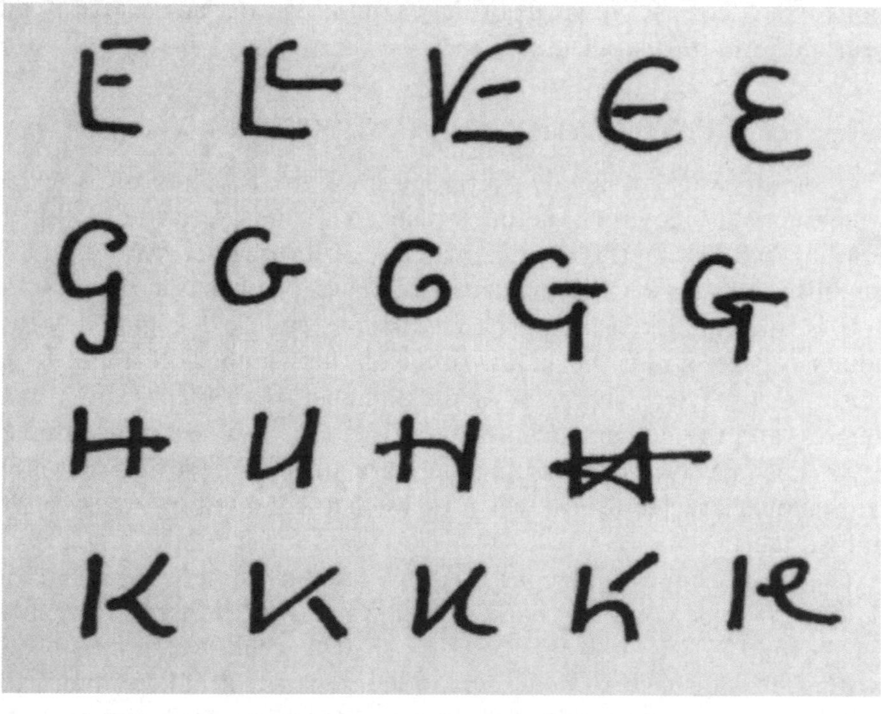

Figure 6.1   *Different constructions of block capital character forms E, G, H and K, as found in UK writings*

'8' can be of crucial importance in distinguishing two different hand-writings. Shape of the individual character form is also an important consideration. Characters may be generally angular or rounded in their formation.

Both the internal proportions of the individual character forms and their relative proportions will be taken into consideration. An individual may consistently introduce a large form of a particular character into handwriting. Similarly only part of a character may be relatively large, such as the bowl of the 'P' or 'R'. Equally important are such features as the point at which a cross-bar is made across a vertical in such characters as 'H' and 'T'.

It is frequently the case that the detailed construction of character forms cannot be determined accurately without magnification of the pen lines. This is particularly the case where it is necessary to determine the order and direction of movement of the pen lines. In some circumstances, the direction of movement of the pen can be used to determine if the writer is left- or right-handed. There are a number of different features of pen lines which can be used to determine the direction in which the pen is travelling.

Ballpoint pen lines show features which are particularly useful in determining direction. They frequently show striations which follow the direction of curves and in addition often deposit ink after a change in the direction of the pen line. Lightening of pressure at the endings of strokes indicates the pen leaving the surface of the paper fluently, whereas a definite spot of ink on the paper can indicate the position of the beginning of the pen stroke. Some writings made with a ballpoint pen showing striation marks are illustrated in Figure 6.2. Other writing instruments produce pen lines which are more difficult to interpret.

The use of a good stereoscopic microscope is essential to study the character forms in detail. A hand lens may provide some assistance in determining structure but is an inadequate tool for detailed analysis since unlike the stereo-microscope the ordinary lens permits no perception of depth in the field of view.

It is essential that handwritings are compared 'like with like'; a character 'a' with a character 'a', the character 'B' with the character 'B'. Similarly, cursive writings must be compared with cursive writings, block capitals with block capitals.

6.3.2 Natural Variation

Very few people write every character in the same manner on every single occasion. All handwritings will exhibit some natural variation

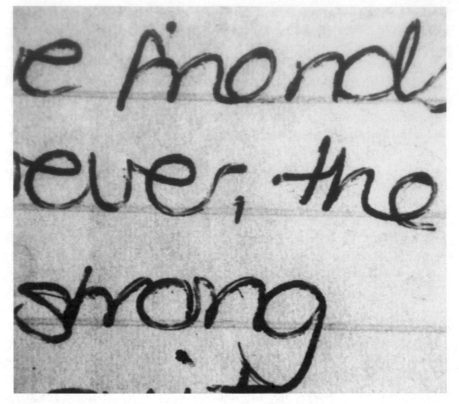

Figure 6.2   *Ballpoint pen lines showing strong striation marks*

most often in the shape and proportions of characters but also occasionally in structure. It is essential that sufficient handwritings are available for examination to enable this range of natural variation to be determined. If insufficient handwritings are available for examination there will always be the problem of not knowing if any differences have arisen because they were written by different people or merely the result of natural variation in one person's handwriting.

6.3.3   Comparison Material

The choice of comparison handwritings may well determine the scope of the final analysis. The best possible handwritings for an examination are those made in the course of day-to-day business as close in time as possible to the writings in question. This is particularly important when the writings of the young and elderly are under consideration. Specimen writings produced specifically for the purpose of comparison are of limited assistance in forensic examinations. It is difficult to produce entirely natural writings under such circumstances. Further-more, writings made all on one occasion will not show the normal range of variation which will occur from day to day. Sufficient comparison material must be examined in order to establish the full range of natural variation in the person's handwriting under examination.

If only specimen handwritings are available great care must be taken in their production to limit the possibility of copying the questioned document or including unnatural features. This is most easily achieved by dictating passages at intervals with breaks in between and removing the specimens from sight as they are produced.

6.3.4 Other Forms of Variation

Handwritings can vary involuntarily for a number of reasons: illness, increasing age, difficult writing positions, use of alcohol or drugs, stress or tiredness. In general these types of problems are evinced by lack of pen control and do not result in fundamental differences in structure which are seen in the writings of two different individuals.

Handwritings can also be disguised. Such handwritings are often detected by their inconsistency: the slope of a piece of handwriting may vary or unusual construction of character forms may appear. Often some characters are formed in eccentric ways or features such as loops and curls are added.

Conversely, an attempt can be made to simulate another person's handwriting. Any attempt at simulation is a compromise between accuracy and fluency. To simulate another person's handwriting it is necessary to suppress one's own natural handwriting characteristics and adopt those of the person whose handwriting is the target of simulation. This is a difficult process and one which cannot be maintained over long periods. Inevitably the simulated handwriting will show character-istics of both the simulator and the target handwriting with the balance changing as time proceeds. The simulation will also be identifiable by differences in detail. Whereas the simulator may achieve a superficial similarity to the target handwriting it is unlikely that the detailed struc-ture of the individual character forms will be correctly perceived or reproduced.

Handwritings undergo the greatest degree of change in adolescence, when the individual style is forming, and old age when pen control becomes increasingly difficult. However, mature handwriting can also undergo developmental changes over relatively long periods. By using this variation it is occasionally possible to place a piece of handwriting within a specific timeframe.

6.3.5 Non-Roman Script

The principles of consistency in construction and natural variation are the same for all handwritings whether they are written in Roman, Arabic, Chinese or any other script. The forensic document examiner trained in the study of UK handwritings will be able to detect significant differences between handwritings in different languages and scripts but will be handicapped in not being familiar with the expected range of variation for any particular problem. In such circumstances the forensic document examiner will be cautious in attributing significance to similarities and differences.

6.3.6 Expression of Handwriting Conclusions

Having completed the comparison of two handwritings, the forensic document examiner will assemble the observations of differences and similarities in the detailed structure, shape and proportions of the component character forms of the handwritings, and from these draw the appropriate conclusions. Unlike fingerprint analysis, which is based on a specific number of points of similarity, the forensic document examiner's conclusions are based upon not only the number of similarities but also their quality. The variations in construction of character form appear in different frequencies in the population. For instance, in the UK population the radial form of the block capital 'K' is far more common than the propped 'V' form.

In all handwriting comparisons the presence of differences is of profound significance. There will always be similarities between handwritings because there are a finite number of character forms which can be used. However, the presence of even a single, consistent difference between handwritings must be explained since it is a strong indicator of different authorship. Where there is a sufficient quantity of both questioned and genuine comparison handwritings, and they match in all respects without any significant differences being detected, then a safe, firm conclusion of common authorship can be given.

There are, however, many occasions when the quantity of handwriting, either questioned or genuine, is restricted. There may also be clear indications that one or other of the handwritings under examination is not natural or has been altered by an outside influence. In these circumstances a qualified conclusion can be given.

The practice of forensic science laboratories in the UK and around the world, is to express conclusions on a qualitative scale describing the strength of the evidence. Typically this is:

Positive		*Negative*
Conclusive evidence		Weak evidence
Very strong evidence	Inconclusive evidence	Strong evidence
Strong evidence		Very strong evidence
Weak evidence		Conclusive evidence

The term 'no evidence' should not be interpreted as 'did not write'. There are, however, some circumstances, such as when considering if a very young or infirm person could have been responsible for a particularly fluent, well-developed piece of handwriting, that exclusion of an individual as a possible writer is acceptable. Nevertheless, because handwriting is, to an extent, a voluntary act and can be consciously changed, it is unusual to be able to say with total confidence that a person was not the author of a particular piece of handwriting.

6.3.7 Copies

Forensic document examiners are often asked to examine copies of documents. The skilled examination of handwriting and signatures requires an analysis of the fine detail of the handwriting including stroke direction and order, crossings between strokes and pressure. These details are lost during the copying process and hence the conclusions that can be drawn from the examination of copies will be restricted. However, where the amount of handwriting under examination is large and its image clear, it is possible to determine sufficient characteristics of the handwriting for strong, and in some cases, very positive conclusions to be reached.

Carbon copies, having been made by direct contact with the writing instrument, frequently show more detail than photocopies and again there are circumstances where positive conclusions regarding authorship can be drawn.

The process of facsimile transmission greatly distorts the handwriting image and in general only qualified conclusions can be draw from their examination.

6.4 THE EXAMINATION OF SIGNATURES

Signatures are very specialised pieces of handwriting and there are specific problems involved in their identification. Our signature is the piece of handwriting which we all use most frequently. Even individuals who write infrequently can produce a consistent signature. Because we use our signature so frequently it becomes more or less unconsciously produced each time it is written.

The greatest problem in identifying signatures as genuine or forgeries lies in the small amount of comparable material which is contained within any signature. This factor, combined with the natural variation which is inherent with any signature, makes signature identification one of the more testing areas of forensic science.

Because of these difficulties it is essential that the original signatures are examined. Photocopies are not adequate substitutes. The amount of available information contained within a signature is so restricted that every available feature must be used. Fluency cannot be assessed accurately from a photocopy nor can pen lifts or guide lines be detected. Only restricted conclusions can be drawn from the examination of copies.

The basis of signature identification is very much the same as that for handwriting. However, since signatures are used for personal identification they are frequently the targets of forgery. Considerable effort will be expended in attempts to simulate another person's signature. This can be achieved in a number of different ways.

6.4.1   Tracing

If a document bearing a genuine signature is placed on a window or a sheet of glass over a light source and overlaid by another document it is possible to trace the outline of the genuine signature onto the forged document. The tracing, often made in pencil can be then inked over. The resulting forgery although superficially similar to the genuine signature will be detectable by its lack of fluency as the simulator laboriously follows the line of the original signature and by mistakes in the detailed construction of the character forms. These will be superficially similar to those in the genuine signature but the simulator is generally only interested in obtaining a pictorial, passable simulation and will not necessarily use the correct number or order of strokes to achieve this.

Guide lines may be left on the document and even if only fragments remain these can be detected by viewing the signature under specialised lighting conditions. Pencil lines remain opaque when viewed in the infrared region of the spectrum whereas many inks can be transparent at these wavelengths.

Guide lines may also be in the form of indented impressions on the target document. These are produced by placing a genuine signature on top of the target document. The genuine signature is written over

Figure 6.3 *Traced simulation of a signature as seen in obliquely directed light*

heavily so that impressions are transferred to the document beneath. These are then inked in to produce the simulation. The impressions can be readily detected in light shone at a shallow oblique angle to the document as seen in Figure 6.3. In addition the signature is likely to lack fluency and contain mistakes in a similar manner to the direct tracing.

Tracing may be more subtle with sections of several signatures being incorporated into a single simulation. Several instances have been seen of simulations produced from templates prepared from genuine signatures.

6.4.2 Freehand Simulation

The freehand simulation is likely to be the most fluent but the simulator still needs to achieve a compromise between accuracy and fluency. Unusual hesitations and pen lifts in the pen line will occur whilst the simulator pauses before embarking on the next section. Often these types of simulation show the greatest degree of deviation from the genuine signatures at their ending when the simulator relaxes concentration and becomes more confident. This usually results in a reversion to the simulator's natural handwriting rather than that of the target signature. Again the simulation is detected by its superficial similarity and its differences in the detailed shape, structure and proportions of its component character forms.

6.4.3 Authorship of Simulation

Simulated signatures are not natural writings. To produce a simulated signature the forger's own natural characteristics of writing must be suppressed and those of the person whose signature is the target of forgery adopted. Hence it is not normally possible to identify the author of a simulated signature by comparing it with natural handwriting.

6.4.4 Self Forgery

Certain signatures are written by individuals with intent to deny them at a later date. These signatures are generally different from normal natural signatures in some very obvious feature, often the form of initials or slope, but the detailed structures of the less obtrusive features are unchanged.

6.4.5 Vulnerable Signatures

Signatures which are short and simple and which contain a number of natural pen lifts are vulnerable since they reduce the requirement of maintaining fluency whilst executing unfamiliar and possibly complex character forms. Signatures which naturally demonstrate a wide range of variation are also vulnerable since it is difficult to assess if differences have arisen merely as a result of this variation or because of simulation by another person.

Signatures of the elderly often naturally contain many of the features which are the hallmark of simulation such as lack of pen control, undue variation and poorly formed character forms. These signatures are, therefore, particularly vulnerable.

6.4.6 Guided Hand Signatures

An infirm, elderly or partially sighted individual may be assisted by another person in signing documents. Guided hand or assisted hand signatures are a particularly difficult area for the forensic document examiner. The characteristics of such a signature may be a mixture of those of the signator and the assistant.

The forensic document examiner is often asked to comment on the degree of influence exerted by the assistant. Clearly if the signature deviates strongly from the genuine signature and contains a predominance of the assistant's signature characteristics a degree of influence has been exerted. However, the forensic document examiner must be cautious in commenting on the intent of the assistant and the person signing. It may be that the signator, whilst unable to exert sufficient pen control to write, may have been completely clear in his or her desire to sign the document in question.

6.4.7   Comparison Material

The requirement for adequate comparison material in signature examinations is similar to that required in all handwriting comparisons. However, it is worth stating again that like can only be compared with like – in other words, signatures are only comparable with other signatures in the same name. 'Smith' cannot be compared with 'Jones' nor can signatures generally be compared with day-to-day handwritings.

The ideal comparison material would consist of genuine signatures made in the course of day-to-day business close in time to the date of the signature in question.

6.4.8   Expression of Signature Conclusions

Given a reasonably stable and mature signature and an adequate range of genuine signatures for comparison, the forensic document examiner can expect to identify a genuine signature or a simulation. However, there will be occasions, particularly in this very testing area, where only a qualified conclusion can be given. These qualified conclusions are generally expressed in a similar manner to those for handwriting examinations.

6.5   THE EXAMINATION OF PHOTOCOPIES

The limitations of examining photocopies have already been explained. However, it is often the case that a photocopy is the only evidence of a pre-existing original. The forensic document examiner is then charged with the task of determining from the copy if it was indeed prepared from an authentic contemporaneous original document or if the copy has been recently manufactured to misrepresent the facts.

The photocopy may be a montage produced partly from genuine documents. The simplest demonstration of this is the image of a genuine signature transferred by photocopying onto a fraudulent document. The image may be transferred intact and without apparent disturbance in which case often the only way of demonstrating the montage is to locate the original signature from which the image was taken. However, often it is necessary to manipulate or retouch the image to make it fit the new document. These processes are detectable as inconsistencies in the image of the pen lines.

It may be important to demonstrate the origin of a particular photocopy document. There are two types of mark which appear on photocopies which can be used to do this: trash marks and drum or mechanism marks.

Trash marks are transient and produced by dust particles or debris on the glass surface of the photocopier. These will appear as a series of dots or marks on the resultant copy. The same configuration of trash marks on a number of documents would indicate that they have a common origin. However, it is important to remember that if a copy bearing trash marks is copied, the daughter copies will also bear the same configuration.

Marks made by the drum or the mechanism of the photocopier will persist for a longer period and can be used to identify a specific machine. Drum marks do not necessarily appear in the same position on every page but can be detected by the constant intervals at which they appear. The interval is directly related to the circumference of the drum and not the dimensions of the document being copied.

6.6   PRINTING AND TYPEWRITING

6.6.1   Modern Office Technology

The replacement of the old-style manual typewriter with fixed typebars, first by electric typewriters containing golf-balls and print-wheels, followed by the dot-matrix printer and today the ubiquitous laser printer, has completely revolutionised the forensic examination of typescript in the last 20 years.

6.6.2   Word Processors

The term word processor actually refers to the computer program used by the keyboard operator to prepare a computer file of the document which is to be printed. The printing of the document from the computer can be made using a number of different types of machine. These include laser printers, ink-jet printers or dot-matrix printers. The appearance of the printed characters which finally appear on the paper is to a large extent governed by the computer program. More primitive word processor systems were connected to electronic typewriters fitted with daisy-wheels. The number and nature of the typestyles which could be prepared from these machines is much more limited.

6.6.3 Laser Printers

The laser printer works very much on the same principle as a photo-copier. It contains a photosensitive drum which becomes charged as the laser light hits it, producing an 'electrostatic negative' of whatever image is to be printed. The drum is sprinkled with negatively charged toner which clings to the charged areas originally scanned by the laser beam. A sheet of paper is passed under the rotating drum and given a charge greater than on the drum allowing transfer of the toner from the drum to the paper where it is then fused in place by heat or pressure. Images produced by laser printers are generally of very high quality. They are also very reproducible and it is extremely difficult to distinguish between the work of two different laser printers. However, it has been established that, on occasions, drum faults may occur on laser printers in the same way as they do on photocopiers.

As difficult as it is to distinguish the work of one photocopier from another, it is similarly difficult to identify any document as having been produced by a particular computer/printer system. The number of fonts available on laser printers and their versatility is immense.

6.6.4 Ink-Jet Printers

In an ink-jet printer the only moving part in the print head is the ink itself. The printer uses a grid of tiny nozzles to which specially formulated ink flows from a reservoir. The computer sends a signal as to which nozzle to fire, following which a minute resistive heating element built into the base of the nozzle heats up. The ink close to the element vaporises almost instantly and creates a tiny bubble. As the bubble expands, the ink in the nozzle is forced outward and the droplet is jected towards the paper. This happens in a fraction of a second, allowing droplets to be ejected at over 300 per square inch. The main difference between the laser and the ink-jet printer is the state of the ink. When the ink from the ink-jet printer hits the paper it spreads into the fibres and gives the print a slightly ragged appearance. Some systems require the use of a specially adsorbent paper.

Although the work of ink-jet printers can readily be identified it is as difficult to identify a particular machine or differentiate between the work of several machines as it is with laser printer copy.

6.6.5 Dot Matrix Printers

Dot matrix printers work by striking the paper through a ribbon with a number of pins. Each time a pin hits the ribbon it transfers a dot of ink from the ribbon to the paper. The characters are formed from a matrix of pins arranged in a head. These printers are relatively slow and noisy when compared with ink-jet and laser printers. However, they are often retained for processes which involve producing several copies of a document all at the same time such as a Letter of Credit or Shipping Manifest.

These types of printers have a relatively large number of moving parts and there is greater scope for examiners to identify the print. Examples have been seen where individual pins have been damaged or have not functioned. However these tend to be relatively rare since the eccentric character forms produced are very noticeable and rapidly repaired.

6.6.6   Single Element Typewriters

There are still a substantial number of documents which are produced on electric typewriters equipped with either a daisy-wheel or, more rarely these days, a golf-ball. These single elements can be easily removed from the machine and replaced using another of a different style. The elements can also be transferred between machines. One of the most important features of these single elements is that they deteriorate with increasing use. Faults develop on the element and these are transferred to the print on the paper. Figure 6.4 shows typing produced with a printwheel on which the underlining bar is beginning to deteriorate.

These faults can be used to distinguish between work from two type-writers, or a particular document can be identified with a particular print element. However, these faults are relatively transient. Once the fault begins to develop on a daisy-wheel in particular, the wheel will

Figure 6.4   *Typing produced using a damaged print wheel*

deteriorate rapidly, revealing a number of different faults until it needs to be replaced. Once it is replaced, these faults no longer exist. This particular feature can be used to identify the date on which documents were produced.

Some early word processor systems used daisy-wheel typewriters as the printing device. Unlike dot matrix, ink-jet and laser printers, these machines are relatively inversatile. The final print is governed by the single element which is inserted into the typewriter. The spacing of the typescript from these machines is also relatively restricted whereas that from the other types of computer-driven printers is extremely versatile.

6.6.7 Fixed Type-Bar Machines

In this type of typewriter each individual character is fixed to the end of a type-bar. The type-bar is raised as a result of depressing the character key and the character is pressed onto the ribbon and copied to the paper. The number of moving parts within these machines is relatively large and in consequence, usage results in the variation of the relative alignment of individual characters. Furthermore, the individual characters can become worn or damaged so that the printed character is imperfect. These faults and imperfections tend to become progressively worse over the period of usage.

The fixed-bar typewriter can only have one typeface and consistent spacing of the characters. Consequently, there is a much higher chance of identifying the work of a fixed type-bar machine than the products of modern office technology. Although the machines are rarely used in the modern office environment there are still many individuals who keep them for personal use. They are also still prevalent in the Third World and developing nations.

6.6.8 Spacing

Typewritings produced by any means will have a regular vertical and horizontal spacing. Older typewriters produced typewritings in which the horizontal space allocated for each character was precisely the same. The advent of the electronic single-element typewriters allowed the introduction of proportional spacing in which the space allocated depended on the width of the character, with the character 'w', for example, occupying more space than the character 'i'. Modern-day, computer-driven printers can provide a variety of spacing, size and boldness of text.

Notwithstanding the method by which the text is produced, a document printed or typed at one time will have regularity at least in margin positioning and in the spacing of the lines. These spacings as well as the inter-character horizontal spacing can be compared using accurately ruled transparent grids. Any irregularities may indicate insertion or manipulation of the text.

6.6.9   Ribbons, Rollers and Correction Facilities

Where older style typewriters are involved, study of the typewriter itself and any carbon paper or correction papers can reveal useful information. The text of documents typed using carbon film ribbons can be detected fairly easily. These machines function by stamping out the carbon image of the character from the ribbon onto the paper. The resulting negative remains on the ribbon and the text of the document can be followed using a moderately powerful magnification system. The process is, however, very tedious especially when the ribbon mechanism involves complex vertical and horizontal movements. A computerised optical system of transcribing ribbons has been developed and is available in a small number of laboratories.

The roller or platen of a typewriter may also contain information. The image of the text may have been inadvertently transferred to the roller. The roller itself may contain physical defects which affect the text of the document.

Careful examination of corrections, correcting ribbons, correcting papers and transfer or correction fluid onto fabric ribbons can all provide additional information to link a questioned document with a particular typewriter.

6.7   THE ORIGIN AND HISTORY OF DOCUMENTS

There are occasions when documents used in criminal activities or in the course of civil litigation are either created specifically for the purposes of deception or altered for similar ends. For example, entries in diaries, or occasionally whole diaries, are prepared to show that a particular sequence of events occurred in the past. Similarly, files can be reconstructed and correspondence back-dated.

The forensic document examiner is asked to assist in determining the authenticity of such documents and entries. The dating of documents and writings is a difficult forensic science problem. Since there are no reliable techniques available which allow the absolute dating of inks, it is necessary to use other features of documents to determine where and when they were created and their history.

6.7.1 The Examination of Inks

The ink used to write particular entries on a document may be of interest for several reasons. Principally the forensic document examiner will be interested in establishing if inks are similar or different. In this way it is possible to determine if entries have been added to or altered.

The examination of inks on questioned documents is generally confined to non-destructive comparative analysis. Inks which appear similar to the unaided eye can be very different when viewed in the infrared region of the spectrum. Similarly illuminating an ink with high energy light at one wavelength can promote it to release light energy at different wavelengths. The degree to which this occurs depends on the chemical nature of the ink. Figure 6.5 shows a set of black ballpoint pen inks as viewed in normal light, in the infrared region of the spectrum and under conditions which promote fluorescence. Several instruments have been developed which allow inks to be viewed under a wide range of different lighting conditions to enhance some of the less visible effects. The fact that many inks can be rendered transparent in the higher wavelengths of the infrared region of the spectrum is an essential tool in the examination of questioned signatures. Pencil markings remain opaque at these wavelengths and can be easily detected under the transparent ink.

Raman Spectroscopy of inks is a largely non-destructive process which has proved useful in distinguishing between inks which cannot be distinguished using more traditional infra-red and fluorescence techniques. The Raman response can be considerably enhanced by a technique called surface enhanced resonance Raman scattering (SERRS) spectroscopy.

This involves the application of a very small quantity of a silver colloid onto the ink line under examination. The extent of any damage is minimal and can be confined to 1–2 mm within the width of the pen line. SERRS spectra obtained from two black biro inks are shown in Figure 6.6.

On some occasions further destructive techniques can be employed to compare inks more rigorously. In fact this destructive analysis can be carried out using relatively tiny amounts of ink which are carefully removed from within the pen line of a single character form. Thin layer or high performance liquid chromatography of these ink samples is used to separate and compare the dye components. In two studies reported in 2000, combining various techniques (thin layer chromatography, infrared and fluorescence techniques and SERRS spectroscopy) allowed over 90% of the ballpoint pen inks under review to be distinguished.

(a)

(b)

(c)

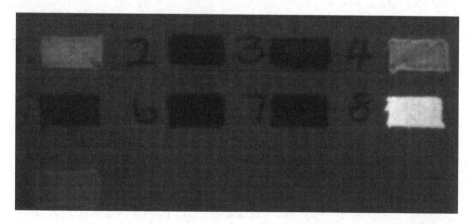

Figure 6.5 *Black ballpoint pen inks as seen in: (a) normal light, (b) infrared light and (c) under conditions which excite fluorescence*

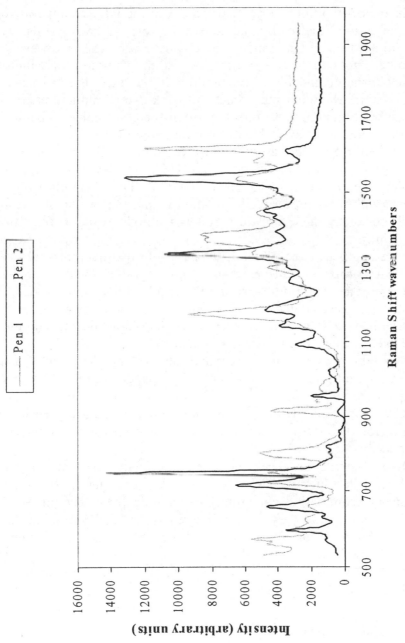

Figure 6.6 *Raman spectra of two different black inks*

The majority of ink analysis carried out in the UK is comparative. If parts of a document can be demonstrated to have been written using two different inks then clearly two pens have been used. However, the opposite is not equally true. If the inks used in two writings are indistinguishable they could have been written either using the same pen or pens containing similar inks. Since most ink dye manufacturers do not change basic ink compositions very often there are liable to be several million similar pens available for any one type of ink.

The Secret Service Laboratory in Washington has maintained an extensive library of inks made in the USA. However, it is rare that absolute identification of an ink is required. This certainly would be necessary if an attempt is made to date the ink. Unfortunately, as indicated previously there are still very few methods available for reliably dating inks.

For a number of years in the 1990s certain manufacturers of inks in the USA co-operated in an ink-tagging programme where rare-earth trace components of fluorescent dyes were added and changed regularly to allow inks to be dated. The number of manufacturers were however, very limited and few laboratories have the sophisticated equipment available for analysis of these components.

Other laboratories claim success in dating inks by comparing the relative extractability of ink components from paper. Again the method is comparative but the inks must be identified to make sure that they are directly comparable. These methods have not attained acceptance in the field of forensic science since they have not been found to be wholly reproducible.

More alarming has been the attempt to date inks using artificial accelerated techniques. There is substantial disquiet in the forensic document examination community concerning these techniques. Current thinking is that they are fundamentally flawed. Where ink dating evidence has been offered in courts of law there has, very frequently, been disagreement between the experts. Although ink dating services are offered by a very small number of laboratories in the USA they are rarely used in other countries.

6.7.2 The Examination of Paper

The size, thickness, density, colour and finish of papers can be compared to determine if they are from a similar or different origin. Optical brightners are incorporated into some papers and give differing reactions when viewed in ultraviolet light. The biological fibres used in a particular paper's manufacture can be identified by microscopy and can be used to give a broad geographical origin of the paper. However, since wood pulp is transported on a global scale, this is often inaccurate.

The origin of a paper is most accurately identified by its watermark. Some paper manufacturers periodically introduce changes into watermarks. Repairs to the dandy roll used in the manufacturing process to produce the watermark may be identifiable in subsequent production runs. However, not all manufacturers keep detailed records of changes made over the years and this has been exacerbated by the take-over and absorption of some of the older brands.

In a similar manner the chemical content and colour of papers can be used to establish when they were manufactured, if the manufacturer can be identified and holds the necessary records.

Pages torn from a pad of paper often show a specific pattern along the top edge where the adhesive has been applied unevenly to the backing sheet. The pattern may vary through the thickness of a pad so that the pages can be identified as being from the same pad, different parts of the same pad or different pads altogether. Similarly, sheets from the same batch of paper may show similar guillotine marks along their edges, although these are more difficult to identify.

6.7.3 Development of Handwriting and Signatures Over Time

As noted previously the handwriting and signatures of individuals vary over the years. If sufficient examples of genuine handwritings and signatures are available to demonstrate the different stages of development, questioned writing and signatures can be matched to a particular period. This type of examination is particularly applicable to the writings of young people and the elderly.

6.7.4 Impressions

Impressions of handwriting may be left on a document as a result of writing on another document which at the time of writing was overlaying the first one. These impressions are rarely visible to the naked eye and need to be detected using specialised equipment and techniques.

Deep impressions can be detected using light directed at a low oblique angle to the document. In this way the surface of the document is thrown into relief and the impressions can often be read. Where the impressions are small and fragmentary the best results are often obtained using a low diffuse light or by viewing the document in natural daylight in the early morning or evening.

Fainter impressions can be detected using the technique of electrostatic detection (ESDA). This technique was developed jointly by the London College of Printing and the Metropolitan Police Forensic Science Laboratory in London in 1979 and revolutionised forensic document examination. The long and tedious hours piecing together fragments of indented impressions in dimly lit rooms was replaced with a relatively fast and highly sensitive technique which could develop impressions made through several layers of paper. Such impressions are too faint to be detected by any other methods.

It is thought that the ESDA process detects areas where the paper fibres have been deformed by the movement of the writing implement or by two documents moving against each other. The technique is not always successful. Impressions cannot be detected on documents which have been treated with solvents. It is therefore imperative that any document which will eventually be examined for fingerprints should first be submitted to the forensic document laboratory before treatment with fingerprint reagents.

Recently a number of reports have been made of 'secondary impressions'. These impressions are caused not by the action of writing but by storing two documents together such that the fairly deep embossing of actual handwriting on one document causes detectable impression marks on the second document. Such 'secondary impressions' can be very confusing particularly if they appear on the same document as those impressions caused by the act of writing.

In the last few years the utility of the ESDA technique has been extended and it can now be used in the determination of the order of writings. The ESDA 'lift' bears not only black traces caused by the development of the impressions but also traces caused by the ink lines present on the document as illustrated in Figure 6.7.

It has been shown that where impressions and ink lines on a document intersect, the appearance of the intersection can be used to determine if the writings or the impressions were made first. Situations where the black impression line can be seen to be continuous over the white ink line indicate that the impressions were made after the writings. The technique can only be used when the ESDA lift is of good quality and there are a substantial number of intersections which are clear, since where impressions have been made after writings on a document only a proportion of the intersections will show the black line continuous over the white ink trace. Experiments where the ink lines of writing were made after impressions on a document resulted in the black impression trace being broken on every occasion. Similar reactions occur between impressions and writing on opposite sides of a piece of paper. The

Figure 6.7 *ESDA 'lift' showing black traces of impressions and white traces caused by writings on the document*

mechanisms of the response are not fully understood and great care must be exercised in interpreting the results of these analyses.

Although ESDA has now been in routine use in forensic document laboratories for nearly 25 years the underlying mechanisms of the process are still not fully understood and the ESDA response can be variable. However, it is one of the most powerful weapons in the armoury of the forensic document examiner.

6.7.5 Folds, Creases and Tears

The physical condition of a document may betray much of its history. The arrangement of folds and creases in a document may indicate if it was folded into an envelope or folded in conjunction with other documents.

Torn edges of a document may be matched to those of another document particularly if they were torn together from the same pad. Portions of torn documents can be matched by studying the mechanical fit of the torn edges.

Envelopes which have been opened and re-sealed may contain fragments of torn paper fibres within the seal. These can be detected by transmitted light through the document or X-ray techniques. The act of re-gluing an envelope flap may require the use of additional adhesive which can be detected by similar techniques.

6.7.6 Staples and Punch Holes

The act of stapling one document to another causes physical damage to the document as the staple ends are punched through. If the staples are removed the staple holes remain obvious. Further damage can be caused to the back of a document where the ends of the staples are forced in, often distorting the paper between the staple holes. The relative size and the positioning of staple holes can be used to identify documents which have been stored together.

The study of staples is also important in considering if sections of documents have been removed, such as in notebooks. Prising open staples often causes distortion to the staple holes which become widened. Microscopic examination of the staple legs will show marks made by pliers or other implements used to prise the staple out. Besides examining the condition of the legs of the staple in the central fold of a book it is important not to neglect the exterior of the spine, where the act of pulling out the staple or replacing it may have left detectable marks on the document.

Punch holes also cause irreparable changes to a document. Punch holes become worn and distorted over a period of time as a result of movement of documents against the holding prongs of a file or clip. Often documents stored together for a period of time will show a similar pattern of damage to the punch hole.

6.7.7 Erasures, Obliterations and Additions

The act of erasing an entry from a document, be it typewritten, handwritten or printed, will damage the document. The extent and nature of the damage will be different depending on whether the erasure was carried out by mechanical or chemical means.

In general, mechanical removal of an entry will physically damage the document. Using an eraser to remove entries results in the paper fibres of the document being disturbed. This disturbance, if large enough, may be detected using obliquely directed light which throws the surface of the document into relief. Less obvious erasures made with a traditional rubber eraser can be detected by the use of a very fine powder of dyed lycopodium spores. The powder will adhere to fragments of rubber left on the surface of a document in the process of erasure. The powder is spread over a document held at an angle and the excess shaken off gently. The powder can then be brushed off the surface of the document leaving it undamaged. However, the action of some plastic erasers cannot be detected in this way. Consequently, the lack of reaction to lycopodium powder cannot be taken as proof that no erasure has taken place.

Chemical erasure involves the use of a solvent to remove an ink entry. This may be comprehensive and leave no trace visible to the unaided eye. The effect of the solvent on the document, however, can be detected by viewing it in ultraviolet light or under conditions which promote infrared fluorescence.

Traces of entries removed by erasure can be enhanced by viewing the document under different optical conditions. Pencil entries are opaque in the infrared region of the spectrum and, therefore, can be enhanced using an instrument which combines this lighting facility with powerful magnifying capabilities. Similarly, traces of an ink which has fluorescence properties can be enhanced by promoting the fluorescence and magnifying the image.

Obliterated entries are detected by separating the original entry from the obliterating material. This can be achieved non-destructively by exploiting the optical characteristics of inks and correcting fluids. A black ink entry obliterated with another black ink can be recovered if the inks react differently, for instance, in infrared light. Entries obliterated with white correcting fluids can be recovered using a combination of lighting conditions including viewing from the back of the document. Today's modern range of instrumentation in forensic document laboratories allows images of such features to be captured electronically and manipulated to give a clear view of the original obliterated or erased entry.

The detection of added ink entries again involves the non-visible properties of the inks. Anyone making a fraudulent addition to a document will make an effort to choose an ink of similar colour. However, these visually similar inks may well have very different properties in infrared light and can easily be demonstrated as having been made with different pens.

There are differences in the manufacture of pencils which can be detected using techniques such as scanning electron microscopy (SEM). However, such investigations have the disadvantage of being destructive and expensive.

Unlike pen tips the point of a pencil alters as writing proceeds. It is only necessary to change the position of a pencil in the hand by 90° to fundamentally change the appearance of the drawn line it produces. In consequence it is extremely difficult to detect tampering of pencil entries.

6.8 PRINTED DOCUMENTS

Most practising forensic document examiners will have a working knowledge of printing techniques sufficient to identify the process by which a document or parts of a document have been produced. However, some fields of printing required a highly specialised knowledge.

High value documents, for example bank drafts and identity documents such as passports, are specially printed using methods to deter alterations or counterfeiting. Documents can be printed on papers which are sensitive to the application of chemicals or mechanical erasures and which change colour. Coloured or fluorescent fibres can be added to the paper. Security backgrounds can be produced with fugitive inks which run on application of a liquid and complex patterns of printing can be produced which are difficult to imitate. Sections of documents can be printed using inks with special spectral properties or which change colour under different lighting conditions. The techniques used for studying alterations to documents can also be used to detect breaches of a document's security.

A knowledge of printing processes is frequently needed in the identification of counterfeits. The counterfeit must be compared directly with the genuine article be it a passport, cheque or perfume packaging. The forensic document examiner needs to be able to identify the differences in printing between the products to show differences in their mode of production.

6.9 PROCEDURES, PROTOCOLS AND QUALITY ASSURANCE

In forensic document examination, as in any forensic science, certain criteria need to be applied to ensure a correct interpretation of the evidence.

Any document, either questioned or of known origin, submitted to a forensic document laboratory needs to be handled properly and securely to ensure it retains its integrity throughout the examination. Documents need to be properly identified, labelled and recorded. The appropriate testing must take place and this should always be the most rigorous available. The results of examinations and tests should be recorded clearly and unambiguously in the laboratory notes along with details of standards and controls.

The forensic document examiner's report should be comprehensive, setting out the examinations employed, results and the conclusions drawn. Forensic document examination is a science but this is no bar to clear, concise and understandable reporting. These processes will at least ensure that a full examination is carried out in a properly equipped laboratory. It is, however, extremely difficult to ensure that the results of an examination are interpreted correctly.

Properly trained forensic document examiners working in fully equipped laboratories can generally be expected to reach broadly similar conclusions from the examination of the same material. Problems arise when two experts are given different material, often in the form of control handwritings or signatures. More serious disagreements arise when inappropriately trained individuals, or those lacking proper examination facilities, are employed as experts. Provided the forensic document examiner has the opportunity to do so it is generally not difficult to refute the evidence of such individuals. Visual presentation of results is the most effective method of presenting the evidence, particularly where there are conflicting views. Indeed it is always wise to work on the premise 'If you can't show it, don't use it'.

6.10 BIBLIOGRAPHY

A.A. Cantu, Analytical Methods for Detecting Fraudulent Documents, *Anal. Chem.*, 1991, **63**, 847–854.

D.M. Ellen, *The Scientific Examination of Documents*, Ellis Horwood Ltd, Chichester, 1989.

W.R. Harrison, *Suspect Documents*, Sweet & Maxwell, London, 1958.

A.S. Osborne, *Questioned Documents*, Boyd, Albany, USA, 1929.

Proceedings of 2nd European Academy of Forensic Science Meeting (European Document Experts Working Group), Cracow, 12–16 September 2000 – papers by C. Neumann & W.D. Mazella, and by E. Wagner.

Computer Based Media

JONATHAN HENRY

7.1   THE CRIME SCENE

The use of computers to commit crimes and computers as the subjects of crime has redefined where and what a crime scene can be. The computer crime scene may well exist as a traditional scene with the evidence clearly visible, but more often the evidence is not in an obvious form and requires the actions of a suitably qualified examiner to preserve it and to interpret its meaning.

Whereas the documentary evidence of a business fraud may previously have been found in the locked filing cabinet of the accountants office, that locked filing cabinet is now more likely to be a folder on their computer hard drive or in the remnants of deleted files thought long gone. The overseas 'hacker' and their computer, penetrating the security of a UK network may be well beyond our physical grasp but the records of the penetration may be available to us and properly preserved and interpreted can lead us to the source. In its most extreme form the crime scene may exist only fleetingly as a conversation in a chat room with unknown traces of the conversation remaining on the participants' computers.

The computer crime scene exists in terms of the traces of actions performed on a computer that remain on the hard disk drive or removable media, a single line in an email header or even as entries in a server log file showing access to files or an email account.

The rule of 'every contact leaves a trace' is just as applicable to computer-based evidence as any other. This applies to the actions of the examiner also. Such is the nature of computer-based evidence, the contact is usually date and time stamped and can prove compelling evidence for the prosecution or the defence. Similarly it is just as fragile as other forensic evidence and without proper treatment can be lost.

In order to understand the forensic evidence available to an examiner it is necessary to have a basic understanding of the methods of data storage on computer media, both in physical terms, *i.e.* the devices used for storage, and logically, *i.e.* the structure and format of the data on those devices. The following information is intended as an introduction to the principles of how data is stored physically on computer media and logically in Windows-based computer systems and how the actions of the user produce these data. The principles behind the physical media are independent of operating systems. Some of the principles relating to the actions of users may be equally applicable to operating systems other than Windows. It is not intended as a guide to performing computer-based evidence examinations, nor as a technical reference on the format and location of data commonly forming the basis of such examinations.

7.2 GUIDANCE ON EXAMINATION OF COMPUTER-BASED EVIDENCE

7.2.1 Principles

There are four overriding principles that form the basis of any examination of computer-based evidence. These have been encapsulated within the Association of Chief Police Officers Good Practice Guide for Computer Based Evidence. Their purpose is to protect the integrity of the evidence, maintain a record of the examination and to protect the rights of the original owner of the evidence. The principles are:

1. No action taken by police or their agents should change data held on a computer or other media that may subsequently be relied upon in court.
2. In exceptional circumstances where a person finds it necessary to access original data held on a target computer that person must be competent to do so and to give evidence explaining the relevance and the implications of their actions.
3. An audit trail or other record of all processes applied to computer-based evidence should be created and preserved. An independent third party should be able to examine those processes and achieve the same result.
4. The officer in charge of the case is responsible for ensuring that the law and these principles are adhered to. This applies to the possession of, and access to, information contained in a computer. They must be satisfied that anyone accessing the computer, or any use of a copying device, complies with these laws and principles.

Combined, these principles mean that examinations of computer-based evidence should only be performed by suitably qualified individuals, in a forensically sound manner, with sufficient records maintained to ensure that another examiner can reconstruct their actions.

7.2.2 Imaging

Principle 1 for the examination of computer-based evidence states that 'No action taken by police or their agents should change data held on a computer or other media that may subsequently be relied upon in court.' An essential step in the examination of computer-based evidence is therefore, where possible, to create a forensically sound and accurate copy or 'image' of the media under examination. This step is commonly referred to as 'imaging'.

The purpose of imaging is to enable the forensic examiner to copy the original media in a forensically sound manner that will not alter the integrity of its contents. Any subsequent examination can then be performed on the copy of the media, leaving the original intact and unchanged. Imaging or Bit Stream Imaging is the process of replicating each individual bit of data from a piece of media. The media could be a hard drive, a floppy drive, a digital camera card or any other storage device.

An image can be made of either a physical disk or a logical volume. It is always preferable to image the physical disk where possible. Unused areas of the disk can and do exist outside those normally visible to the user for a variety of reasons. They may not always have been unused and may have previously contained data that we are interested in. For this reason unused areas should be included in an examination.

7.2.3 Examinations

The actions taken by the forensic examiner after imaging are largely determined by the type of investigations they carry out and the forensic software used. Commonly the 'image' of the media exists as a series of files produced by the imaging software which contains all of the data from the imaged media. Dependent on the software used, the image can then either be examined directly within the forensic software as a 'virtual drive' or a 'clone' of the original media made for the purposes of the examination. In this way the examiner is then working on an exact replica of the imaged media and the original remains intact and unchanged by the examination process.

The procedures to be followed throughout the forensic examination of computer-based evidence are not discussed here as there is a myriad of software applications available to the forensic examiner. They range from utilities specifically designed for the imaging and analysis of computer-based evidence to utilities designed for other purposes *e.g.* recovery of lost or accidentally deleted data, which lend themselves to the needs of forensic examiners. The procedures to be followed for each piece of software, or hardware for that matter, are generally unique to that equipment.

An examination process can be set out technically for the use of any piece of software or hardware, but there is more to the examination than 'ticking the boxes'. The methodology differs for each individual case and depends on many factors: the type of investigation, what is required to be proven, the physical hardware or software applications encountered and the timescale for the examination. Computer forensic examiners rarely exist as examiners alone. They are investigators with specialist technical skills who play an integral part in the entire investigation process.

7.3 STORAGE DEVICES

7.3.1 Ones, Zeroes, Bits and Bytes

Information is stored and processed by computers in terms of binary data, *i.e.* 1s or 0s. Each 1 or 0 is referred to as a 'bit' of information or data. Eight bits of information together make up a 'byte'. 1024 bytes make up a kilobyte (KB). Similarly, 1024 kilobytes (1,048,576 bytes) make up a megabyte (MB). Logically, 1024 megabytes (1,073,741824 bytes) make up a gigabyte (GB) and so forth for terabytes (TB) *etc.*

That is not to say that computer hard drives, CD-ROMs and disks are full of 1s and 0s. The 1s and 0s are a mathematical representation of a two-state or binary system *i.e.* 'on' or 'off'. This is how data is physically stored on hard drives, floppy disks or CD-ROMs. This is also referred to as digital data. On hard drives or floppy disks changing magnetic orientations encoded on a magnetically responsive material represent the on/off states. CD-ROMs represent this by the interference patterns of a beam of laser light bounced off the surface of a disc caused by pits on the disc surface. All computer media, whatever its physical form, stores data in two states. This two-state information is interpreted physically by the hardware, and logically by the operating system, as 1s and 0s.

Storage devices are generally based on magnetic media, optical media or a combination of both technologies. I intend to describe the most commonly encountered storage devices for each category. The examples are by no means exhaustive.

7.3.2 Magnetic Media

7.3.2.1 Hard Disks. Hard disk drives are the heart of data storage in modern computer systems. The computer operating system resides on the hard drive. Every program installed by the user is stored there. Details of the system configuration are found in files saved on the hard disk by the operating system or particular programs. Most importantly it is the most commonly used location for the user to store files they create. From an evidential view the hard disk drive is an extremely rich source of information. It may contain not only the evidence of offences that have been committed, but also how they were committed which may reveal the intentions of the user.

Physically the hard disk drive, or HDD, is a sealed unit usually roughly 3.5 " (modern drives) to 5.25 " (older drives) across, or in the case of a laptop drive 2.5 ". The only visible component is the printed circuit board attached to the outside of the sealed unit. In modern hard drives this contains the hard disk controller which handles conversations between the HDD and the processor, interpreting instructions from the processor and converting information from the HDD into a form the processor can understand and *vice versa*. Contained within the sealed unit is one or more disks mounted centrally on a rotating spindle, each of which has two read/write heads sliding over their surface on arms, much like the stylus on a record player.

Each rotating disk is made up of a thin layer of magnetically responsive material laid or 'spun' onto a rigid substrate of glass or aluminium. It is within this magnetically responsive layer that the data is stored. In modern hard drives to increase capacity this may actually be a number of layers each capable of storing data independently. Each disk is referred to as a 'platter'. The magnetic media layer is coated with an extremely thin layer of carbon to protect it from the read/write heads and a lubricating layer to help the heads slide over the surface.

During operation the platters spin at extremely high speeds, 3,600–12,000 rpm, depending on the HDD used. As they do so the read/write heads glide fractionally (millionths of an inch) above the surface, held up by the pressure of the airflow from the spinning disc. The spinning motion combined with the side-to-side movement of the heads on their arms allows any area of the platter to be accessed very quickly.

Data is written to and read from the disc by the read/write heads. Each head is an electromagnetic induction coil. The read/write head works in two modes. In the 'write' mode data is encoded into electrical pulses that are passed through the coil in the head causing a change in the local magnetic field. Because the head is only a microscopic distance above the platter this change in magnetic field is sufficient to reorient the magnetic alignment of a tiny area of the surface. The affected area retains this orientation until a subsequent write operation occurs. These areas are of a finite size.

In the 'read' mode the head passively passes over the surface of the platter 'sensing' the changes in the magnetic alignment. These changes induce tiny electrical currents in the head that are converted back into data by the appropriate encoding algorithm.

We can therefore see that the basic principle is to write the data, not in 1s or 0s, but in changing magnetic orientations encoded on the surface of the platter which we can then interpret as a two-state system of 1s of 0s.

As the disk rotates the position of an individual head describes a circular 'track'. There are at least two heads in a hard drive and their combined motions describe a 'cylinder'. If we then consider that there are only so many areas in a track that can be individually written to before the magnetic orientations become so compressed as to interfere with each other, we can segregate the track into 'sectors'. By describing the physical structure of the hard drive in this way we can 'address' the location any data storage area in coordinates of Cylinder, Head and Sector. This is known as CHS addressing. As part of the physical preparation of disks by the manufacturer they are physically formatted, *i.e.* arranged into discrete sectors for data storage. The individual sectors have a structure within themselves containing a 'header' noting their physical location (CHS) followed by error correction, the data area and finally more error correction information.

Modern hard drives use Logical Block Addressing (LBA) where each sector is assigned a sequential number, starting at zero, and the hard disk controller remembers its physical location on the disk. Modern drives may still report the capacity of the drive in terms of the CHS but this is a legacy restriction imposed by the architecture and operating systems of older computers to ensure compatibility. Each time the hard disk controller receives a request to read or write data to or from a particular location on the drive it simply performs the appropriate translation to access that data.

7.3.2.2   Floppy Disks.   As the hard disk drive is named for the rigidity of its platters, so the floppy disk is named for the flexibility of its disk. Instead of a rigid substrate, the magnetic media is layered onto a single flexible plastic disk and encased within a rigid plastic case to provide protection. In the centre of the base of the disk is a circular piece of aluminium that provides for alignment of the disk in the floppy drive. The sliding aluminium 'door' on the top of the disk provides access for the read/write heads. As with hard disk drives the data is written on both sides of the disk. Rather than floating above the disk, the heads contact with the magnetic media to read the information. This necessitates a reduction in the speed at which the disk can spin without the heads stripping the magnetic material from the disk. Floppy disks commonly spin at 300–360 rpm. The heads are also fairly crude devices compared with those on hard disk drives and so fewer tracks and sectors can be fitted onto the surface of the platter. This results in a much lower data capacity than a hard disk, but greater mobility and tolerance to damage.

The floppy disk format most people are familiar with is the standard 3.5 " 1.44 MB floppy disk, usually unlabelled, that clutters their desk and drawers. Floppy disks do come in a number of different sizes and capacities, *e.g.* 5.25 ", 3.5 ", 2 MB, 1.44 MB and 720 KB, amongst others.

3.5 " floppy disks have a physical write-protection switch built into their casing. If this switch is in the closed position the floppy drive cannot write to the disk and works in read-only mode. The only safe way to examine a floppy disk in a forensically sound manner is to ensure all disks are 'write-protected' prior to examination. If disks are not write-protected it is possible that the examination process itself may interfere with the forensic integrity of the data on the disk by updating file 'Last Accessed' dates. This is discussed later in Section 7.4.

7.3.2.3 Zip Disks. Zip disks look like a slightly larger, chunkier version of a floppy disk. They contain a single magnetically responsive disk mounted on a spindle within a rigid plastic case. Access to the platter is *via* a small slot with a sliding aluminium cover on the top edge of the case. They have a much greater storage capacity than floppy disks with a range from 100 MB to 750 MB available. They require a proprietary hardware drive that can be internally mounted on the computer or attached externally.

Zip disks have no hardware write-protection facility. Therefore precautions must be taken during the examination to prevent alteration of the media, *e.g.* disabling anti-virus software or imaging in a read-only operating system environment.

7.3.2.4 Digital Magnetic Tape. Magnetic tape is commonly used as a backup media, for archive purposes or for system recovery after failure. Access to the data is linear, *i.e.* the tape must be read from the start to the finish, and is therefore relatively slow compared to disk-based media. Difficulties arise in the examination of digital tape as it exists in many proprietary physical and logical formats and often requires the use of proprietary hardware or software to access the data.

Digital tape media come in all sizes and capacities, from the large 'dinner plate' tape spools to the DDS tape, similar in appearance to a small 8 mm video or DV tape. Modern operating systems, *e.g.* Windows 2000, sometimes provide their own native tape backup utility.

7.3.2.5 Solid-State Storage. Solid-state storage is based on semiconductor technology. There are no moving parts. The media is rewritable and the materials used retain the ability to reorient through hundreds of thousands of write cycles. There is currently a wide variety of solid-state media available for data storage. They range in shape, size and capacity depending on their use. They are commonly used in digital cameras, camcorders, and MP3 players, as an alternative to floppy disks/CDs in the form of small USB 'thumb' drives and as solid-state disk drives.

7.3.3 Optical Media

7.3.3.1 Compact Disk. The most common optical format is the CD, which in itself comes in numerous varieties – CD-ROM, CD-R, and CD-RW being the most relevant for our purposes. Physically a CD consists of an internal data storage layer or layers, sandwiched between two protective plastic layers in a 12 cm diameter disk. Data is stored in a series of pits arranged in a spiral around the centre of the disk. Data is read from the disk by a low power laser passed over the surface and reflected back to a photosensor. When the laser beam encounters a pit on the surface an interference pattern is created in the beam and is interpreted by the photosensor and drive hardware. CDs have a physical capacity ranging from 650 MB to 700 MB.

7.3.3.2 CD-ROM. CD-ROMs are commercially produced read-only media. The recording layer is physically pressed by the manufacturer before being coated with the reflective metal layer that gives them their distinctive silver appearance. Given their origins, the data contained within the disk is generally not of forensic interest other than in cases of suspected software counterfeiting on a commercial scale. The presence of disks containing applications relevant to the offence in question may, however, be of evidential interest to an investigator.

7.3.3.3 CD-R. CD-Rs are a write-once media. They enable any user with a CD writer to create their own discs. A laser in the CD writer 'burns' the pits into the recording layer, which in this case is an organic dye. The green, blue or gold colour of the dye characterises CD-Rs. As the data is recorded onto the disk the appearance of the dye changes and the recorded portion of the disk is visible around the centre when viewed at a slight angle.

7.3.3.4 CD-RW. CD-RWs are write-erase-write or rewritable media. They require a CD rewrite-capable drive. The recording layer is a metal alloy into which pits are burned by a relatively high-powered laser. The data is read as previously but there is the added functionality to 'melt' the recording layer back into its original form and delete any data present. This data is irrecoverable. The read-write process may be repeated many times (up to 1000) before the disc starts to fail. CD-RWs are similar in appearance to CD-ROMs, but with a smoky glaze over the silver layer and usually an absence of commercial artwork on the top of the disc.

7.3.3.5 Data Layout. Data is stored in sectors, normally containing 2048 bytes of data, arranged in a spiral track starting at the centre of the disk. The user can write to the disk in a single 'track' or a series of tracks up to the capacity of the disk. A 'session' is a collection of one or more tracks. When a session is completed, it may be left 'open', *i.e.* available for further data storage, or 'closed', preventing further writing to that session. The disk itself may also be left open for future use or closed preventing further writing. Data may therefore be recorded to the disk over a period of time in a number of sessions, each session containing one or more tracks. Each session appends data to the disk, rather than replacing the data already present. It is quite common to find CD-Rs/CD-RWs containing data placed there on different dates.

A further method of writing to the disk is 'packet writing'. Using appropriate software the disk can be used similarly to a large floppy disk. The disk is formatted prior to use and files can then be 'dragged and dropped' onto it. If a CD-RW is used, data can even be erased from the disk and the space reused. The user adds data to the disk in sessions as previously. After adding data the user may again leave the session open or closed or even close the entire disk.

7.3.3.6 Forensic Implications. Difficulties can and do arise in the examination of CD-Rs and CD-RWs. The major difficulty is incompatibility between the disk and drive formats in general and also in differences in quality of disks and drives. CD-R and CD-RW disks are potentially volatile media and thus the safest environment for their examination is a read-only CD-ROM drive.

Modern CD-ROM drives will read multisession CD-R disks, but older CD-ROM drives may not 'see' all of the sessions. CD-ROM drives will also not read CD-Rs created with the UDF packet writing applications unless the particular session has been 'closed'. A session left 'open' for further use is only visible to CD writer or CD rewriter drives with the appropriate UDF software installed. In this environment there is the potential for changes in the content of the media, and great care must be taken during the examination.

CD-RW disks may be viewed in some modern CD-ROM drives, but not necessarily all. UDF packet CD-RWs may be read with some modern CD-ROM drives, but only if UDF software is available. Similarly, some CD writer drives will view the contents of CD-RW disks. All formats of CD-RW will be readable in a CD rewriter drive, but again this is an environment where changes could potentially be made to the media. In these circumstances the examiner should be guided by Principle 2. for the handling of computer-based evidence.

The contents of CD-Rs/CD-RWs can be examined thoroughly with the aid of dedicated forensic software that allows examination of the track and session structure.

7.3.3.7 DVD Formats. DVDs are the successors to CD-ROMs and are almost identical in appearance. Their data storage capacity is much greater and this provides for their use as an entertainment film media. The technology behind DVDs is the same as that of CDs. Data is arranged in a spiral track and recorded in a series of pits that are interpreted by reflected laser light. Reduced spacing between pits allowing greater data density provides their increased data capacity. Further capacity can be gained by using both sides of the disk and multiple data storage layers. As for CDs there are a number of types of DVD.

7.3.3.8 DVD-ROM. DVD-ROMs are commercially produced read-only media. As for CD-ROMs these are of limited forensic interest unless their content is such to relate directly to an offence. Their capacity ranges from 4.7 GB to 17 GB.

7.3.3.9 DVD-R. DVD-Rs are a write-once media with a capacity of 4.7 GB per side. It exists in two types of its own: DVD-R(A) for professional DVD authoring and DVD-R(G) for consumer use.

7.3.3.10 DVD-RW. DVD-RW is a furtherance of the DVD-R media with read-write capability. DVD-RW disks have a capacity of 4.7 GB per side, and similarly to CD-Rs can be rewritten approximately 1000 times.

7.3.3.11 DVD-RAM. DVD-RAM is a rewritable media using the same phase change technology as CD-RWs. DVD-RAM operates on the same principle as a very large floppy, allowing data to be dragged and dropped to a disk and even erased with up to 100,000 rewrite cycles. Their capacity ranges from 2.7 GB to double-sided media with capacity of 9.4 GB.

7.3.3.12 DVD+RW/DVD+R. DVD+RW is yet another rewritable format of the DVD with a capacity of 4.7 GB per side. DVD+R is a write-once implementation of the same format with a capacity of 4.7 GB.

7.3.3.13 Forensic Implications. Again the difficulties arise through incompatibility. Any modern DVD-ROM drive can read most DVD media with the exception of DVD-RAM (DVD-R+RW is fully compatible with all DVD-ROM drives, DVD-RW may not be). They are therefore relatively easy to examine in a forensically sound manner. DVD-RAM disks require the use of a DVD-RAM drive, which obviously presents the danger of writing to the disk. As for all other volatile media, adequate precautions should be taken in the examination of such media to ensure forensic integrity.

7.3.4 Magneto Optical Media

7.3.4.1 Magneto Optical Disks. Magneto optical disks are a combination of the technologies used in magnetic and optical storage devices. These disks and the technology that support them are more expensive than other storage options and tend only to be used by larger commercial concerns for the storage and backup of important information. The media is based on magnetically responsive disks and strong magnetic fields are required to affect the orientation of the media, thus providing protection against accidental corruption or loss, together with a long data life (up to 390 years is claimed). In order to write to the disk, a laser beam is used to heat the magnetically responsive material of the disc while it is exposed to a small magnetic field. This allows the relatively weak magnetic field to reorient the heated area of the disc, which retains this orientation after cooling.

Magneto optical discs are available in both rewritable and write-once read many times (WORM) forms. They exist in a number of physical formats and capacities: 5.25 ", 12 ", 1.2 GB to 30 GB. The media can only be read in a magneto optical drive matching their physical characteristics.

7.3.4.2 'Super' Floppies. A cheaper form of magneto optical disk is the LS120 super floppy technology. Based on a 3.5 " floppy disk this media uses normal floppy disk read/write technology with the head alignment accuracy improved by the use of a laser beam. The disk requires the use of a proprietary hardware drive and has a capacity of 120 MB. The super floppy incorporates a write-protection switch similar to a normal floppy.

7.4 LOGICAL STRUCTURE

7.4.1 Partitions and Logical Drives

We have seen how data is physically stored on media, but how does this relate to the operating system and its 'logical' data structure. Logically, we talk in terms of 'drives' rather than physical disks. A physical hard disk drive, as illustrated in Figure 7.1, may be 'partitioned' into logically discrete data storage areas within the hard disk drive, each containing their own independent file structure and format. Partitions can contain one or more logical drives, or even another partition. Removable media such as floppy disks, CDs, *etc.* do not normally contain more than one partition.

At the start of the physical hard disk is the master boot record which contains information relating to the number of partitions on the disk, where they start and finish, the type of file system they use (*e.g.* File Allocation Table (FAT) or New Technology File System (NTFS)) and which partition the operating system resides in. The logical volumes within these partitions are referred to as drives in common usage and are assigned letters by the operating system to identify them. At the start of each logical drive is another boot sector containing the partition table for that partition. Letters 'A:' and 'B:' are historically reserved for the floppy drives. The initial boot drive is normally the 'C:' drive.

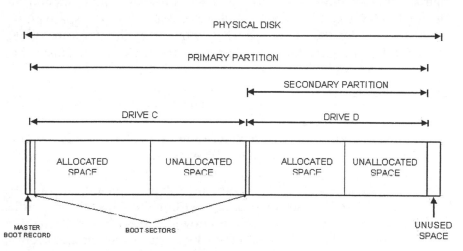

Figure 7.1 *Physical and logical disk structure*

Further hard disks (if any) are assigned drive letters after 'C:' followed by any further logical volumes. CD-ROMs, DVDs, zip drives *etc.*, are normally assigned letters following the drives contained on the physical hard disk *e.g.* 'D:' 'E:' *etc.*

7.4.2 Directory Structure

In Windows operating systems the data in a drive is arranged in a tree or directory structure, with a 'root' folder off which branch folders or files. The 'branch' folders themselves contain further folders or files. As an analogy we can think of a drive as a filing cabinet. The filing cabinet is the root folder. The drawers in the cabinet represent the folders branching off the root folder. Contained within the drawers are further folders and files. A hard disk drive has a huge data storage capacity and as mentioned is often partitioned into smaller discrete areas or drives. To further our analogy, a hard disk drive can be thought of as a room containing one or more filing cabinets. Each cabinet then represents a drive on the hard disk and has its own directory structure of drawers with their folders and subfolders.

7.4.3 File Allocation Table and Master File Table

To keep track of the data within a drive the operating system makes use of an 'index', usually found at the start of the drive. In systems such as Windows 98 this is known as the file allocation table (FAT). Each Windows directory contains a record (directory entry) for each file or folder in that directory detailing their name (including file extension), their starting locations, their size, their relevant dates and times and the file attributes *i.e.* system, hidden, archive or read-only. The space physically occupied by the file is recorded within the file allocation table. The FAT has an entry for each cluster on the drive. If the cluster has been allocated for storage, the FAT entry indicates whether the file allocated that cluster resides there solely and if not, points to the next cluster occupied by that file and so on. The concept of file structure and clusters is discussed later.

In NTFS operating systems such as Windows NT or Windows 2000, the situation is more complex. Each file or folder (as folders are essentially just another type of file) has its own separate entry in the 'Master File Table' which records all relevant information relating to the file. In fact, an entire file can exist solely within the Master File Table, which is in itself a file. A further file known as the 'Cluster Bitmap', a representation of the volume showing which clusters are in use, keeps a record of which clusters have been allocated for data storage.

7.4.4 Allocated and Unallocated Space

In discussing the logical structure of the disk the concept of space being 'allocated' to a file was introduced. This is a convenient way to think of how data exists on the disk or in fact on any media. Any file that is stored on the media has an area that it is allocated for storage. The combined allocated areas of all the files on the logical drive are often referred to collectively as 'allocated space'. Allocated space is therefore the areas of the disk currently assigned by the operating system for the storage of files. Obviously, unless the logical drive is full, not all of the available space is allocated for the storage of data. The remainder is known as 'unallocated space' (see Figure 7.1).

The amounts of allocated and unallocated space are dynamic and change as files are created and deleted, but always total to the size of the logical drive. The physical disk may be larger that the combined sizes of the logical drive or drives contained within it, leaving 'unused space' at the end of the drive. This can be due to a number of reasons including the fact that the sizes of logical drives are entirely user-definable whereas the manufacturer defines the physical disk size at the time of production. It is possible and conceivable that the user may wish to define their logical drives or partitions to be smaller than the physical constraints of the disk. Unused space can also be created by mathematical constraints on the way space is allocated.

7.4.5 File Structure

The structure of a file is roughly illustrated in Figure 7.2.

7.4.5.1 Header. Included in the directory entry for a file is its file extension, *e.g.* 'txt' for a text file, 'jpg' for a picture file, *etc.*, which identifies the file by type to the operating system and any software application. In addition to this, within the body of the file itself is a sequence of bytes characteristic to the particular file type. This is known as the 'header'. Each type of file has its own header signature that

Figure 7.2 *File structure*

uniquely identifies it. This signature is contained in the first few bytes of the file and it is this code that the operating system or application recognises. These signatures can easily be seen with a hex viewer application. For example, Figure 7.3 shows the header signature for a simple JPG picture file.

The first two bytes read hexadecimal 'FF D8'. This is the characteristic header signature of a JPG file. Other information is clearly visible within the file, *e.g.* 'Photoshop 3.0', suggesting this application was used to create the image. Whilst not forming part of the file structure this sort of information is commonly included in files.

7.4.5.2 Data Block. The layout of the data block depends on the type of file. Some file formats are universal standards common to numerous applications *e.g.* JPG files, whilst others, such as ART picture files, are proprietary. The data contents of some types of files *e.g.* Microsoft Word files, can be clearly seen in plain text. The layout of the document is contained in formatting information that surrounds the plain text, and so the actual appearance of the document cannot be immediately ascertained. As will be discussed later it is possible to search allocated and unallocated Space for this plaintext. It is this that allows us to recover even partial fragments of deleted files from unallocated space. An example of such data content is shown in Figure 7.4, again as viewed with a hexadecimal viewer.

The content of the document is clearly visible in plaintext. As well as formatting information, details such as the author and the full path of where the document was stored are also included within the file, as shown in Figure 7.5.

The value of such information as author details must not be over-stated. The author details are most likely those that were input when the software was installed and registered by the user. They are not necessarily those of the user who created the document. Information such as file paths may be of great evidential and investigative importance, particularly if they point towards removable media, such as zip or floppy disks, which may contain the complete file. The example given is specific to MS Word documents but the principle may apply equally to other file formats, while still more may appear incomprehensible when examined.

7.4.5.3 Footer. The file footer is the signature that indicates the end of the file to the software application, which then knows to read no further data for this file. Returning to our example of the JPG picture file, Figure 7.6 shows the end of the file.

```
          0  1  2  3  4  5  6  7  8  9  A  B  C  D  E  F
00000000  FF D8 FF E0 00 10 4A 46 49 46 00 01 02 01 00 F0   yØyà..JFIF.....8
00000010  00 F0 00 00 FF ED 0C 1C 50 68 6F 74 6F 73 68 6F   .8..yi..Photosho
00000020  70 20 33 2E 30 00 38 42 49 4D 04 04 00 00 00 00   p 3.0.8BIM.....
00000030  00 1C 43 02 1C 02 00 02 78 04 1C 02 74 0C 59 6F   .C......x..Yo
00000040  73 65 6D 69 74 65 2C 20 43 41 00 74 0C 39 26 4A   semite, CA..t.&J
00000050  61 79 20 54 6F 72 62 6F 72 67 20 31 39 39 39 00   ay Torborg 1999
00000060  28 77 77 77 2E 74 6F 72 62 6F 72 67 2E 63 6F 6D   (www.torborg.com
00000070  2F 6A 61 79 29 00 38 42 49 4D 03 ED 00 69 F0 DD   /jay).8BIM.i....
00000080  00 01 38 00 04 00 00 00 00 00 04 00 0A 00 04 00   ..8......8.....
00000090  00 01 38 42 49 4D 4D 0D                           ..8BIM....
000000A0  00 78 38 42 49 4D 3B 4D 49 49 00 4D               .x8BIM.6......
000000B0  00 01 00 38 42 49 4D 27                           ....8BIM....
000000C0  00 01 00 00 38 42 49 4D 27                         ...8BIM'......
000000D0  00 00 00 0A DA                                      ....
```

Figure 7.3 *JPG header signature*

```
Offset      0  1  2  3  4  5  6  7  8  9 10 11 12 13 14 15   ASCII
00001456   00 6D 05 00 00 00 00 00 00 00 AA 01 00 00 00 00   .m.......à.....
00001472   00 6C 01 00 00 00 00 00 00 00 8C 00 00 12 00 00   .l....%........
00001488   00 CE 00 00 00 0E 00 00 00 00 A8 00 00 00 00 00   .i............
00001504   00 A8 00 00 00 00 00 00 00 00 A8 00 00 00 00 00   ..............
00001520   0D A8 00 00 00 00 00 00 00 00 02 00 D9 00 00 00   ....Ù.........
00001536   0D 0A 05 48 65 6C 6C 6F 2E 20 54 68 69 73 20 69   ..Hello. This i
00001552   73 20 61 20 65 78 61 6D 70 6C 65 20 64 6F 63 75   s a example docu
00001568   6D 65 6E 74 20 74 6F 20 69 6C 6C 75 73 74 72 61   ment to illustra
00001584   74 65 20 66 69 6C 65 20 73 74 72 75 63 74 75 72   te file structur
00001600   65 2E 0D 00 00 00 00 00 00 00 00 00 00 00 00 00   e.............
00001616   00 00 00 00 00 00 00 00 00 00 00 00 00 00 00 00   ..............
00001632   00 00 00 00 00 00 00 00 00 00 00 00 00 00 00 00   ..............
00001648   00 00 00 00 00 00 00 00 00 00 00 00 00 00 00 00   ..............
00001664   00 00 00 00 00 00 00 00 00 00 00 00 00 00 00 00   ..............
00001680   00 00 00 00 00 00 00 00 00 00 00 00 00 00 00 00   ..............
```

Figure 7.4 *Data content within a file*

	0	1	2	3	4	5	6	7	8	9	A	B	C	D	E	F	
00001330	0E	00	4A	00	6F	00	6E	00	61	00	74	00	68	00	61	00	..J.o.n. . . . .
00001340	6E	00	20	00	48	00	65	00	6E	00	72	00	79	00	1C	00	. .H.e.n.r.y...
00001350	43	00	3A	00	5C	00	4D	00	79	00	20	00	44	00	6F	00	C.:.\.M.y. .D.o.
00001360	63	00	75	00	6D	00	65	00	6E	00	74	00	73	00	5C	00	c.u.m.e.n.t.s.\.
00001370	74	00	65	00	73	00	74	00	20	00	64	00	6F	00	63	00	t.e.s.t. .d.o.c.
00001380	2E	00	64	00	6F	00	63	00	FF	40	03	80	01	00	18	00	..d.o.c.y@.□....
00001390	00	00	18	00	00	00	30	98	8F	02	01	00	01	00	18	00	......0~□.......
000013A0	00	00	C0	00	0C	00	17	00	00	00	00	00	00	00	02	10	....
000013B0	00	00	C0	00	0C	00	00	19	00	00	00	90	00	00	08	00	...........□....
000013C0	40	00	C0	FF	FF	01	00	00	00	07	00	55	00	6E	00	6B	@..yy......U.n.k
000013D0	00	6E	C0	6F	0C	77	00	6E	00	FF	FF	01	00	0B	00	00	.n.o.w.n.yy.....
000013E0	00	00	C0	00	0C	00	00	00	00	FF	FF	01	00	00	00	00	.........yy.....
000013F0	00	FF	FF	00	0C	02	00	FF	FF	00	00	00	00	FF	FF	00	.yy....yy....yy.
00001400	00	02	00	FF	FF	00	00	00	00	03	00	00	00	47	16	90	...yy........G.□
00001410	01	00	00	02	02	06	03	05	04	05	02	03	04	87	3A	00	.............‡:.
00001420	00	00	00	00	00	00	00	00	00	00	00	3C	00	FF	00	00	.............y..
00001430	00	00	00	00	00	54	00	69	00	6D	00	65	00	73	00	20	.....T.i.m.e.s.
00001440	00	4E	00	65	00	77	00	20	00	52	00	6F	00	6D	00	61	.N.e.w. .R.o.m.a
00001450	00	6E	00	00	00	35	16	90	01	02	00	05	05	01	02	01	.n...5.□........
00001460	07	06	02	05	07	00	00	00	00	00	00	00	10	00	00	00	....
00001470	00	00	00	00	00	00	00	00	80	00	00	00	00	53	00	79	........□....S.y
00001480	00	6D	00	62	00	6F	00	6C	00	00	00	33	26	90	01	00	.m.b.o.l...3&□..
00001490	00	02	0B	06	04	02	02	02	02	02	04	87	3A	00	00	00	...........‡:...
000014A0	00	00	00	00	00	00	00	00	00	00	00	FF	00	00	00	00	...........y....
000014B0	00	00	00	41	00	72	00	69	00	61	00	6C	00	00	00	22	...A.r.i.a.l..."

Figure 7.5 *Typical information within an MS Word file*

```
           0  1  2  3  4  5  6  7  8  9  A  B  C  D  E  F

0001E910   F4 EB 40 60 18 D0 FD 0D 54 10 0D 33 B4 DF CA 80   δë@`..Đý..T..3´ßÊ□
0001E920   81 3E 07 C3 4A A1 82 47 43 7F 2A C8 EB F8 1A 08   □>.ÃJ¡‚GC□*ÈëΦ..
0001E930   82 46 87 E9 57 41 4F 91 A2 38 DE D0 68 04 CF 81   ‚F‡éWAO'¢8ÞĐh.Ï□
0001E940   FA 1A 8A 0B 12 FA C1 83 24 88 D6 A2 3F FF D9      ú.Ø...úÁf$ˆÖ¢?ÿÙ
```

Figure 7.6 *Footer signature for a JPG file*

The last two bytes contain the hexadecimal characters 'FF D9'. This is the footer signature for a JPG file. It should be noted that not all files have a common footer signature.

Above we have noted the information that can be found within the file. What cannot normally be determined from examining the file only is information such as the file name and its relevant date and time stamps. This information is not normally contained within the file itself but in its directory entry (though there are exceptions). Such information may, however, be inferred from the contents of the file where *e.g.* a date is used in a letter.

7.4.6 Dates and Times

Contained within the directory entry for Windows files are two sets of dates and times and one date. These represent respectively the 'file created' date and time, the 'file last modified' date and time and the 'last accessed' date. They are collectively referred to as date and time stamps. For Windows NTFS operating systems the last accessed date and time are recorded, along with a fourth date and time representing the date and time the MFT entry for that file was modified. In order to understand the significance of the dates it is first necessary to have an understanding of what they represent.

7.4.6.1   File Created Date and Time.   This is set when the file initially comes into existence on a particular logical drive regardless of its source *e.g.* created using an application or downloaded from the Internet. It is not a permanent record of the date and time a file was created, only when it initially existed on this logical drive. Within a logical drive files may be 'moved' between directories without changing the file created date and time. The term 'move' means using 'drag and drop' or 'cut and paste' for single or multiple files. However, if a file is 'copied' from one directory to another using 'copy and paste' or 'drag and drop' with the Ctrl key depressed then the file created date and time will be updated. This makes sense as effectively the copy command brings into existence a second version of the original that was created at the time of copying. If a file is moved or copied between two logical drives, normally a second copy of the file is created on the target drive leaving the original on the source drive and the file created date and time of the new file will therefore be updated. To further complicate matters if a file is dragged and dropped between two logical drives with the shift key depressed or using cut and paste then the file is removed from the source drive and placed on the target drive and the file created date and time remain unchanged.

7.4.6.2   File Last Modified Date and Time.   The file last modified date and time represents the date and time that the system last modified the contents of the file and wrote the changes to the disk. Logically this is initially when the file was created and changes every time the file is modified and the changes saved. Viewing the contents of a file or changing its name do not affect the last modified date and time, as these are changes to the directory entry and not the data contained within the file itself. When a file is moved within or between logical drives the last modified date and time is not affected. The contents of the file have not after all changed, only its location.

7.4.6.3   File Last Accessed Date.   The last accessed date for a file is simply the date on which the user last accessed it. It may be thought of as a 'hair trigger' date as it is updated each time the file is accessed and each access overwrites the previous date. No record exists of any previous access, only the most recent.

7.4.6.4   Forensic Implications.   Interpretation of file dates can provide information on what the user may have known or done with files. For example, an apparent anomaly often arises where examination shows that the created date and time for a file is after the file was last modified by the system. Interpreted literally this suggests that the file was some- how changed before it ever existed. This can obviously never be the case. However, an understanding of how the dates and times work suggests that the file may have been initially created on an unknown date, prior to its last modified date, and copied to its present location on the date now stated as its file created date. This does not tell us much about the origin of the file, it may have been simply copied from another directory or even extracted from a zip archive file and the result would be similar. What it may infer however, is knowledge on the part of the user of the files existence in that they have performed actions in relation to it.

Individual software applications may affect date and time stamps differently. Compression software is a good example of this. Winzip, a very common application, maintains the last modified date only when it compresses a file. When uncompressed the file created date and time is exactly that of the last modified date and time. Other compression programs however, *e.g.* EnZip similarly maintain only the last modified date on compression, but on decompression give the file created date and time as per the current date and time, but maintain the last modified date and time as per the original file.

The dates and times used by the operating system are obtained from an electronic clock on the motherboard with its own small independent power supply. As the computer starts up, the operating system reads the current date and time from this clock and applies this to itself. Similarly, if the user changes the date and time in the operating system environment, this updates the system clock. The impact of this is that the dates and times associated with files are dependant on the accuracy of the system clock and are subject to manipulation by the user. It is an essential part of any examination of a computer to establish the date and time as recorded in the system BIOS and note any discrepancy.

7.4.7 Sectors and Clusters

Previously we have noted that the sector is the smallest physically addressable area on a disk. Logically a sector corresponds to 512 bytes of storage capacity (with the exception of CDs normally with a sector size of 2048 bytes). Historically, the origin of hard disks and modern storage media is the floppy disk. For small capacity media such as the floppy disk, data can be effectively and efficiently written to the disk sector by sector. The capacity of storage devices has increased rapidly since floppy disks and to access huge numbers of sectors individually is neither effective nor efficient. To overcome this difficulty the operating system may not allocate data to individual sectors but to blocks or 'clusters' of sectors. The number of sectors in a cluster is determined by the size of the disk and how it is logically formatted *i.e.* FAT12, FAT16, FAT32 or NTFS. Whatever the logical size of a file, it is allocated disk space in a single cluster or multiples of clusters. If a file is larger that a single cluster, a second cluster is assigned for the excess and so on. The sectors within a cluster are contiguous, that is to say they occupy areas of the disk that are adjacent to each other. Multiples of clusters forming the same file however may not be contiguous but 'fragmented' *i.e.* occupying areas throughout the disk.

7.4.7.1 Slack Space. Whilst data is allocated to files in cluster-sized blocks, it is written to the disk in sector-sized chunks of 512 bytes at a time. A whole cluster, or multiple of clusters is assigned for the storage of a particular file and that file only, regardless of how little of the cluster it occupies. The forensically interesting thing about this is contained within the final cluster (or at the end of its only cluster) for each file. The whole cluster is allocated to the file but the file may not be large enough to 'fill' it completely.

Consider the theoretical situation as illustrated in Figure 7.7. The operating system is allocating space to files in four sector for cluster blocks *i.e.* 2048 bytes. The file 'Hello.txt', a simple text file, is 800 bytes

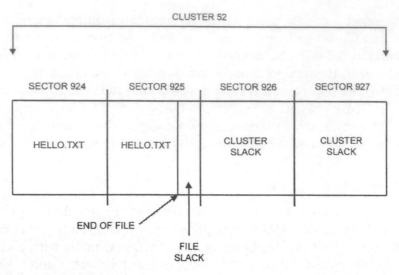

Figure 7.7   *Slack space*

in size, but is allocated one cluster (2048 bytes) of storage space. The operating system can only write to the disk in sector-sized 512 byte chunks, so it writes 512 bytes to each of the first two sectors in the cluster and leaves the remainder untouched, but still allocated to 'Hello.txt'. This raises two interesting situations:

1. The file is only 800 bytes in size, but the operating system has written 1024 bytes of data to the first two sectors. Where did the extra 224 bytes of data come from? The answer is from whatever surplus data the operating system has lying around in its data buffers or registers from previous operations it was performing. Thus, data that the user has been working on may be found in the 'slack space' between the end of the file and the end of the last sector the file was written to. This will obviously only be small amounts of data but potentially could include user names or fragments of documents. This type of slack space on the disk is referred to as 'buffer' or 'file slack'.

2. The remaining two sectors in the cluster are untouched by the file being written to the disk. Their contents are quite simply whatever was last written to those sectors by the operating system. This is due to the manner in which files are deleted which will be discussed later. Therefore the sectors may contain the remnants of whatever file was last allocated to this cluster but not overwritten by 'Hello.txt'. This type of slack space is referred to as 'cluster slack'.

The example used of four sectors for cluster is artificial: ratios of 32 or 64 sectors for cluster are more likely, giving us a possible cluster slack space approaching 32 KB in size. This is a considerable amount of storage space and could contain a great deal of evidential information. As slack space forms part of the area assigned to a file it will remain untouched unless the file 'grows' larger, when it will start to overwrite its own slack space. To the normal user of the system, slack space is invisible. Slack space is part of what we have considered to be allocated space, even though it contains the remnants of deleted files that once formed part of unallocated space. It should be noted that when a file is moved or copied from one logical drive to another it does not retain the slack space of the original file, either file slack or cluster slack.

7.5 CONTENTS OF ALLOCATED SPACE

All files that are 'live' on the media exist in allocated space, *i.e.* the space assigned by the operating system for storage of those files. If a file exists on a piece of media it must have been created by a user, by an application as a part of its operating processes or received from an outside source *e.g.* the Internet or email. While the content of any file in allocated space may be of an evidential nature, the following are examples of files of particular forensic interest. Details of Internet activity as discussed in Section 7.7 may also fall within the boundaries of allocated space.

7.5.1 Link Files

Link files are 'shortcuts' created by the Windows operating system for ease of use by computer operators. They are the shortcut icons on the Windows Desktop, the list of recently opened documents we see when we click on 'Start, Documents', the contents of the 'Start Menu' and the list of 'Send To' options presented when we right-click on a file. Link files have the file extension '.lnk'.

7.5.1.1 Desktop. The desktop icons we look at each time we use Windows are contained within the 'Windows\Desktop' folder. Their purpose is to provide quick and easy access to installed programs or a favourite file or folder. The presence of link files on the desktop can indicate which applications or files the user wishes to access frequently or quickly, *e.g.* an encrypted disk area that the user can simply drag and drop files into. Further they may indicate a degree of knowledge on the part of the user that a relevant application was installed or a file present on their computer.

7.5.1.2 Recent Documents. The list we see in the 'Start, Documents' menu is of link files contained in the 'Windows\Recent' folder relating to the most recent files accessed by the user. These are particularly valuable links. They may point to now deleted files, files the user accessed over a network and never saved locally, files stored on removable media or even files in encrypted volumes. The presence of the link file proves that the user, or other person using the computer, has accessed these files. It is not enough that the user has right-clicked, moved or deleted a file; in order to create a link file in the 'Recent' folder they must have opened the file.

7.5.1.3 Start Menu. The contents of the start menu are link files within the 'Windows\Start Menu' folder. As previously they may prove user knowledge that applications relevant to particular offences were installed on the system.

7.5.1.4 Send To. When we right-click on a file or folder, we are presented with a series of options including 'send to'. The link files within the 'Windows\Send To' folder provide these links to applications or physical locations. These links can provide very valuable indications to the examiner that the suspect is utilising removable media for data storage, file encryption software or even secure deletion applications.

Link files are extremely valuable sources of information about the files they relate to. A link file is just like any other file, with date and time information of its own. However, contained within the link file itself is not only the name and location of the target file, but also the date and time stamps of the target file. This information would, for example, assist an examiner in connecting a file found on removable media to a link file found on the computer, by examination of date and time stamps, thus proving access on the suspect computer. The format of the information stored within the link files is such that all the data content is not viewable in plaintext. Forensic utilities are available that allow examiners to extract information from these types of file.

7.5.2 System Swap File

When running applications, computers make use of their random access memory (RAM) to hold short-term data necessary to perform tasks. There is only a finite amount of RAM available to use so the operating system also uses temporary space on the hard disk as memory. This operation is known as 'paging'. The area used for paging is referred to commonly as the 'swap' file. In Windows operating systems the swap file is a single system file whose size is dynamic and controlled by the operating system (by default) or set to a fixed size by the user. In Windows 95/98 systems this file is called 'win386.swp' and resides in the root folder of the operating system partition. In Windows NT and 2000 the swap file is called 'pagefile.sys' and again resides in the root folder of the operating system partition.

7.5.2.1 Forensic Implications. The contents of the swap file can literally be any part or all of any file, be that a document, image or other, that was created, accessed or edited, either locally, *i.e.* on this machine, or remotely *via* a network. Obviously this material is of great interest in any investigation. The swap file is a hidden system file and its operation is invisible to the user. Its contents can, however, be examined with any file viewing utility *e.g.* Quick View Plus, *etc.*

As noted, the size of the swap file can be fixed or dynamic, defaulting to dynamic unless fixed by the user. A dynamic swap file increases and decreases in size automatically as required by the operating system and can reach a considerable size (100s of MB). A great deal of information may therefore be contained within it. Its contents are continually written and overwritten and as its size decreases the excess becomes part of unallocated space. Unallocated space is further discussed in Section 7.8. A fixed swap file will not change size and its contents, as they are written and overwritten, will remain in allocated space.

7.5.3 Digital Cameras

The use of digital cameras by commercial and private photographers is now widespread. An integral part of digital photography is the use of computers for image enhancement or printing.

7.5.3.1 Storage Formats. Digital cameras store their images in internal solid-state memory or removable media, *e.g.* solid-state 'cards' or microdrives. Most modern cameras use the latter. There are a number of different physical solid-state formats available including Smartmedia, Compact Flash and Secure Digital (SD) cards, and the Sony Memory Stick. Card media are currently available to around 512 MB capacity. A further high capacity storage method, the 'microdrive', employing a miniaturised hard drive, is available for some cameras with up to 1 GB of capacity currently available. Data format on the media is in a FAT format with a sector size of 512 bytes. Images are stored in any of a number of different graphical formats (JPEG, TIFF, *etc.*) depending on the size of the images and the quality required by the user.

The media is rewritable and the data contained on them is therefore volatile. Write-protection is available on most of the media formats and should be used where present to ensure integrity of evidence. Users access the images either directly from the camera using a proprietary cable and software provided by the manufacturer or make use of a hardware card-reading device. The media is simply treated by the operating system as a removable disk.

7.5.3.2 Exif Data. All images on the media have the date and time information associated with any normal file recorded within their directory entry. In addition, when an image is taken with a digital camera it is very common for technical data about the image, known as 'exif' data, to be recorded. This data is recorded within the body of the image by the camera. Exif data normally includes information about exposure, aperture, ISO speed, *etc.* However, it may also contain information about the date and time the image was originally taken, the make, model and possibly even the serial number of the camera used. These may be preserved when the image is transferred to the hard disk drive or CD-ROM for storage or archiving, providing very valuable information about its source.

Whether or not the exif data is preserved depends on number of factors including the software used to transfer the image from the camera, the format chosen by the user for storage, *e.g.* JPEG or TIFF, and the use of image manipulation software. The exif data is visible to many utilities including the more common image manipulation packages or utilities designed specifically for reading exif data such as 'Exifread'. Note, the dates and times recorded in exif information are set by the camera clock and are dependant on these settings being correct or even having been set at all. They are therefore subject to manipulation by the user.

7.5.3.3 Deleted Images. The deletion process is the same as for any other file. The unallocated space on the disk will therefore contain the data from any deleted images. Due to the limited space available on the storage media, it is likely that images will be overwritten very quickly after deletion if the media is in constant use.

7.6 CONTENTS OF UNALLOCATED SPACE

The contents of unallocated space (if anything at all) are the remnants of data that must once have existed on the system as 'live' files and have subsequently been deleted. The files could have come to be on the system in any of the ways listed for the contents of allocated space. What is important is that the files must have existed as live files on the media in order for them to now reside in unallocated space.

Unallocated space exists on any piece of storage media that it not filled to capacity. Examination of the contents of unallocated space can provide valuable evidence or suggest further directions of enquiry to the examiner.

The contents of unallocated space are invisible to the user accessing the media through their operating system. However, forensic tools from complex imaging and examination tools to simple hexadecimal viewers allow examiners to see the physical data layout on the media, *i.e.* to look beyond the logical data structure as shown by the operating system and examine what data lies outside the boundaries of live files.

The following is a description of the general process of deletion and examples of the type of processes that normally take place in the general use of a computer that may leave traces of this activity in unallocated space. They are only a few examples; the list of potential processes leaving traces in unallocated space is endless.

7.6.1 Deleted Files

As explained eariler files are allocated storage space on the logical drive by the operating system and their locations are tracked by means of their directory entry and file allocation table or *via* the master file table. But what happens when a file is deleted? Contrary to what is commonly believed data is not removed from the drive.

7.6.1.1 The Deletion Process. The detailed mechanics of the deletion process are not vital to understanding the forensic implications. Windows FAT and NTFS systems handle file deletion in different ways. They both make use of a 'recycle bin'. For FAT systems there is only one recycle bin whereas for NTFS systems a folder called 'recycler' exists containing a recycle bin for each user profile. When the user initially deletes a file, Windows operating systems effectively 'place' the file in the recycle bin from where it can either be deleted (by emptying the recycle bin) or restored. When the recycle bin is emptied the operating system marks the area previously occupied by the deleted file as available for reuse, so that the file is no longer 'addressed' by the operating system. The data remains in its original location and form until the operating system uses that same space again to store another file. What it is important to understand is that regardless of the method of deletion used by the operating system, the data is in no way affected by the process.

7.6.1.2 The Recycle Bin. Examination of the recycle bin entries can provide valuable information for the examiner. The naming of files in the recycle bin is significant as it adheres to a strict protocol. Files are named according to the convention 'D<Original drive letter> <number>.original file extension' *e.g.* 'DC1.doc'. This may point the examiner or investigator towards files in removable media of which they were previously unaware. A special system file called the INFO2 file also resides in the recycle bin. This file is hidden from the normal user but can be seen on an imaged disk just as any other file. Its purpose is to record the names and locations of files that have been deleted and the date and time of their deletion to allow them to be restored if necessary. The exact format of this file is dependant on the variant of Windows installed, but the names and original location of files are in plain text and the dates and times recorded in Windows date and time format. When the recycle bin is emptied, the file entries and the INFO2 file are deleted.

7.6.1.3   Forensic Implications.   The essentials of the above are that on deletion, the data contained in a file is not 'gone', it is merely 'hidden' from the operating system and the space it occupies made available for reuse. While the data is deleted, it still resides in the space previously allocated to it unless overwritten. Thus deleted data may be overwritten, either completely or partially, and become corrupted. Large files often exist in multiple fragmented clusters and after deletion become individual fragments of the file, each of which may be overwritten or remain separately intact in unallocated space.

It is possible to 'undelete' a file even after the recycle bin has been emptied. This can be done manually, but realistically is best left to software designed for this purpose. When files are undeleted the examiner has all of the information associated with the original file, such as dates and times, available to them. The only information that cannot be recovered is the first letter of the file name, which unfortunately is overwritten in the deletion process.

Even if a file cannot be successfully undeleted, it may still be intact in unallocated space. By searching unallocated space for the header and footer signatures of particular file types we can effectively 'recover' its contents. Again, it is possible to perform this task manually, but the facility is built into many modern forensic examination tools. Even if the file has been partially overwritten, some of its contents may still be visible, as shown in Section 7.4.5.2. While these do not form complete files they may still contain valuable evidence.

There are evidential difficulties with files recovered from unallocated space. We cannot state the date and time attributes of even a complete file found in unallocated space, as we have no directory entry or Master

File Table entry relating to the file. It is possible that there may be information contained within the file from which we may infer an approximate date of creation *e.g.* the contents of a letter may contain a date of writing or refer to incidents whose timing is known.

We have previously referred to slack space as part of allocated space, which it correctly is. However the contents of the cluster slack were in unallocated space until the cluster was reallocated, so we are really looking at a piece of unallocated space. It is possible to infer date and time information for data contained in cluster slack not only from content but also from the date and time information of the file that overwrote it, *i.e.* the data in cluster slack must have existed prior to the file that overwrote it.

7.6.1.4 Wiping. Deletion is often confused with other actions such as wiping or formatting the drive. Wiping is better described as secure deletion. It differs from deletion in that wiped files have their directory entries and their allocated space physically overwritten by a random or even user-definable series of characters. The overwriting is often repeated to allow for deviations in the movement of the disk read/write heads. Free space or unallocated space wiping is also often available where the contents of unallocated space are overwritten as above. If files are properly wiped they are irrecoverable by normal forensic means. That is not to say there will be no evidence of their having been present at some time, *e.g.* link files.

7.6.1.5 Formatting. Formatting a drive or disk is normally done initially to prepare the drive for use. Reformatting a drive or disk that has been in use will destroy the FAT/MFT and mark all space on the drive as available. This effectively deletes all the pointers that mark the start of files but leaves the data that formed the files intact in their original locations.

7.6.2 Word Processed Documents

7.6.2.1 Temporary Files. Word processing applications are an essential part of most business and personal computer software libraries. Users may create or edit their own documents, or those created by others, or simply view the contents of a document previously created on another computer. In all of these situations it is possible that all or part of the document may be left in unallocated space long after the user has viewed, printed, deleted or saved the document to removable media. This is due to the way in which applications of this type operate.

Perhaps the most commonly used word processing application is Microsoft Word. MS Word makes extensive use of temporary files to allow normal operation of the computer to continue unhindered while the user works on documents. Temporary files are also created to ensure that data is not lost when the application performs certain tasks *e.g.* when saving a file. When MS Word has completed the task for which the temporary file was created, it is deleted. There are a large number of temporary files used by MS Word and their interactions can be complex. However, a few deserve particular mention, such as those created during the normal save process and as automatic saves performed by the application to recover the file if the application is unexpectedly terminated *e.g.* by a system crash. These temporary files are a copy of the original file and contain all the information normally associated with that file *e.g.* content, author details, full path, *etc.* When the application has completed the task for which the temporary file was created, it is deleted and enters unallocated space.

7.6.2.2   Locations of Documents.   The locations of user created files in general can be of significant evidential value. If a user creates a new folder to store documents or images they have created, Windows will by default name that folder 'New Folder'. A second new folder in the same directory will be named 'New Folder 2' and so on. If the user should, for example, rename the new folder 'Young' and store indecent images of children in it, the implication is that the user created the folder for that specific purpose and was aware of the presence of such images and their contents. The principle is equally relevant to documents.

The default save location for MS Word files is the 'My Documents' folder, though the user can reconfigure this. If an MS Word document is saved elsewhere it must have been placed there by the user. For Windows NTFS systems, a separate profile exists for each user with their own 'My Documents' folder.

7.6.2.3   Naming of Documents.   MS Word not only defines a default location for saving documents but also automatically suggests a name for the document to be saved based on the contents of the first line or sentence of the document.

7.6.2.4   Summary.   The above are specifics for MS Word but the general principles may be equally applicable to any word processing application. The details are dependant on the individual application used. In relation to the production of documents in general, it is possible that while a document was being produced a portion of the document, particularly if it is a large document, may have been placed in the swap file and thus entered unallocated space. Whilst it is maybe not feasible to search unallocated space for all coherent text fragments, it is certainly possible to link a document suspected of being produced on a computer to that computer by means of a keyword search for relevant or unusual terms contained within the text. This would identify any of the text fragments left behind by auto saves, deleted copies of the document or contents of the swap file.

7.6.3 Printed Documents

The printing methods employed by the various Windows operating system variants have differences in detail, but the concept is the same throughout. In order to print the contents of a file, be that a document or an image, Windows makes use of two temporary files.

The purpose of the first temporary file is to convert the data contained within the file to a format that the printer can understand. To do this Windows creates a temporary file in either Windows Enhanced Metafile (EMF) or in a RAW graphical format. These are both graphical formats and in simple terms the printer receives a 'picture' of the document to be printed.

In addition to the 'picture' of the document, Windows also creates another temporary file containing the information needed to complete the print job, *e.g.* the name and path of the file to be printed, the printer to use, *etc*. These files are only temporary files. After printing is complete they are deleted by the operating system. Their contents then become part of unallocated space.

7.6.4 Summary

As we can see, the contents of unallocated space can be a very rich source of information. Files that the user has deleted and thought gone may still reside there. Searching unallocated space with suitable forensic tools for the content of documents, names or even particular types of files can produce valuable evidence. As previously noted in relation to the swap file, literally any file accessed on the computer can leave traces of its passing in unallocated space. The rate at which the contents of unallocated space are overwritten is determined by a number of factors including the size of the hard disk, the size of the area currently unallocated and the amount the computer is used. Obviously data will be overwritten at a slower rate on an infrequently used computer equipped with a large hard drive and only a small amount of allocated space.

7.7 INTERNET ACTIVITY

7.7.1 The Internet

Firstly, what is the Internet? It would be an oversimplification to describe the Internet as a huge network of computers, but it does consist of literally millions of computers, either stand-alone or in networks of their own, with the ability to communicate with each other by means of a common protocol known as the TCP/IP protocol.

These millions of computers are not directly connected to each other all the time. When a computer or 'host' is connected to the Internet it is assigned an 'address' in the form of an IP number. The IP number system uniquely addresses each computer and allows the host to send and receive data in small parts called 'packets' to and from another host. Error checking and acknowledgement facilities are included to ensure the data arrives intact at the correct location. The packets of data travel from host to host independently of each other by whatever is the most viable route available to them at that time.

The parts of the Internet that we are familiar with such as email, web browsing, *etc.* are utilities that make use of this connectivity between computers.

7.7.1.1 Connecting to the Internet. The majority of personal users access the Internet *via* a dial-up connection to their 'Internet Service Provider' (ISP) whereby the user connects to their ISP's internet server *via* their telephone line, are assigned an IP address by the ISP and then proceed however they choose. Larger organisations such as large businesses or academic organisations connect to the Internet through their own servers with direct connections *via* 'leased lines' *i.e.* high-speed telephone lines. The physical links that provide the connections between computers are leased lines, satellite links, microwave transmitters and receivers. In its travels from one computer to another, a packet of data may well make use of all of these types of connection.

7.7.2 Internet Protocol (IP) Numbers

An IP number takes the form of four numbers separated by a period (.), each number in the range of 0 to 255. This number forms a unique address or identifier for each host. However, the situation is slightly more complex. Large organisations such as Internet Service Providers are assigned a 'block' of IP addresses for their exclusive use. What normally happens is that they have more customers than IP numbers, so each customer cannot be assigned an IP number for their exclusive use. The ISP overcomes this by dynamically allocating IP addresses to each customer as they go online. When the customer goes offline and no longer requires the IP number it is placed back into the pool and can then be reallocated to another customer. This is the case for the majority of dial-up customers. The implication of dynamic IP allocation is that an IP number uniquely identifies a customer for the date and time that IP number was assigned to them. Forensically, IP numbers are the fingerprints of users when they are online.

7.7.2.1 Domain Name System. In order to make the Internet more human friendly, a system of converting the IP numbers of hosts to names was established. This is known as the 'Domain Name System' or DNS. Domain names are registered to particular companies or organisations. When a user attempts to connect to a host *e.g.* a web page or sending an email, they enter the domain name of the host they wish to connect to, rather than the IP number. This is then converted to an IP number by referring to one of a number of worldwide servers that maintain details of the registered domain names and their corresponding IP numbers. The connection can then proceed based on the IP numbers of both parties. Similarly, given an IP number it is possible to 'look up' the various registries and establish to which domain this IP number belongs.

The allocation of IP numbers and domain names is handled by ICANN, the Internet Corporation for Assigned Names and Numbers. Regional internet registries detailing the allocation of IP number blocks and domain names are maintained by a number of regional organisations worldwide. They are: 'RIPE' – Reseaux IP Europeens; 'ARIN' – American Registry for Internet Numbers and 'APNIC' – Asia Pacific Network Information Centre.

7.7.3 World Wide Web (www)

The World Wide Web is commonly misconceived as being the Internet. It is a vast collection of 'web pages', produced using a common language (Hypertext Mark-up Language or HTML) with multimedia-based content. The pages are graphical in nature and highly interactive. Each web page has a unique address known as a uniform resource locator (URL) with the form *e.g.* 'http://www.microsoft.com/downloads.html'. The URL consists of the protocol used to access the page (http:// or hypertext transfer protocol), the domain (www.microsoft.com) and the filename (downloads.html). The URL may not point directly to a filename but merely the domain of the site, in which case a default index page for the site is automatically loaded.

To give an idea of scale, Google, one of the most popular search engines for the World Wide Web, currently searches 2,469,940,685 web pages (September 2002).

7.7.3.1   Web Browser Applications.   The World Wide Web is accessed using web browsing software applications. There are a large number of web browser applications available, from commercial products (MS Internet Explorer, AOL, Netscape Navigator) to free shareware utilities developed by users for their own needs. Browsers interpret the contents of web pages and present it to the user in visual and even audio format. Regardless of the particular browser used, the concept of how they work is as follows.

7.7.3.2   Web Page Structure.   The structure of a web page is contained within its HTML code. This records the layout of the page and where its contents, *e.g.* the images on the page, are to be found on its host server. The user views the contents of a web page by copying it to their own computer. This is known as 'downloading'. The user's web browser downloads the HTML code for the page requested by the user along with any content, *e.g.* pictures, as indicated within the HTML code. The browser then interprets the data and displays it to the user. The web page is not simply 'viewed' on the computer monitor; a copy of the web page is made on the hard drive of the user's computer.

7.7.3.3   Internet Cache.   The copy files stored by the browser are generally known as 'internet cache' files or 'temporary internet files'. Their location on the hard drive is dependant on the web browser used. Most browsers have a default location that may be user configurable. Unless otherwise instructed by the user the web browser will store the cached files in case the user wishes to visit the page again, in which case the browser will refer to its cached copy of the web page to speed access. The browser normally retains these files for a set period of time that can be reconfigured by the user.

To keep track of which files relate to which web page and when that page was visited *etc.*, the browser may create further files for its own use. Microsoft Internet Explorer is perhaps the most popular web browsing application in use today. Internet Explorer stores information such as this in a number of files named 'index.dat' files. These are found in default locations and record information including the URLs of every site visited, along with the date and time of the visit, files accessed, *etc.* The format of the 'index.dat' files, and indeed such files for other browsers, is proprietary to the application. As such they can only be interpreted fully either by their parent application or by the use of specialised software designed specifically for this purpose.

7.7.3.4 Forensic Implications. The content of these files can be extremely valuable. Included within the URLs will be the name of any file accessed *i.e.* downloaded, using the browser. The URLs also contain information regarding the use of common search engines. When the user submits a query to the search engine, the results are displayed in a web page whose URL may contain the search terms. For example, a search on Google for 'computer forensics' will return a search page whose URL is: 'http://www.google.com/search?hl=en&lr=&ie=UTF–8&oe=UTF–8&q=computer+forensics'. Information such as this can obviously be indicative of a suspect's intentions when browsing.

7.7.4 Email

Email is undoubtedly the most popular and widely used Internet application. Put simply, it is a means of communicating electronically with a single person or group of persons. Email messages take the form of a text message with the facility to send any other type of file with the message including graphics, multimedia, executable files, *etc*. The email message consists of two parts – the message 'body' and the message 'headers'. Email addresses also contain two parts, the domain name of the recipient and the recipient's user account in that domain. For example, 'jon.smith@fireserve.com', where 'fireserve.com' is the domain name for the recipient and 'jon.smith' is the particular user in that domain.

7.7.4.1 Headers. The headers are the information that helps the message to reach its destination and allow the recipient to reply. The fields visible at the top of every email message, *i.e.* the 'From:', 'To:', 'Date:' and 'Subject:' lines, are part of the headers but do not show the full information. These are normally referred to as the simple headers. The full header information is much more detailed and can be viewed by accessing the properties of an individual message. The message headers contain a record of each stage of the journey from sender to recipient, tracking their origin and the location of the computers they passed through en route by means of IP numbers.

In the first stage in the journey, the sender composes the message using their email application software (*e.g.* Pegasus Mail, Outlook express, *etc.*), connects to the Internet *via* their ISP and sends the message. The email application adds the first set of headers to the message, the simple 'From:', 'To:', 'Date:' and 'Subject:' lines, and passes the message to an server belonging to their ISP dedicated to dealing with email, known as a 'mailserver'. The ISP mailserver determines the location of the intended recipient's mailserver by examining the simple headers and converting the domain name portion of the recipient's email address into an IP number using the Domain Name System (DNS). It then directs the message onwards to the mailserver of the intended recipient.

The message does not usually pass directly from the sender's mailserver to the recipient's. Normally it is routed through one or more servers on the way. Each time the message is received by another server another line of information is added to the top of the headers in the form of a 'Received From:' field. This line details:

1. The domain name and IP number of the computer the message was received from.
2. The domain name of the receiving computer.
3. The date and time (including time zone or deviation from GMT) of the transfer.
4. A unique ID number assigned by the receiving computer for its own handling purposes.
5. The intended recipient of the message (by the recipient's mailserver on the final transfer only).

For example:

'Received from: mail.fireserve.com (mail.fireserve.com [198.212.34.56]) by server8. messlabs.com with SMTP; id 183uhG-0004vH-00 12 Oct 2001 08:47:29 –0000,
   i.e. a message received at 08:47:29 on 12th October 2001 (–0000 hours from GMT) by 'server3.messlabs.com' and assigned ID '183uhG-0004vH-00', from 'mail.fireserve. com' whose IP address is '198.212.34.56'.

The recipient's mailserver determines which of their customers the email is for and places it in their 'mailbox', awaiting collection. When the recipient next checks their mailbox the message is passed to their computer and removed from the mailserver.

7.7.4.2 Forensic Implications.   The exact content of the email headers will vary from message to message. Some applications or servers may add additional lines to the headers for their own purposes. It is normal to trace the route of a message by reading from the top of the header to the bottom, as a line or layer is added to the top of the header each time a transfer takes place. The IP address of the sender is normally the last IP number before the message body.

Examination of the contents of email message headers is obviously a valuable source of information about the identity of the sender.

Unfortunately, some of the information may not be reliable. All fields attached to the message by the sender's computer are suspect in that the sender can change them using simple tools or techniques in an attempt to mask their true identity. This would possibly include information such as the sender's email address or date and time of the message. However, once the message is in transit it is beyond their control. This means that the information contained within the 'Received From' lines is independent and reliable. Tools used for mass mailing or 'spamming' often insert spurious 'Received From' lines into the message header, but careful examination and verification of the details should identify these types of entry.

7.7.4.3 Body. The message body contains the 'data' of the email – the text message (if there is one) and any attached files. The protocol that handles email (simple mail transfer protocol or SMTP) is text-based, therefore any graphics *etc.* attached to the message are 'encoded' by the sender and 'decoded' by the recipient. This is done automatically by the mail handling software and is invisible to the user.

7.7.4.4 Storage. Depending on the email application, the user will normally have an 'Inbox' folder for incoming/received messages, a 'Sent Items' folder for messages sent by them, a 'Deleted Items' folder much like a recycle bin, a 'Drafts' folder for messages written but not yet sent and any user created and named folders.

As previously stated the protocol used by email messages is text-based. For the forensic examiner this has the benefit that the textual content of the messages, including all of the date and time information in the message headers, may be visible using a file viewer. The matter is complicated by the fact that messages may not be stored individually, but grouped together in an 'archive' file. The content of the messages in an archive file may be visible with a file viewing application, but this is dependant on the application. For example, Outlook Express messages in their archive files are clearly visible in clear text when examined with a file viewer, but Outlook messages which are all stored in a single archive file are not. Regardless, unless the messages are viewed with the email application that created the archive file, generally they cannot be seen as they appeared to the sender or recipient. In particular, any files, such as pictures, sent with the message are text-encoded and must be decoded by the email application to be properly viewed. The location and format of the archive files is dependant on the application used.

As a final note, the protocol used by email applications to send messages is independent of the individual application. Users of AOL software can send and receive messages from Outlook users without difficulty.

7.7.5 Webmail

Webmail can be thought of as where email meets the World Wide Web. Whereas email requires the use of an application such as Outlook Express to send and receive messages, webmail allows the user to perform the same tasks using their normal web browsing software. The user simply accesses the website providing the webmail service *e.g.* Hotmail, Yahoo, *etc.*, and after entering their username and password, can compose and send messages in email format or read any email messages sent to them. Messages are still sent and received using the SMTP protocol and so are compatible with ordinary email applications. The difference is in how the messages are 'retrieved' from the server. Users browse their email messages and view their contents without downloading them to their own computer. Services such as this are also provided by most ISPs thereby allowing their subscribers to access their email account from computers other than their own.

7.7.5.1 Forensic Implications. Email messages received *via* webmail are not stored on the user's computer unless deliberately saved. The messages remain on a server belonging to the webmail or service provider, where the user can view, delete or save them. The messages are simply 'viewed' as the contents of a web page. They may therefore be found in the Internet cache, in unallocated space or in print spool files if printed. Note that the Internet cache stores web pages downloaded to the computer (*i.e.* received messages), but not information sent to the server by the user (*i.e.* sent messages). Therefore it may be possible to recover messages the user has read, but possibly not those they have sent. However, different webmail providers operate in their own way and it is not uncommon for the user to see a preview web page showing a message they have typed and asking for confirmation to send the message. In this way messages the user has sent may be cached on their computer and be recoverable.

Webmail and email operate under the same protocols and the information recorded in the headers of an email message is similarly recorded in the headers of a webmail message. Again, normally only simple headers are shown at the top of the message, but options are available to show the complete header information. Some webmail providers now include an extra line in the header information by default, clearly stating the 'Originating IP' of the sender in an attempt to discourage abuse of the system.

7.7.6 File Transfer Protocol and Peer-to-Peer Applications

The File Transfer Protocol (FTP) is a set of common rules that allows computers to connect to each other directly and transfer data. Modern peer-to-peer applications are a new incarnation of this existing Internet technology. The first well-publicised peer-to-peer application was 'Napster', which allowed users to connect to each other and share MP3 music files.

Napster has been succeeded by applications such as Morpheus, Kazaa, WinMX and many others. The idea behind the applications is that individual users choose to share certain files on their computers with other persons generally unknown to them. The application software handles the communication and transfer of files between users. Users choose which files they wish to share (if any), generally by placing them in a default 'shared' folder. The location of the shared folder is dependant on the application used, but is normally within the program folder created on installation of the application.

Downloading files using peer-to-peer applications differs from downloading from a web site in that the file transfer is directly between two or more users. There is no central repository of all the files on a server as with a web site. Each user is in effect making their own computer available as a server for others to download files from. To download from other people, the user can choose to view all available content of a certain type, *e.g.* music, video or still pictures, or enter search criteria based on filename, file type or content. The application software then establishes which other users are online currently sharing files which meet the requested criteria. The user is then presented with a list of files available for download including details such as the file name, the user making the file available, the size of the file, the speed of the internet connection available for that user, *etc.* The user chooses which files they wish to download and the download then commences direct from one or more users who possess that file and have chosen to share it. It is even possible to view the contents of another user's shared folder directly and choose what material to download. As the user is downloading files, other users may be simultaneously downloading that file or other files from the user's computer. The application software manages the entire process. The content of the files downloaded is generally not visible to the user until the download is complete.

7.7.6.1 Forensic Implications. Certain applications maintain a database file logging files that have been downloaded including information such as file names, date and time information or even the user the file was downloaded from. The format of these files is proprietary to the application. They may contain information in plain text, but the full content is visible only with specialist forensic software.

The applications mentioned are all highly configurable by the user. Users can set their own preferences for any number of the application features: the location of the shared/download folder, whether or not other users can download from them, whether or not files are to be viewed immediately on completion of the download, and many more. The preferences of the individual user may be recorded within a configuration file forming part of the program installation or within the Windows Registry entry for the application, again usually in a format proprietary to the application. The contents of these configuration files can obviously be of benefit in determining how the user obtained material, whether or not it was shared, viewed immediately, *etc.*

7.7.7 Newsgroups

7.7.7.1 Introduction. Newsgroups are the precursors of chat rooms but do not provide 'real-time' conversations. News servers throughout the world contain thousands of newsgroups, each created for discussion on a particular topic. Users wishing to share files, information or just their opinion can 'post', *i.e.* upload, a message to a newsgroup. Similarly, any user viewing the newsgroup can read (*i.e.* download) the contents of a posted message, which may include attached files such as pictures.

7.7.7.2 Newsgroup Applications. Newsgroup reading applications are many and varied, as are the records they maintain. Normally, however, a user must do two things before viewing a newsgroup. Firstly, they must download a list of the newsgroups available on that server. Secondly, they must 'subscribe' to the newsgroup whose contents they wish to view. Contents of particular postings in that newsgroup can then be downloaded. Both of these actions involve downloading of data from the news server. Obviously, it would be very cumbersome for the application to download the same information each time the user wished to access a newsgroup, so a record is usually maintained by the application of the newsgroups available on that server and the newsgroups to which the user has subscribed, possibly including the current postings the user has downloaded to enable offline viewing of the material. The location and format of these records is dependant on the application used.

The user cannot see the contents of a newsgroup posting until it has been downloaded to their computer. The user normally sees a list of messages posted to that group showing a subject line indicating content, the date of posting, the name of the sender and an indication that a file is attached to the posting. Applications are available to automatically download all postings to particular newsgroups without the user being present. The user defines the newsgroups whose contents are to be downloaded and the application takes over. The user can then view the postings at a later time and decide what to keep and what to discard. Such applications are commonly used to download large numbers of picture files. The records stored are again dependant on the application used, but may include details of the newsgroups marked for download by the user.

7.7.8   Chat Rooms and Applications

7.7.8.1   Chat Rooms.   'Chat rooms' are virtual rooms where computer users can engage in real-time 'conversation' in text, audio or even video. Text based chat rooms are by far the most common. As a user types a message it is visible to everyone in the room at that time. Other users can reply to the message in a similar fashion. New users entering the room can view the conversation from the time of entry onwards. There are a plethora of chat rooms on the Internet, each dedicated to a particular subject or group of persons *e.g.* 'scuba diving' or 'Thirtysomethings'.

Chat rooms are generally hosted on a server and communication between all persons in the chat room takes place *via* that server. It is also possible for users in the room to have a 'private' conversation with another user and possibly even exchange files with them. In this case the conversation takes place directly between the two users and is not visible to other people in the room.

7.7.8.2   Chat Applications.   Access to chat rooms requires either the downloading of a chat program (*e.g.* Microsoft Internet Relay Chat, MIRC) or is by access to a World Wide Web-based chat room.

Other chat applications are available that allow conversations to take place solely between two people privately *e.g.* Instant Messenger or ICQ. Users log into the service when they go online and are notified when any 'friends' also come online. A chat session can then take place directly between the two users, with no server involved. Some applications also allow users to exchange files.

7.7.8.3   Records.   There are many chat applications. The records maintained on the users' computers are dependant on the application and the settings applied by the user. For example, Microsoft Internet Relay Chat (MIRC) is a long-standing and still popular application that allows users to 'text chat' in 'channels' hosted in servers worldwide, engage in direct conversation privately and exchange files of any type.

Logs of MIRC chat sessions are created at the user's discretion in the form of plain text files of the chat session conversations. The location of the log files is within the program group for the application or another location as determined by the user.

For all applications, even if 'logging' is not switched on or simply not available, it may be possible to recover fragments or entire conversations from the swap file or unallocated space using keyword searches, or from print spool files if a printed record of the session has been made. Similarly, when webcams are used for video links, the video is made up of a series of still images, some of which it may be possible to recover, again from unallocated space or the swap file. For web-based chat rooms, Internet cache files may indicate visits to the web site and unallocated space may contain parts of chat sessions as mentioned previously.

In some chat applications it is possible to use other utilities from within the chat room, *e.g.* to identify the IP number of other people in the room or to set up a file server to allow automated file transfers (using a file server utility such as 'fserve'). Logs may be maintained by these utilities that record the details of other people in the room or files transferred. The location, format and content of these logs will be dependant on the utility used.

7.8 CONCLUSION

Computing is a rapidly developing technology in terms of both hardware and software. Advances in storage technology are providing users with ever increasing amounts of storage capacity in their computer systems, forcing the development of more powerful forensic tools with faster imaging speeds and greater analytical functions. Development of software operating systems and applications is a constant, with each new variant requiring further knowledge on the part of the examiner.

7.9 BIBLIOGRAPHY

F. Clark & K. Diliberto, *Investigating Computer Crime*, CRC Press, Florida, 1996.
S.H. James & J.J. Nordby, *Forensic Science – An Introduction to Scientific and Investigative Techniques*, CRC Press, 2003.
D.B. Parker, *Fighting Computer Crime*, John Wiley & Sons, New York, 1998.
P. Stephenson, *Investigating Computer-Related Crime*, CRC Press, Florida, 1999.

CHAPTER 8

Fire Investigation

ROGER IDE

8.1 INTRODUCTION

Fire investigation is one of the most difficult studies undertaken routinely by the forensic scientist. Not only are flammable items in the vicinity destroyed by the fire, but materials and surfaces at some distance from the seat of a fire are either affected by the heat or coated with deposits of soot.

Fire-fighting techniques must necessarily be intrepid and enterprising and as a result, items at the scene may have been damaged, displaced or removed by fire fighters. However, the greatest loss of evidence may occur after the fire, during the cleaning up and salvage procedures. At this time too, unscientific preliminary investigations into the fire cause can result in the loss of irreplaceable evidence.

Despite all of these problems, it can be demonstrated that significant evidence remains after even the most destructive of fires. Evidence of directional heating effects, smoke records, temperature indications, debris layer sequences and implicative trace evidence may remain. Such evidence may not be immediately recognised and it is essential therefore that evidence should be preserved after even the most simple fire. Identification of the cause, whether accidental or deliberate, is recognised as being increasingly significant in the prevention of further fires and in bringing offenders to justice.

8.2 THE NATURE OF FIRE

The three main requirements for a fire to occur are heat, oxygen and fuel. These, for no very logical reason, are normally represented in the form of a triangle. It is self-evident that methods of fire-fighting normally depend upon the removal of one or more of the three components in order to extinguish the fire.

For a fire to start, an appropriate fuel must be heated in the presence of oxygen, to a temperature sufficient to initiate a chemical reaction. Most commonly encountered fuels contain carbon and hydrogen and in most cases the oxygen is supplied from the air. Once the reaction has been initiated sufficient heat must be produced for the fire to continue.

8.2.1 The Burning of Methane

When methane (the major component of natural gas) burns, its reaction with the oxygen in the air is as follows:

$$CH_4 + 2O_2 \rightarrow CO_2 + 2H_2O + Heat$$

One molecule of methane reacts with two molecules of oxygen forming one molecule of carbon dioxide and two molecules of water. Since the volume of a gas is proportional to the number of molecules, it follows that the perfect mixture for an explosion of methane in oxygen would be one volume of methane mixed with two volumes of oxygen.

Mixtures of chemicals in their exact reaction ratios are known as 'stoichiometric'. However, oxygen comprises only 21% of the composition of air, and for this reason the stoichiometric mixture of methane in air is approximately 9.5% methane in 90.5% air. The most destructive methane explosions occur when the mixture has this approximate composition.

8.2.2 Flammability Limits

It is self-evident that there must be a concentration below which a flammable gas cannot burn in air In the case of methane, this concentration is 5.3%. There is also an upper limit above which methane cannot burn in air. This concentration is approximately 14%. These two concentrations are known as the lower and upper flammability limits for methane, (abbreviated to FL1 and FLu).

As a general approximation, one kilogram of a hydrocarbon will react with approximately 15 kilograms of air and the lower flammability limit for a gaseous hydrocarbon is likely to be very approximately half of the stoichiometric concentration. The lower and upper flammability limits for a number of commonly encountered fuels are listed in Table 8.1.

It can be seen that there is a wide variation in the flammability ranges of different fuels. Fuel gases having a wide flammability range, such as

Table 8.1 *The stoichiometric concentrations and lower and upper flammability limits for some common fuels*

Fuel	Stoichiometric Concentration (%)	Lower Flammability Limit (%)	Upper Flammability Limit (%)
Methane	9.50	5.3	14.0
Butane	3.10	1.5	8.5
Kerosene	0.90	0.7	6.0
Ethanol	6.50	3.3	19.0
Petrol	1.65	1.4	5.9
Hydrogen	29.58	4.9	75.0
Acetylene	8.40	2.5	82.0

hydrogen and acetylene pose a significantly greater threat to safety than those having a narrow range.

8.2.3 Pyrolysis Products

Most solid materials which become involved in accidental fires are either natural products such as wood, paper and cotton, or synthetic polymers such as plastics, rubber and paint. When any of these materials become involved in fire, the mechanism is likely to be roughly the same. Radiant heat from the flames causes the material to decompose (pyrolyse), forming flammable, low molecular weight, volatile compounds known as 'pyrolysis products'. These compounds will not at first be produced in sufficiently high concentrations for them to reach their lower flammability limits and they cannot therefore burn at this stage. If the material is subjected to higher temperatures, then the volatile compounds can be produced in sufficient quantities to exceed the lower flammability limit. In effect, a plume of flammable gas is being produced by the solid material. If this is ignited, then it will burn as a flame, which will itself continue to heat the solid material, resulting in the production of more flammable pyrolysis products. It can be seen that most solid materials do not themselves burn but act as generators of fuel gases. For a normal solid fuel material to burn it must therefore be subjected to sufficient heat to cause the evolution of pyrolysis products in concentrations within their flammability limits before a flame can result.

8.2.4 Flash Points

The burning of a flammable liquid is in many ways analogous to the pyrolysis of solid fuels. The flash point of a liquid is defined as 'the

Figure 8.1   *A fire ignited in a building with substantial quantities of petrol may result in a destructive explosion followed by a fire. This had been a two-storey brick-built shop. The remains of two 5L plastic petrol containers were found in the debris*

lowest temperature at which the vapour produced by the liquid can be momentarily ignited by the application of a small flame'. At a slightly higher temperature, the fire point, sufficient vapour will be produced so that continuous burning of the vapour will result after ignition. Flash and fire points are measured in specially designed apparatus and there may be slight variations in the results depending upon the type of equipment used. The flash point of a liquid is therefore the temperature at which the vapour above the liquid reaches its lower flammability limit.

Certain volatile liquids such as petrol have flash points well below ambient temperatures and as such present significant fire and explosion hazards. Liquids, such as kerosene, which have flash points significantly above normal temperatures cannot normally be ignited in bulk unless they have been heated to a temperature above their fire point. The presence of an absorbent material which can act as a wick serves to immobilise the liquid in the vicinity of the flame so that it can be heated locally to a high temperature.

8.2.5 Smouldering Combustion

Smouldering is a form of flameless combustion which can occur in cellulosic material or substances which can form a solid char. The reaction takes place at the surfaces within the material and can occur at very low oxygen concentrations. The rate of propagation of smouldering can vary depending upon the oxygen concentration and the materials involved. Although temperatures of approximately 650 °C are typical, smouldering can also occur at temperatures significantly below this.

It is possible for smouldering material to develop into flames, particularly if greater quantities of oxygen become available, for example if there is a draught or if structural components fail allowing greater quantities of oxygen into the vicinity. Flaming combustion can revert to smouldering combustion, a widely observed example being when most of the wood on a fire is consumed leaving just glowing embers. During the smouldering of normal organic materials in a room, large quantities of flammable pyrolysis products may be produced which can ignite explosively if a door or window is opened. This phenomenon, known as 'ventilation induced flashover' or 'backdraft', has been the cause of loss of life in a number of cases.

8.3   FACTORS AFFECTING FLAME PROPAGATION

The rate of flame propagation depends primarily upon the availability of fuel in the gaseous form. Flames can propagate with explosive violence through mixtures of flammable gases with air. The vapours of low flash point liquids such as petrol behave in a similar manner.

In the normal situation where solid fuels are involved, the rate of flame propagation depends largely upon the rate at which flammable pyrolysis products can be released. It is easy to understand that solid materials, whose chemical instability allows them to decompose at relatively low temperatures, can cause rapid spread of flames. However, the physical properties of the material are equally significant. Materials which are good thermal insulators tend to heat up more rapidly in fires and therefore quickly attain sufficiently high surface temperatures for decomposition to take place. For this reason flammable materials having a low thermal conductivity, low density and low specific heat are inherently far more likely to ignite when subjected to radiant heat.

Foamed plastics present a greater fire hazard than solid blocks of the same material. Thin sheets of material are more easily heated to a temperature at which they can decompose than thick blocks. It is common experience that wood shavings are more easily ignited by a match than thick pieces of wood. Even the shape of the material is likely to have an effect. Sharp corners allow heat to penetrate the intervening portion of material from both sides, increasing the opportunity for it to reach its decomposition temperature.

8.3.1 Orientation

It is well known that fires spread more rapidly upwards than downwards. If a cotton curtain is ignited at the bottom, the flames produced by the burning of the lower fabric, heat up the regions above. This causes the cotton to decompose producing more flammable pyrolysis products which, in their turn, ignite and produce flames higher up. If the same curtains were ignited at a top corner but remained suspended, most of the heat produced by the burning of the cotton would be lost and the lower regions of the curtain would be subjected to only limited amounts of radiant heat. For this reason the flames would spread only slowly down the curtain if at all, although significant lateral spread might be expected.

8.3.2 Flashover

When a fire occurs within a closed room there is normally insufficient ventilation for unlimited burning to take place. In extreme cases, the fire may self-extinguish for lack of oxygen. However, where there is limited ventilation, the flames may increase in height because the limited supply of oxygen does not allow the flammable pyrolysis products to

Figure 8.2   *When there is the possibility of vertical spread of flame, for example in goods stored on racking, fires initiated at low level develop very rapidly. The flammability of the stored goods may be of little relevance because the fire can be propagated rapidly through packing materials*

burn completely as quickly as would have occurred had the fire been out of doors. This flame elongation effect is more pronounced when the fire occurs near to the wall of a room. In this case, air is available from only one side and the effect may be to push the flame against the wall with very considerable elongation. As fire develops, the flame may eventually reach the ceiling and when this occurs horizontal elongation of the flame results. Under the ceiling there is likely to be a considerable amount of smoke and very much greater oxygen depletion. As a result, the horizontal elongation of the flame occurs with dramatic speed and within a very few seconds the whole of the ceiling may be enveloped in flame. When this occurs, considerable amounts of heat are radiated downwards and this heat is sufficient to cause all flammable material in the room to begin to decompose. Within a few seconds all of the flammable items in the room catch fire. Within the next few minutes, windows break, doors burn through and the room is likely to be gutted. This phenomenon is known as 'radiation induced flashover'. Under the worst conditions a room, such as a domestic lounge, can be totally destroyed within a few minutes of ignition.

Flashover is more likely to occur if:

1. The first item to be ignited allows vertical spread of flame.
2. The item ignited is near a wall, or more particularly, in the corner of a room
3. There is a low ceiling.
4. The compartment is of a small size.
5. The articles of furniture are less than a metre apart or the floor covering is flammable.

8.3.3 Ignition Temperatures

The 'ignition temperatures' of commonly encountered materials are widely quoted in the literature. They normally represent the lowest temperatures at which a sample of the material has been found to ignite when heated under experimental conditions. These quoted temperatures should be treated with caution and should not be interpreted literally. They may, however, provide guidance and may assist in the interpretation of hypotheses. A more significant temperature is the spontaneous ignition temperature (SIT) which can be measured by heating a sample of the material in an oven, raising its temperature gradually and measuring the temperature within the bulk of the sample. At the SIT the temperature of the sample will rise above the temperature of the surrounding oven. However, variations will occur in the measurement of the SIT depending upon the bulk of the sample, the amount of air

Table 8.2 *Spontaneous ignition temperature (SIT) values for some common materials*

Material	SIT (°C)
Coal	125–130
Hay	172
Sawdust	192–220
Cotton	228

present, the surface area of the particles and other factors. Table 8.2 lists the SIT values for a number of common substances.

Rather than considering the temperatures at which materials ignite, it is sometimes more convenient to consider the amount of heat in the form of radiation incident upon the surface of the material. Experiments have shown that wood can ignite when subjected to a radiant heat flux of approximately 40 kW/M². However, wood decomposes producing flammable pyrolysis products which can be ignited by a small pilot flame when subjected to about only half of this level of radiation.

8.3.4 Spontaneous Combustion

It is well known that such materials as hay, sawdust and oil-soaked rags can spontaneously ignite under certain circumstances. The mechanism for this phenomenon is fairly simple.

Many chemical reactions, particularly oxidation reactions, are exothermic, that is, they generate heat during the process of the reaction. Normally, most of this heat is lost to the surroundings by radiation, convection and conduction. Many chemical reactions accelerate as the temperature rises, with the effect that they produce heat more rapidly. Most of the heat is lost to the surroundings and as a result a reacting system tends to settle down at a temperature slightly above the ambient temperature and reacts slightly more rapidly than it would have otherwise done at the lower temperature.

When large quantities of materials react, the total mass of reacting material is greater in proportion to the surface area than when small quantities react. Since the heat can only be lost from the external surface, greater quantities of heat may be produced than can be lost and the temperature rises; this causes the rate of reaction to accelerate and the rate of heat production to increase accordingly. At the higher temperature the heat losses are considerably greater and the system may equilibrate at this particular temperature. However, above a critical mass for the particular system, more heat is generated than can be lost and the temperature rises until ignition occurs.

Since heat losses occur from the outer surfaces of the bulk of the aterial, the highest temperatures are attained in the centre. One characteristic of fires which have started as a result of spontaneous combustion is that the fire may be seen to have originated within the bulk of the material.

Spontaneous combustion can occur in reactive porous materials such as sawdust, coal, hay and oil-soaked rags. When spontaneous combustion occurs in an industrial process, it is normally as a result of a change in procedure or circumstances. Fires can be triggered by an increase in the quantity of reacting materials, an increase in the ambient temperatures, improved insulation of a reacting mass, or an improvement in the formulation of the reactants.

8.4 THE INVESTIGATION

Fires are investigated in order to detect deliberate ignition, to identify dangerous appliances or materials, to establish liability and to provide data for future policy decisions. Some of the processes used in the investigation are likely to be destructive and it is important that anything of evidential significance should be recorded and preserved or recovered. Unscientific investigations into the causes of fires can result in the loss of irreplaceable evidence. Significant evidence remains even after the most destructive of fires and it is important that this evidence should be preserved and recognised.

8.4.1 Sequence of Events

Because of the wide diversity of situations likely to be encountered in fire scene investigations, it is not possible to formulate one single procedure for the approach. However in general the investigation may follow the general pattern:

1. *Identify Objectives.* A common requirement is to establish the cause of a fire. However, the investigator may be called upon for some other reason, such as to explain an unusual mode of fire spread, to offer an opinion regarding whether life was placed at risk or to comment upon the accuracy of statements made by the owner or suspect.
2. *Investigate Background.* If documentary evidence, witness statements and photographs are available before the scene investigation, then it will be possible for the investigator to direct enquiries more effectively.

3. *Consider Safety.* Certain destructive fires in the past have caused such serious damage to the buildings that they could only be investigated after demolition of the majority of the structure.
4. *Locate Seat.* Many techniques exist for the location of seats of fire. The accuracy with which this can be achieved depends to a large extent upon the rate of development and size of the fire.
5. *Consider Possible Ignition Sources.* In most fire investigations there are normally several possible accidental ignition sources, together with the ever present possibility that the fire had been deliberately ignited. All ignition processes, even those which are only remotely possible, must be considered.
6. *Excavate Seat.* The seat of the fire should be excavated with archaeological care, seeking possible accidental or deliberate ignition sources and noting the relative positions of items as they are recovered.
7. *Take Samples.* Samples of debris may be taken for analysis for volatile fire accelerants. Items of possible evidential significance may be tested to establish whether they could have caused the fire.
8. *Formulate Hypotheses.* Throughout the investigation the investigator will have been formulating hypotheses, rejecting some of them as impracticable. By this stage it is probable that only a limited number of plausible hypotheses remain; it may be possible to devise ways of testing the likelihood of each.
9. *Report Conclusion.* The results of the examination should be reported fully, accurately and objectively. Adequate information should be given. If recognised as an expert, the investigator may offer opinions. These must be distinguished from factual information in the report.

8.4.2   Witness Evidence

It is normally possible to acquire some background information prior to the physical examination of the fire scene. Such information may be available from the police, fire brigade, eye witnesses and from the last legitimate visitor to the premises. Whilst not all of this information may be available initially, the investigator should take steps to obtain and record the evidence as it becomes available.

Fire-fighters may provide information regarding the security of the building, the apparent position of the seat of fire, any unusual circumstances, the time of the initial call and the time that the fire was brought under control.

Other eye-witnesses may provide information regarding the position of the seat of the fire at an earlier stage, colour of flames or smoke and the incidence of explosions. Eye-witnesses are notoriously unreliable and it is likely that the most reliable accounts will be obtained if the witnesses are questioned objectively by personnel experienced in interviewing techniques.

8.4.3 Background Information

It is important at some stage to acquire information from the owner of the premises or the last legitimate occupant. Enquiries should be made into the security of the premises, the positions of items prior to the fire, the presence of stored flammable substances, the status of electrical and other appliances and the history of previous fires. In addition it may be possible to acquire plans of the building, photographs taken prior to or during the fire and video recordings. In the case of industrial premises it may be possible to examine undamaged equipment similar to that which has been destroyed in the fire and to take samples of unreacted process chemicals and solvents.

8.4.4 Recording of Information

The information obtained during the investigation should be recorded either in writing or on audio tape. One convenient method of recording

Figure 8.3 *Photographs taken during the progress of a fire may help to establish the exact seat and may in addition show the face of the person who deliberately started the fire, watching from the midst of a crowd*

the information is by the use of pre-prepared forms which list the information required in the investigation of a typical fire scene. One advantage of this approach is that the form itself acts as an *aide memoire*, reducing the possibility that the investigator may inadvertently neglect to obtain important relevant information. It is often the case, however, that the information volunteered by witnesses and others does not follow the sequence of questions dictated by the form. It may be more convenient to make long hand notes on writing paper and to transfer the information, where appropriate, to the form if required. A tape recorder can be used in the same way, provided that a transcript can be produced for perusal shortly after the scene visit.

It is normally desirable to produce a plan of the whole premises or at least that part of the premises where the fire is believed to have started. Whilst it is preferable that the plan should be to scale, practical difficulties in the fire damaged buildings may compromise accuracy. It must be recognised that the positions of items at the time of the examination may not be those occupied by the items at the time of the fire because of disturbances caused during fire-fighting. In addition, in the case of deliberate fires, items may have been moved from their legitimate positions by the fire setter. It may therefore be necessary to draw more than one plan to illustrate the positions of items at various stages during the incident.

Photography, too, is a valuable method of recording information at fire scenes, and it is desirable to take photographs not only of items and areas considered relevant, but also of other parts of the building. After the investigator has left the scene questions may arise which had not previously been considered. Examination of relevant photographs may provide information which would not otherwise have been available. One of the greatest difficulties experienced at a fire scene is in seeing sufficient detail with the limited illumination available. Automatic cameras with electronic flash often produce high quality images which may allow features which had hitherto been unobserved to be identified in the photographs.

Camcorders may also be used to record information at fire scenes. They have a particular value as a training medium and with the increasing light sensitivity of video cameras, reasonable quality recordings can be obtained. It is possible also for the investigator to provide some sort of commentary during the filming, although there is a danger that unexpected comments and noises from other sources may be recorded.

With all methods of information recording it is necessary to preserve the original notes, even if fresh notes and plans are made from the originals. In the case of tape recordings and video recordings, the original tapes must be retained. Where there is a possibility that a criminal prosecution may result, every single item of paperwork relating to the case must be preserved for disclosure to the defence.

8.4.5 External Examination

It is desirable to examine the outside of the building before entering and carrying out a detailed examination. The external examination may provide evidence of points of entry and may give indications as to the location of the seat of the fire. It is also important to consider the safety of the building before entering.

8.4.6 Point of Entry

It is normal for fire-fighters to effect entry by forcing doors. Criminals may also force doors but commonly enter buildings by breaking a window, reaching through and releasing a window catch. The point of forcible criminal entry to a building need not necessarily be the place where the criminal has started the fire and it is possible that evidence at the point of entry has not been destroyed by the fire. This evidence should be preserved, bearing in mind the possibility that there may be fingerprints, footprints, instrument marks, fibres or even blood from the criminal at the point of entry.

Windows may also be broken by fire and during fire-fighting activities. It is also feasible that the breakage of a window at an early stage in the fire may have allowed the deposition of dense plumes of soot on the exterior wall above the broken window. The presence of unsooted broken glass in the vicinity may also be an indication that a particular window had been broken prior to, or during, the early stages of a fire.

Analogous effects may assist in the interpretation of instrument marks. Deep damage caused during fire-fighting may result in the removal of soot-coated paint, whilst instrument marks caused prior to the fire are likely to be coated uniformly with soot.

The examination of locks found in the debris or attached to doors may indicate whether they had been in the secured or in the unlocked position and may demonstrate whether they had been forced prior to the fire.

8.4.7 Safety

It is important that investigators should take precautions to minimise risk to themselves and to others at the scene. During the external examination of the building it may become apparent that certain walls or parts of the roof are unsafe.

Figure 8.4 *The collapse of lintels or other supporting structures can result in the survival of large areas of unsupported brickwork. As the bricks cool the forces holding them in place become less strong and sudden catastrophic collapse may occur. Dangerous areas should be identified at an early stage in the investigation and subsequently avoided*

Walls may be seriously cracked or distorted. The destruction of interior structural members may leave large areas of wall unsupported. Roof components may be manifestly insecure and there may be evidence that tiles or slates have recently fallen. The investigator should wear adequate protective clothing including a helmet, armoured boots, strong gloves and, where appropriate, a respirator. Advice and opinion should be sought from all relevant authorities including the fire-fighters. Whilst structural dangers may be readily apparent, there is also the possibility of the presence of hazardous materials such as asbestos, beryllium oxide and toxic pyrolysis products. It may be necessary for the investigator to regard certain parts of the structure as unsafe and confine the investigations to those regions of the building which can be entered in relative safety.

8.5 LOCATION OF POINT OF IGNITION

To establish the cause of the fire and the circumstances relating to its development, it is normally essential to establish the point of ignition with as much precision as possible. At least 40 different techniques have been employed in the location of points of ignition, although some of these are of questionable value. The number and diversity of techniques employed is in itself evidence that there is no single infallible location technique.

Many methods employed depend on the principle that the fire is likely to have burned longer and to have developed higher temperatures at the point of ignition. Clearly this assumption is not always valid. Techniques based upon this principle are known as 'time temperature dependant'.

8.5.1 Time Temperature Dependant Techniques

8.5.1.1 Measurement of Depth of Char. Exposed wood chars at a rate which is dependent upon the amount of radiant heat flux incident upon the surface and upon the time that the wood was subjected to this heat. Structural woodwork which has been subjected to high temperatures for a long period of time will have suffered greater charring than wood-work in other regions. The widespread belief that wood in fires chars at a constant rate of 1/40 of an inch per minute is, however, a myth. There is a steep thermal gradient in most rooms involved in fires. For this reason woodwork at a high level is likely to char more rapidly than that at a low level. Comparisons should be made between similar types of wood at the same level.

8.5.1.2 Spalling of Plaster. When subjected to heat, plaster may spall from the underlying brickwork. Regions where this has occurred may be supposed to have been subjected to high temperatures or a sudden rise in temperature. However, variations in the quality of plaster in a particular room and the sudden cooling effect of fire-fighting jets can provide anomalous results. Plaster which has been removed by heat may be indicated by the presence of smoke staining on the underlying brickwork. Brickwork exposed by the removal of plaster during fire-fighting is unlikely to show smoke staining evidence because smoke does not normally deposit once water has been applied.

8.5.1.3 Distortion of Rolled Steel Joints. Rolled steel joints (RSJs) used in the construction of some industrial buildings may become distorted as the result of the fire. The amount of distortion will depend upon a number of factors, including the temperatures to which they have been subjected, the load which they are carrying, their geometrical orientation and the time that they have been exposed to heat. Regions where significant distortion has occurred may indicate where the fire started.

8.5.1.4   Melting of Glass.   Glass does not have a fixed melting point and tends to become less viscous with elevated temperatures. At the temperatures experienced in normal fires, glass can become sufficiently mobile to distort significantly. Normal soda glass, used for glazing windows, shows evidence of distortion at approximately 700 °C. At 800 °C there is considerable distortion and rounding of sharp edges and at 850 °C the glass flows, trickles and takes on the appearance of frozen treacle.

Many other time temperature dependant techniques are available to the investigator. These techniques all share the disadvantages that similar effects will be caused by regions of high fire loading, good ventilation or a delay by the fire-fighters in extinguishing the fire.

8.5.2   Geometrical Techniques

Due to the buoyancy of the hot gaseous combustion products, fires tend to spread in an upward direction. Destruction of items during the fire causes them to collapse under the action of gravity and, as a result, there are many directional indications (particularly in the vertical alignment) which may give evidence of the position of the point of ignition. Geometrical techniques cannot be fully divorced from the problems associated with the time temperature dependant methods.

8.5.2.1   Low Burning.   Since there is a tendency for fire to spread upwards, it might be thought that a point of low burning is likely to be the point of ignition. This is not necessarily true because fires can pread slowly downwards. Burning material can also drop from high positions causing secondary fires which may develop more rapidly than the original fire seat. In certain cases the lowest point of burning includes a hole in the floor. This may sometimes give rise to difficulty in interpreting at which level the fire in the building started.

8.5.2.2   Characteristic Structural Collapse.   Buildings may collapse in a manner which indicates the region where the first failure of structural members occurred. Caution must be used in interpreting such evidence because the expansion of structural members may be masked by the softening or weakening effects of elevated temperatures. The rate of rise of temperature may be a key factor in determining the way that a particular type of structure may fail.

8.5.2.3   Thermal Direction Indicators.   Any material which suffers a change in appearance as a result of elevated temperatures may provide evidence to indicate the direction from which it has been subjected

Figure 8.5 *Fire spreads in an inverted conical pattern. Many materials show evidence of heat damage either by colour change or by some other effect. The V-shaped burning pattern indicates that the seat of fire had been towards the middle of this room*

to heat. The effect is likely to be most marked in items having a low thermal conductivity. If the item can be shown to have been in a particular position at the time of the fire, then the direction of heating can be deduced by examination. The examination of a number of items showing directional indications can be used in conjunction with one another to provide evidence which can be recorded graphically on a plan. Because this technique depends upon the attainment of high temperatures, it is subject to similar limitations to those suffered by time temperature dependant methods.

8.5.2.4 Operation of Alarms. Many smoke detectors and intruder alarms are addressable and may record which detector operated first. This information may be retained by the equipment for only a limited period of time, which may depend on the efficiency of a back-up battery. If it is known that an addressable system was in operation then immediate steps should be taken to recover the information from the equipment.

8.5.3 Development Techniques

Changes occur during a fire as a result of the elevated temperatures, increased oxidation and displacement of items. Consideration of the changes which occur throughout the development of the fire may provide evidence which could assist in the location of the seat.

8.5.3.1   Smoke Records.   Smoke tends to deposit preferentially upon cool rather than hot surfaces and rough rather than smooth surfaces. The manner of its deposition may depend upon the direction of movement of the smoke, its temperature, buoyancy and composition. In a hot oxidising environment, such as may occur near windows, previously deposited smoke may burn away during later stages of the fire. The nature and position of smoke deposits may not only give evidence to assist the location of the seat of the fire, but may provide information relating to the original positions of items such as doors, locks and switches. Smoke produced by a smouldering fire is likely to be chemically and physically different from smoke produced during a free burning, flaming fire.

Figure 8.6   *Smoke deposits can reveal the positions that have been occupied by items at the time of the fire. In this case an attempt had been made to open the safe and impressions, including those of keys, are clearly visible*

8.5.3.2 Enhanced Ventilation. Windows and roof panels are likely to fail first in regions near the seat of the fire. Their failure may result in significantly enhanced ventilation in the vicinity. Areas where there has been enhanced ventilation are likely to attain higher temperatures and sometimes show effects characteristic of more efficient oxidation.

8.5.4 Human Indications

8.5.4.1 Eye Witnesses. Eye witnesses are notoriously unreliable but they may provide the only substantial evidence available to the investigator. It is essential that they should be questioned in a manner which does not distort or degrade the information which they could provide. Lay witnesses have a limited understanding of the mode of fire spread in buildings and inaccuracies may arise as they try to rationalise phenomena which they do not understand.

It must be borne in mind that the witness volunteering information may in fact have started the fire deliberately. It is sometimes necessary to consider the evidence of an eye witness and form a judgement as to whether it is scientifically possible for events to have taken place in the manner described, or whether the witness is confused or lying.

Eye witnesses amongst fire-fighters are likely to be very much more reliable, although they would have had many distractions demanding their attention which may affect the accuracy of their recollections.

8.5.4.2 Position of Bodies. People who died in the building will have been eye witnesses to the fire. They will have been trying to escape or extinguish the fire or they may have been involved in its ignition. It is not necessarily valid to assume that the person was attempting to escape from the fire in a logical manner. He or she may have been affected by carbon monoxide and confused by the limited visibility caused by the smoke. For this reason it is possible that the deceased had been crawling towards the fire rather than away from it.

8.5.4.3 Examination of Photographs and Videos. Members of the public increasingly carry cameras and camcorders with them on a wide variety of occasions. It is relatively common for photographs to be taken of fires in progress. Sometimes, the photographs have been taken before the arrival of the fire brigade and may show evidence indicating where the fire started. In addition to this evidence, there may be pictures of members of the public which might include the fire setter. The person who took the photographs or video recording should not be eliminated as a possible suspect.

8.5.5 Diagnostic Indications

Certain information present at some fire scenes may help to indicate the location of the point of ignition, but may also provide some evidence of the cause.

8.5.5.1 Presence of Incendiary Devices. Fires are sometimes started by the use of incendiary devices. Such devices may be totally destroyed by the fire but recognisable remains can survive. Burnt remains of an incendiary device will indicate one seat of fire and demonstrate that the fire has been started deliberately. It is possible that other such devices had also been distributed around the burnt premises and relatively undamaged unsuccessful devices may be found in the unburnt part of the premises.

8.5.5.2 Presence of Extraneous Flammable Liquids. Evidence of pool burns and smells of unburnt flammable liquids used as accelerants may provide evidence of location and the cause of fire. However, flammable liquids may be accounted for innocently and it is entirely possible that containers of particular liquids had been legitimately present in the building prior to the fire. It is important therefore to investigate all possible innocent explanations for the presence of such accelerants.

Figure 8.7   *Hard edge pool burns may indicate that a liquid fire accelerant has been used. However, fires involving melted plastics and certain other materials may mimic this effect. The presence or absence of liquid fire accelerants can only be proved by laboratory analysis*

8.5.5.3 Forcible Entry to Specific Areas. Evidence that an intruder had forced entry into a building, but had only attained access to certain specific regions, may suggest that the fire had been ignited by the intruder and may provide limits as to the exact location of the seat.

8.5.5.4 History of Accidental Fires. A history of previous fires in the same building, particularly an industrial building, may indicate that there has been a particular procedural fault or danger and it may indicate a particular region where the fire is likely to have started. However, fires previously thought to have been accidental may have in fact been deliberate. It is also possible that a sequence of accidental fires may incite a fire setter to start a deliberate fire in the same building.

8.5.6 Confidence Perimeter

No matter how effectively the location of the seat of the fire may have been investigated, it will not normally be possible for it to be specified exactly. It should, however, be possible to define a boundary or confidence perimeter within which the fire must have started. In the case of small, slow developing fires this boundary could be a circle with a radius of perhaps only 100 millimetres or even less. In many building fires, the radius of the confidence perimeter is likely to be one or more metres, whilst in a large, fast developing fire the radius may be tens of metres. In the worst possible case, the whole building may be included within the boundary. In most cases it is possible to establish the room of origin with certainty and to demonstrate conclusively that the fire had originated within a small fraction of that room. When searching for the remains of the original ignition source, it will be necessary to examine the whole of the area defined by the estimated confidence perimeter.

8.6 EXCAVATION

In most fires of any magnitude it is possible that the original region where the fire was ignited will be buried under layers of debris. Meticulous removal of successive layers of debris may provide evidence not only of the ignition source but also of the sequence of events which occurred during the development of the fire.

It is by no means certain that any physical evidence of the ignition source will be found, because it may have been destroyed by the fire and, if the fire was ignited deliberately, the ignition source may have been removed by the fire setter. In important cases it is necessary to remove layers with the care of an archaeological excavation, recording the relative positions and orientation of items as they are removed. In many cases, however, limited resources dictate that the excavation must be brief.

8.6.1 Sampling

In most investigations of suspicious fires, samples are taken for subsequent analysis for liquid fire accelerants, such as petrol or paraffin. Analyses of samples from the vicinity of the point of ignition are likely to provide evidence as to whether such accelerants have been used, even after severe and prolonged burning. Certain fire accelerants can diffuse through normal polyethylene bags and for this reason samples must be taken in impervious plastic bags, glass jars or metal containers. When plastic bags are the chosen packaging material, the sample, typically one to two kilograms, should be placed in a large bag leaving a volume of head space air above the sample. The neck of the bag should be twisted tightly, swan necked and secured, preferably by tying with string and labelled. If control samples of debris are taken they should be packaged and transported in an identical manner to the questioned samples.

Stringent precautions should be taken to avoid contamination when sampling, transporting and storing samples. Gloves and sampling tools which have previously come into contact with high concentrations of volatile fire accelerants are likely to cause contamination of samples taken from the seat of the fire. Sharp items, such as fragments of glass nails and lengths of wire should preferably be removed from the sample before it is placed in the bag. Samples smelling strongly of fire accelerant and containers of liquid fire accelerant should be kept separate from fire seat samples at all times. When samples are taken they should be labelled recording their original location, the name of the person first taking possession and the date. Each sample should be given a unique alpha numeric reference.

8.6.2 Suspected Accidental Ignition Sources

During the excavation, the investigator may encounter items legitimately present in the premises, which could have accidentally caused the fire. Lighting, cooking and heating appliances, materials capable of self-heating, electrical equipment and residues of smoking materials may be found. In some cases it may be necessary to seize these items or materials for laboratory examinations and tests. Unless volatile fire accelerants are suspected, it will not be necessary to use impervious sample bags to package such materials. Any form of packaging which protects and contains the item or sample will be adequate. In common with samples for analysis, adequate labelling is essential.

8.6.3 Incendiary Devices

The investigator may, on rare occasions, encounter devices which have caused delayed ignition. These devices will incorporate some form of timing mechanism such as a clock, together with a form of igniter, often electrical. The laboratory examination of such items may provide evidence to link the device with known terrorist organisations and it may be possible to establish the period of delay and likely effectiveness of the device.

It is possible for a fire setter to construct incendiary devices which are likely to be totally consumed by the fire. A number of unsuccessful devices may be found in the vicinity of a small fire which has been caused by a successful device. Unsuccessful incendiary devices should not be handled or approached by individuals unaccustomed to the correct procedures involved in their investigation.

8.7 LABORATORY EXAMINATION

8.7.1 Analysis of Debris

Samples of debris taken from the fire scene can be analysed by a number of techniques, the most common of which is gas chromatography. Gas chromatographic techniques can be devised to separate and identify any volatile material. The most commonly encountered fire accelerants are readily available fuels such as petrol and premium paraffin. The normal techniques of analysis have been devised with the objective of detecting these accelerants effectively. However, it is also important that other fire accelerants such as diesel fuel, turpentine substitute, methylated spirits and various organic solvents should also be detected by the standard technique.

Most hydrocarbon fuels consist of a mixture of flammable compounds produced from crude oil by distillation and other processes. The gas chromatography technique separates the individual components of a fuel such as petrol, detecting and estimating their quantities and showing the results graphically in the form of a chromatogram.

The analysis of debris samples is generally carried out by taking head space samples from the container of debris and in some way introducing the volatiles from the head space onto the gas chromatograph. The techniques used vary, but an effective method is to heat the debris and draw head space air from the sample through a tube containing an absorbent such as Tenax. The absorbed volatile compounds can subsequently be thermally desorbed from the Tenax when the sample tube has been introduced into the gas chromatograph. The breakdown of

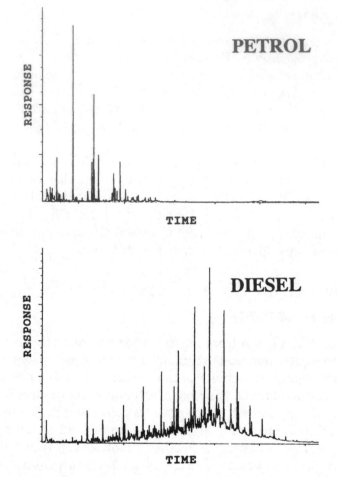

Figure 8.8 *Gas chromatograph traces of petrol and diesel fuel. In general the lower boiling point compounds are eluted at the earlier stages and appear as peaks on the left hand side of the trace. The constituents of petrol are more volatile than most of the constituents of diesel fuel*

plastics and natural materials as a result of the heating effect of the fire can produce volatile compounds which may mask the accelerants present, but it is generally possible to recognise the characteristic peak patterns of common fire accelerants, even in the presence of volatile pyrolysis products.

8.7.2 Examination of Clothing

Clothing can be analysed in a similar manner to debris samples and normally the absence of pyrolysis peaks renders identification of accelerant more simple. However, it is possible for perspiration to undergo chemical reaction in the presence of certain fabrics producing a characteristic series of volatile compounds. These compounds are unlikely to be confused with fire accelerants by any experienced scientist.

The presence of volatile fire accelerants on clothing may subsequently prove to have an innocent explanation. Additional evidence may be provided by the presence of characteristic heat damage which may provide convincing evidence that the owner of the clothing has been subjected to a heat flash, such as might be caused by ignition of a cloud of gas or vapour. Synthetic fibres such as nylon, acrilan and polyester may melt in a manner which can be recognised by microscopic examination. Small holes may have been burnt into the clothing by sparks, and certain fabric dyes undergo colour changes when they are heated. However, some fabric treatments and chemical spillages can cause similar effects and consideration should always be given to innocent explanations for these phenomena.

8.7.3 Examination of Electrical Equipment

Many fires are attributed to electrical causes and it is often necessary to make an examination of electrical appliances, switches and fuses. Any electrical appliance which has genuinely caused a fire is likely to have suffered considerable heat damage and limited evidence may remain. It is however, often possible to establish whether switches had been in the 'on' or 'off' position and whether fuses had failed as a result of the heat, overload or a short circuit. Radiographic examination can often provide the required information, but it is sometimes necessary to embed the item in resin and to examine it in section.

Detailed examination of a piece of equipment, particularly if it can be compared with an undamaged item of the same type, may provide strong evidence to indicate the cause of any failure.

8.7.4 Examination of Heating Appliances

Many fires are caused by paraffin and gas fuelled heating appliances, and examination of such appliances recovered from fire scenes may provide evidence that an individual appliance has been defective or misused. Examples of misuse include improper use of the appliance for airing clothes, the use of incorrect fuels and lack of maintenance. Recently acquired heating appliances are sometimes improperly set up and traces of packaging material intended by the manufacturer to be removed before use may still be present.

8.8 QUALITY MANAGEMENT

A person required to investigate fires for anything more than statistical purposes should have a thorough grounding in the scientific background of fire dynamics and combustion chemistry. Whilst the cause of a fire may appear obvious, there may be misleading phenomena which, if not recognised, could lead to an incorrect conclusion. Above all, the investigator must show objectivity and impartiality in the investigation. It is essential that the investigator should be capable of justifying his or her conclusions in a scientific manner.

Courts are increasingly recognising the value of NAMAS and BSI accreditation for laboratories which undertake analyses and examinations. Fire investigations are likely in some cases to result in the imprisonment of supposed fire setters. It is essential that the highest standards should be maintained in order to ensure that no miscarriage of justice occurs.

8.9 BIBLIOGRAPHY

R.A. Cooke and R.H. Ide, *Principles of Fire Investigation*, Institution of Fire Engineers, Leicester, 1992.
J.D. DeHaan, *Kirks Fire Investigation*, 5th Edition, Brady, New Jersey, 2002.
D.D. Drysdale, *An Investigation into Fire Dynamics*, John Wiley & Sons, 1985.
D.M. Wharry and R. Hirst, *Fire Technology, Chemistry and Combustion*, Institution of Fire Engineers, Leicester, 1974.

CHAPTER 9

Explosions

LINDA JONES and MAURICE MARSHALL

9.1 INTRODUCTION

The investigation of accidental or illegal explosions, and scientific analysis of their causes, has a long history which, in the United Kingdom, started formally in 1871. An explosion at a factory in Stowmarket making guncotton led the then Home Secretary to instigate an inquiry. A Royal Engineer, Captain Vivian Majendie, who was an expert on explosives, led the investigation. He recruited a chemist, Dr A. Dupre, to assist. The arrangement proved so successful that the Home Office decided to continue it, leading to the present day Forensic Explosives Laboratory. One of the first fruits of their collaboration was the 1875 Explosives Act, which embraces various aspects of explosives including their manufacture and storage. Subsequent bombing outrages led Parliament to enact the 1883 Explosive Substances Act which deals with the criminal use of explosives and devices; this Act was intended specifically to deal only with the most serious offences, hence the special provision made requiring the fiat of the Attorney General for any prosecution under it.

Many of the questions which Captain Majendie and Dr Dupre were asked to address are still relevant to today's forensic explosives investigators, including the following:

1. Was it an explosion?
2. Was it an accident or a bomb?
3. Is this an explosive?
4. Was this a viable device, or a hoax?
5. Are these items or materials intended for making explosives or bombs?
6. Has this person been in contact with explosives?

241

7. Have explosives been stored in this place?
8. Are there similarities between these items or incidents that link them together?
9. Could the items have an innocent use?

Actual cases involve a variety of circumstances and the questions that are appropriate will vary accordingly.

In order to address the above points, the forensic explosives scientist requires a sound grounding in the requirements of the judicial system and the ethical principles underlying it, together with a detailed knowledge of the science and technology of explosives, practical experience in the construction of all types of explosive devices and a clear understanding of their effects.

9.2 EXPLOSIVES TECHNOLOGY

9.2.1   What is an Explosion?

A convenient working definition is 'a sudden and violent release of physical or chemical energy, often accompanied by the emission of light, heat and sound'.

To the human observer an explosion seems instantaneous; however, the chemical reaction actually proceeds at a finite, albeit high, speed, progressing through the material as a definite 'front' or 'wave'. Two types of event may be defined: a 'deflagration' being an event where the decomposition within the explosive occurs at a speed equal to or less than the velocity of sound within the material, and a 'detonation', being an event where decomposition occurs at a speed greater than the velocity of sound within the material. Sometimes deflagrations are referred to as 'low-order explosions' and there is a mistaken tendency to assume that they are in some way less serious than detonations. This is an error: many of the most devastating accidents with explosives have involved deflagrations rather than detonations.

Typically deflagrations occur with velocities of less than 2000 m s^{-1} while detonations may reach velocities of 6000 to 8000 m s^{-1} for certain high performance military explosives.

9.2.2   Types of Explosion

Explosions may be characterised by the source of their energy, *i.e.* physical or chemical, and also by their locus, *i.e.* whether dispersed or condensed phase.

Examples of the various classes are:

1. Physical – an exploding pressure vessel, for example an overheated gas cylinder.
2. Chemical – explosion of a mass of blackpowder, or as it is more commonly known, gunpowder.
3. Dispersed – detonation of a cloud of flour in air.
4. Condensed – detonation of a stick of dynamite.

Each of these types can be further classified according to the speed and duration of the explosive event and this can be linked to the type of practical effect observed, enabling the skilled investigator to draw some meaningful conclusions from examination of the types of macroscopic and microscopic damage found at the scene of an explosion. Thus high velocity detonations cause shattering and cutting of metal, whilst low velocity events result in tearing and heaving.

9.2.3 Types of Explosives

Although dispersed phase explosions can be of great power, as can physical explosions, the vast majority of practical explosives used for either commercial or military purposes are condensed phase chemical explosives. These energetic materials are best considered according to their function. Thus we have:

1. Pyrotechnics – used for the production of heat, light, sound or smoke, for instance in fireworks or signalling flares.
2. Blasting explosives – for example to break up rock in quarrying operations.
3. Initiatories – used to transform a small mechanical or thermal impetus into a violent shock wave capable of causing detonation of less sensitive energetic materials such as blasting explosives. Initiatories are also referred to as primary explosives indicating their role at the start of a chain of explosive events. Analogously the less sensitive explosives used for the main charge are referred to as secondary explosives.
4. Propellants – energetic materials which deflagrate in a controlled fashion to allow their energy to be used, for example in propelling rockets or projectiles from guns.

Apart from the above scientific definitions, the forensic scientist also needs to be familiar with the legal definitions; for example, the United Kingdom's 1875 Explosives Act Section 3 states:

'The term "explosive" in this Act –

1. Means gunpowder, nitro-glycerine, dynamite, guncotton, blasting powders, fulminate of mercury or of other metals, coloured fires, and every other substance, whether similar to those above mentioned or not, used or manufactured with a view to produce a practical effect by explosion or a pyrotechnic effect; and
2. Includes fog-signals, fireworks, fuses, rockets, percussion caps, detonators, cartridges, ammunition of all descriptions, and every adaptation or preparation of an explosive as above defined.'

Other countries have their own legal definitions.

9.2.4 Chemistry of Explosives

The requirements are simple: the ingredients of the explosive, an oxidant and a fuel, need to undergo some very rapid chemical reaction liberating large amounts of energy. In practice the most difficult part is achieving control of this process so that the explosive only reacts when required and not otherwise.

The earliest known explosive was gunpowder, more correctly known as blackpowder, first developed by the Chinese, and being an intimate mixture of charcoal and sulphur (the fuel) and potassium nitrate (the oxidant). The chemical reactions which occur are in fact highly complex, resulting in the formation of oxides of carbon, sulphur and potassium, together with potassium/sulphur compounds, all in a range of oxidation states. Blackpowder illustrates the way in which the behaviour of explosives can be tailored to a particular application. Thus if spread out in a thin layer in the open air it will merely burn violently, while if confined in a pressure-tight container it will readily detonate if ignited. Under light or partial confinement blackpowder deflagrates and can be usefully employed as a propellant, for example in the guns of Lord Nelson's *HMS Victory* at Trafalgar or in modern day display rockets on Guy Fawkes night. This versatile material is also very widely used in current military munitions as an igniter powder; the fiery particles of molten inorganic slag which form on its decomposition make it particularly suitable for this purpose.

In broad chemical terms blackpowder is an intimate mixture of fuels and oxidants. Other inorganic and organic materials can also be combined to produce explosive mixtures, the commonest example being the use of ammonium nitrate/fuel oil mixtures (ANFO) in quarrying. Inorganic compounds are also used widely in pyrotechnic mixtures, for example, chlorates, perchlorates and nitrates of alkali metals are common oxidants.

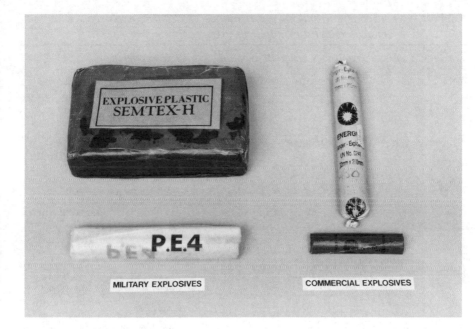

Figure 9.1   *Some military and commercial explosives*

The other approach, which can yield even higher explosive performance, is to combine the fuel and oxidant in the same molecule. Although some organo-fluorine explosives are known, these have not been generally adopted and the majority of organic explosives are organic nitro compounds. The earliest example was nitroglycerine (glyceryl trinitrate), commercialised by Nobel as 'Dynamite' or 'Gelignite'; others include the nitro-aromatics such as 2,4,6-trinitrotoluene (TNT) or the nitramines, for example RDX. Some commercial and military explosives are shown in Figure 9.1, together with their chemical names and properties which are presented in Table 9.1

9.2.5   Initiation and Detonation of Explosives

As mentioned above, the most important practical problem in explosives technology is the safe control of initiation. This has led to the widespread use of the concept of the 'explosives train'; a small quantity of a very sensitive explosive is used to receive an initial stimulus in the form of thermal, mechanical or electrical energy, and amplify this so as to start a reaction in a larger mass of less sensitive material.

Table 9.1 *The common and chemical names of organic explosives with their chemical structures and properties*

Common Name	Chemical Name and Structure	Properties
TNT	2,4,6-trinitrotoluene	Low melting solid; can be conveniently cast into bombs and shells; dissolves in organic solvents
Nitroglycerine	Glyceryl trinitrate	Viscous liquid; toxic vapour; neat liquid is sensitive and unstable
PETN	Pentaerythritol tetranitrate	White crystalline solid
RDX	Cyclotrimethylenetrinitramine	White crystalline solid; high density; high detonation velocity; moderately soluble in acetone
HMX	Cyclotetramethylenetetranitramine	White crystalline solid; high density; high detonation velocity; high melting point (275°C); very stable; moderately soluble in acetone
Tetryl	2,4,6-Trinitrophenylmethylnitramine	Yellow solid

This leads on to the concept of the 'detonator' invented by Nobel. Commercially, two main types are produced: 'plain detonators' and 'electric detonators'. Figure 9.2 shows the essentials of the latter type, together with some examples of the wide variety available.

A sealed pencil-shaped metal tube contains a metal wire filament embedded in a heat sensitive match composition (the 'fuse head'); when an electric current is passed through the wire, the resulting heat from the filament ignites the match composition. This in turn ignites the next element in the detonator. In practice this may be a delay composition (the 'delay element'), used to provide greater control in the sequencing of industrial blasting operations, or the ignition may pass directly to the small charge of primary explosive of initiatory, (the 'priming charge'), which once ignited, burns to detonation. The detonation of the primary explosive in turn causes detonation of a slightly larger quantity of a secondary explosive (the 'base charge') which acts as a 'booster', amplifying the effect several-fold. In a plain detonator the wire filament is replaced by an igniferous fuse.

Table 9.2 gives some examples of the energetic materials used in detonators for each of the functions shown above.

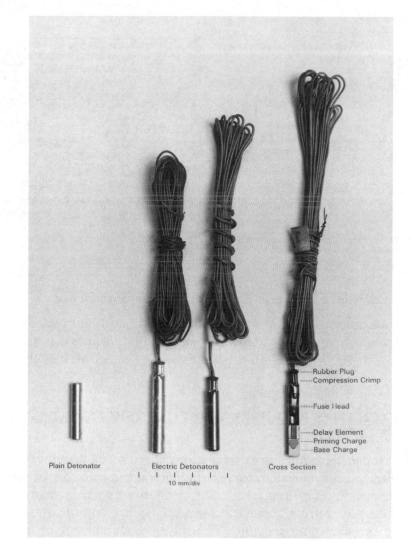

Plain Detonator   Electric Detonators   Cross Section
10 mm/div

Rubber Plug
Compression Crimp
Fuse Head
Delay Element
Priming Charge
Base Charge

Figure 9.2   *Detonators*

Table 9.2   *Some explosive materials used in detonators*

Common Name	Function
Mercury fulminate	Primary explosive
Lead azide	Primary explosive
Lead styphnate	Primary explosive
Potassium chlorate/lead thiocyanate	Match composition
Barium peroxide/selenium	Delay element
PETN	Secondary explosive
Tetryl	Secondary explosive

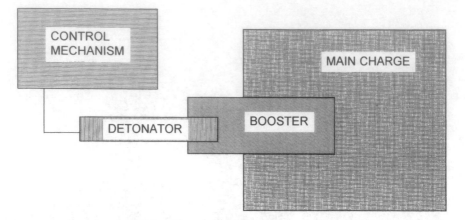

Figure 9.3   *Elements of an improvised explosive device*

9.2.6   Essential Elements of an Improvised Explosive Device

Great ingenuity has been displayed in the creation of improvised explosive devices. However, common elements can be identified in terms of basic functions. Figure 9.3 shows the essential elements in schematic form.

9.3   FACILITIES REQUIRED FOR FORENSIC EXPLOSIVES EXAMINATIONS

In the United Kingdom, strict regulations apply to the keeping and use of explosives and no laboratory or individual may possess explosives without appropriate licences. Similar regulations exist in many other countries. Apart from these legal aspects, safety considerations also need to be taken into account and explosives should not be handled without adequate training and facilities for the purpose.

9.3.1   Safety

Explosives explode. This simple rule needs to be borne in mind at all times when handling energetic materials of any kind; all processes and equipment should be designed on the basis that at some time an unintended explosion will, in fact, occur and therefore the safety procedures must be such as to minimise the harmful effects. Most nations have detailed explosives regulations covering the safe storage, transport and use of commercial and military explosives. However, these often cannot be applied to the case of illegal materials, and scientists investigating such matters must be careful to minimise hazards both to themselves and others. In particular the possibility that improvised explosive devices may incorporate concealed anti-handling features should always be addressed.

Explosive devices should always be examined at the scene by properly qualified bomb disposal personnel before being disturbed in any way whatsoever; the bomb disposal personnel will also be able to provide the necessary advice on safe packaging and transport arrangements. Practical experience shows that the most dangerous devices are often those made by amateur experimenters, for example schoolchildren.

Specialist facilities must allow for four activities:

1. Receipt.
2. Storage.
3. Examination.
4. Disposal.

9.3.2 Receipt

The arrangements for receipt of items have some features in common with those generally used in forensic work, for example the possibility of cross contamination between submitted items must be rigorously and demonstrably precluded. The general requirements for preserving the integrity of trace and contact evidence are discussed in Chapters 2 and 3. In addition, the system must always preserve the chain of custody of evidence. Explosives laboratories also need preplanned procedures for dealing with devices which may deteriorate into an unsafe condition or are suspected of incorporating anti-handling features. Thus immediate access to bomb disposal equipment and personnel is essential, together with radiographic facilities designed to allow safe examination of explosive devices and provide details of the internal construction of such items before disassembly.

The mechanics of receipt have to cater for a range of items which can cover anything from a small, sealed package of hand swabs for trace examination to a blown-up motor vehicle or several tonnes of bomb scene debris. In practice it is convenient to have different physical arrangements for the various streams of evidence types whether bulk or trace, explosive or non-explosive, clean or dirty.

9.3.3 Storage

Detailed safety guidelines and regulations cover the design of explosives storehouses (magazines). These require, for example, separation of detonators and bulk explosives; they also include limits on the mass of explosives which may be kept in particular designs of store and also specify security features to prevent unauthorised access.

A typical laboratory magazine is strongly constructed of reinforced concrete, provided with spark-proof electrical lighting and lightning protection, and fitted with high security doors and locks. The whole is enclosed within a substantial barricade intended (i) to protect the magazine from an explosion in any other adjacent building, and (ii) to deflect upwards the blast wave which would occur if the contents of the magazine itself were to explode.

The laboratory will also need to store non-explosive exhibits ranging from items for explosives trace analysis to bulk debris in a way which meets all the requirements for preservation of the chain of custody and protection of exhibit integrity.

9.3.4   Examination

The items to be examined can be grouped as follows:

9.3.4.1   Examination of bulk explosives.   The design of facilities for bulk explosives is governed by both the type and mass of explosive being handled. In general, different work areas need to be provided for examination of detonators and primary explosives, pyrotechnics and incendiaries, and secondary explosives. Where practicable it is also wise to segregate activities involving large and small quantities.

For example, it is often convenient to break down devices containing large explosive masses in a separate building, take small samples and then examine these in more detail in the main laboratory. Buildings for explosives work require appropriate electrical fittings to prevent spark and dust hazards, together with lightning protection and antistatic precautions. Particular attention needs to be paid to fire prevention and escape in the building design.

In a laboratory for examining detonators and primary explosives, the floor and benches are normally covered in conducting rubber sheet which is earthed to reduce hazards from sparks due to static electricity, as is all the laboratory equipment, and all personnel entering the laboratory have to check themselves on a special meter to ensure that they are properly earthed. In dry climates it is wise to humidify the laboratory atmosphere to reduce hazards from static electricity. The detonators and primary explosives are manipulated behind stout armoured screens which need to be of a design which has actually been type-tested to prove that they really do withstand the effects of an explosion of the quantity likely to be involved. It is not sufficient to rely on normal laboratory safety screens, which are most unlikely to contain the explosive effects of a typical commercial detonator. Similarly, fume cupboards used for chemical treatment of primary explosives (which are generally toxic) should be suitably armoured with arrangements to vent blast pressure waves due to any detonation.

Special machinery is essential if detonators are to be opened to allow examination; lathes, cutters, manipulators and screens must all be type-tested to prove their suitability and safety before use. The superficial simplicity of the process conceals many potential hazards and it most definitely should not be attempted without proper facilities and training.

9.3.4.2 Examination of Samples for Traces of Explosives. This can include swabs or other samples taken from people's hands or property, or samples obtained from items which have been involved in an explosion. As with all other contact trace evidence, where the objective is to detect very small quantities of the analytical species, it is essential to carry out the examination in a controlled environment where it can be demonstrated by objective measurement that the results obtained are meaningful and not the consequence of contamination either from the general environment or from other items at the laboratory.

In general this will require precautions such as the use of clean disposable overclothing, separate work areas for different trace operations, regular monitoring and cleaning and control of access to ensure that people and items that enter the trace area are free from explosives.

9.3.4.3 Examination of Non-Explosive items. Bomb components. Live explosives will often be accompanied by non-explosive items either as part of a device or in a collection of suspected bomb-making equipment. This includes containers, timing devices, power supplies, arming mechanisms and switchgear. Facilities for mechanical examination and disassembly, measurement, photography and microscopy will be needed.

Post-explosion debris. For example, the remains of an exploded firework, or more generally an exploded device, a bomb-damaged motor car or tonnes of debris from a bombed building. Each of these evidence types requires a different approach and different facilities. Conventional, well lit, clean laboratory benches covered with fresh disposable paper will suffice for examination of the remains of a small exploded device. Handling bomb damaged motor vehicles requires special trolleys and lifting gear, particularly since such vehicles are likely to be structurally unsafe and may easily collapse on nearby people if lifted without properly designed equipment operated by trained personnel. Examination of bulk debris from bombed buildings requires a system for the mechanical handling of the material, together with facilities for drying, sieving and searching.

Debris from bomb scenes can present a range of hazards: there may be materials such as asbestos present in building debris which is also liable to be contaminated with sewage; damaged vehicles may contain flammable liquids; and all are liable to be contaminated with biohazard materials, particularly from events where people have been injured or killed. Thus laboratory facilities for this type of work need appropriate mechanical ventilation for dust and vapour extraction, supplemented as necessary by personal breathing apparatus. Vaccination of staff against a range of diseases is also advisable.

9.3.5　Disposal

A major consideration in explosives work of any kind is the need to provide for the safe and environmentally acceptable disposal of all waste and unwanted material. It is not permissible to simply mix explosives waste with normal garbage and most particularly not with waste chemicals. Waste explosives and explosives removed from improvised devices are likely to be less stable than newly manufactured material as a result of rough handling, inadequate storage and contamination, and should therefore be treated with especial care. Material may be destroyed chemically or more commonly by burning at an appropriately licensed facility.

9.3.6　Reference Collections and Databases

Reference collections are commonly used in forensic science; they are of great value in explosives work. Thus samples with different compositions, and examples of packaging and labelling are all most useful in the identification of individual commercial explosives. Whilst the identification of an intact cartridge of explosive bearing the manufacturer's label may appear to be a trivial task, identification of the same item from a few small damaged fragments of wrapper and a milligram of undetonated composition is a more challenging problem. Likewise, specimens of items such as detonators and blasting accessories are useful for physical comparison. Such collections of physical hardware can advantageously be supplemented with a library of commercial literature which can prove especially helpful in identifying unusual or old items.

A set of well characterised samples of known explosives are essential as comparison and calibration standards for chemical analysis by techniques such as GC or HPLC. Collections of reference spectra are also helpful, but the possibility of errors in commercially published spectral libraries must be borne in mind and it is wise to establish the authenticity of spectra used in critical work.

A carefully indexed database of devices should be maintained to facilitate correlation between incidents, yielding pointers to individuals and organisations involved in series of outrages.

9.4 FORENSIC QUESTIONS

9.4.1 Was it an Explosion?

This can involve examination of the damage at the scene either at first hand or, less satisfactorily, by viewing photographs, reviewing witnesses' accounts of the event and laboratory examination of debris from the scene. The occurrence of sudden loud reports, flashes, violent projection of debris, smashing, tearing or rupturing of structural materials and the formation of craters at the seat of a supposed explosion are all useful indicators.

9.4.2 Was it an Accident, or a Bomb?

This has to be a combined process, both of elimination of possible accidental causes such as leaks of flammable gases or the presence of flammable dust clouds, and, wherever possible, the positive identification of physical or chemical evidence resulting from the bomb.

Physical evidence can involve recovering fragments of the bomb from the debris; contrary to popular misconception explosions do not always vaporise everything in the immediate vicinity. Rather it is more accurate to say that 'bombs shatter and scatter. Thus teams of trained and diligent searchers were able to recover evidentially significant fragments of the device in many of the cases which occurred in Great Britain between 1990 and 1996. Such fragments may include pieces of the bomb container, the waterproof plug from the detonator and parts of any timing mechanism such as clockwork, electronic circuitry, batteries and wires. Fragments of the device are likely to be found lodged in comparatively soft objects near the seat of the explosion, for example in vehicle tyres and seats in the case of under-car booby traps. The bodies of victims are also potentially good receptors for bomb fragments; medical staff should routinely be asked to X-ray victims and pass on any fragments recovered for scientific examination.

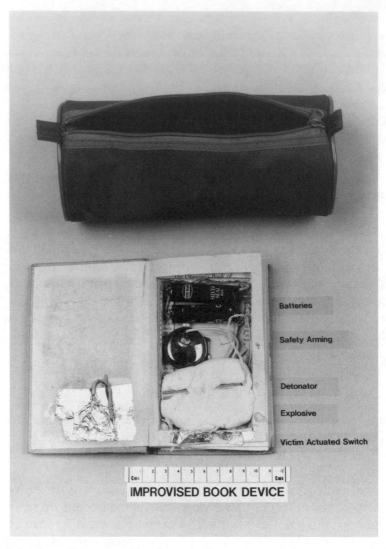

Figure 9.4 *A small improvised bomb in a book*

A small victim-actuated bomb in a book is shown in Figure 9.4, together with the accompanying bag in which it was transported. Such an item would be intended to explode when the book was opened.

Although only containing some 25 g of plastic explosive, such a device would be likely to maim or kill anyone opening it. The fragments recovered after explosion can be seen in Figure 9.5; information of considerable investigational and evidential value might potentially be obtained from such material. The metal fragments, for example, might yield chemical residues of the explosive; the battery fragments, type and

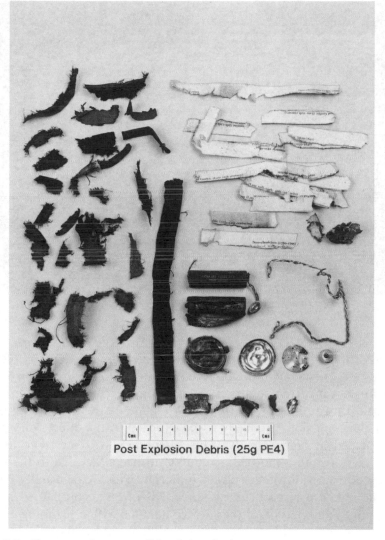

Post Explosion Debris (25g PE4)

Figure 9.5 *Fragments from a small bomb in a book*

batch numbers; the watch might have toolmarks or manufacturer's marks; and it might be possible to link the wires with similar wires from a suspect's premises. Likewise, the fragments of the book and the bag might provide links to a suspect.

Incendiary devices which have partially functioned often produce a fused mass of charred and burnt material, particularly if the device was in a plastic container. In such cases radiography can often be used to reveal details of the construction of the device from otherwise intractable evidence.

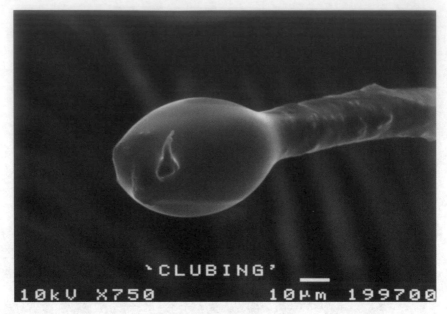

Figure 9.6   *Clubbing on a nylon fibre as seen in the SEM*

In addition, explosions cause characteristic damage to many receptor surfaces as a result of the unique combination of high velocity shock and transient high temperatures. Surfaces near to the seat of an explosion, at a range of a few centimetres to some metres depending on the size of the explosion, are likely to be exposed to the effects of very hot gases ejected at high velocity. This phenomenon, referred to as 'gaswash', gives rise to a mottled, irregular surface effect due to the partial melting and erosion of the target material, which can usually be seen with the naked eye, particularly since it is often accompanied by soot deposition. Examination of the surface damage with a scanning electron microscope can be highly informative. Nylon textiles may display characteristic 'clubbing' or 'toffee apple' damage; the momentary high temperature causes the polymer to melt and then re-solidify to produce a small spheroid on the fibre end which is readily visible in the SEM. An example is shown in Figure 9.6.

Characteristic effects also occur on metals; explosive cladding of one metal upon another can occur and characteristic microcraters can be formed, particularly in aluminium alloys. Figure 9.7 is a scanning electron micrograph of the surface of a piece of aluminium which had been in contact with a small explosive charge, showing a typical microcrater formed in an explosive event. The rolled edges due to flow of the molten aluminium are particularly characteristic.

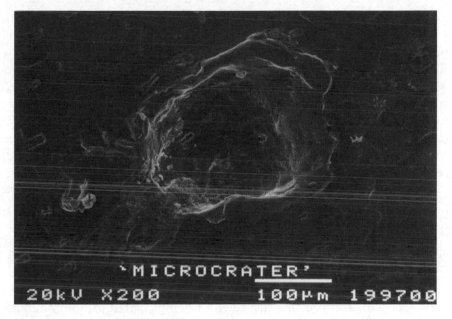

Figure 9.7   *A microcrater in aluminium revealed in the SEM*

Many explosives also leave deposits of soot on cooler adjacent surfaces where partially combusted material and reaction products condense. Large metal fittings and items such as metal window frames and railings often act as receptors for soot in this way.

Chemical residues of explosives or their decomposition products may also be sought, and if found, provide useful evidence. In the case of uncommon chemicals with no non-explosive uses (*e.g.* tetryl or RDX), the detection of residual traces is strong evidence.

The gross physical damage at the scene of a suspected bombing is also informative, since careful examination will enable the scientist to distinguish between a dispersed phase explosion and a condensed phase explosion, and in the latter case to locate the seat of the explosion. Structural damage, crater depth and breakage of window glazing are all useful in assessing the mass of explosive involved. However, it must be realised that calculations of the size of bombs from observations of scene damage are subject to large margins of error.

Particular difficulty can be experienced in determining whether an aircraft crash is due to the explosion of a bomb or some other cause. In this type of case it is necessary to distinguish between what may be very extensive mechanical damage due to the crash and damage which could only be due to an explosion. Where reconstruction of the crashed aircraft is feasible, this can be extremely helpful in locating the position of an explosion. Tracing of the path of shrapnel from the bomb can sometimes be accomplished by examination of the aircraft seats, *e.g.* by threading stiff wires through the fragment penetration holes, thereby enabling location of the site of the bomb. Where a bomb has been placed in hold baggage, careful examination of the recovered baggage can sometimes enable identification of which items of baggage were adjacent to the bomb and which specific item of baggage contained the bomb. This can provide exceptionally valuable assistance to the investigator, enabling possible links back to the individual responsible for checking the baggage onto the aircraft.

9.4.3   Is this an Explosive?

The forensic scientist is likely to encounter explosives samples in five main ways:

1. As a bulk of an unknown substance, whether in isolation or together with components suspected of being related to bomb-making activities.
2. From a suspect device which has been rendered safe by bomb disposal personnel. Apart from suspect explosive, other components of the device such as batteries, timepieces and igniters are also likely to be submitted.
3. As traces on items of post-explosion debris.
4. As traces on items suspected of having been in contact with explosives.
5. As traces on people suspected of having handled explosives.

9.4.3.1   Bulk Explosives.   Although some organic compounds are well known explosives, *e.g.* TNT (2,4,6-trinitrotoluene), there are many mixtures of ingredients which may be explosives depending on the precise formulation and circumstances. Thus it is necessary to make use of both chemical analysis to identify suspect materials and also a variety of physical tests to demonstrate both the energetic nature and detonability of samples.

The very first step is of course to look at the sample; if it happens to bear a manufacturer's label identifying it, this can be quite helpful! However, it is still necessary to confirm the true identity of labelled items since the label may not be accurate. Visual examination with a low power microscope is of considerable value, particularly for recognising materials such as homemade blackpowder.

The simplest test is to observe the behaviour of a small (tens of milligrams) quantity of the material when it is ignited. This test must be donc behind a substantial armoured glass screen, preferably within an armoured fume cupboard since some explosives produce toxic vapours on decomposition, *e.g.* lead salts. Points to note are:

1. Does it burn? Smoulder? Melt? Char?
2. How does the flame spread – are there flashes or sparks?
3. What is the appearance of any residue?
4. Did the burning material emit a characteristic odour?

A more elaborate test is the Lidstone cartridge test, which assesses the detonability of small quantities of suspect materials. A 2 g sample of the suspect material is packed in a 0.303 " brass cartridge case, a standard detonator fitted in a sleeve in the open neck of the case and the assembly fired in a tank filled with wood pellets. The wood pellets are then sieved and the remnants of the cartridge case recovered. Comparison with firings of known explosives allows a rough assessment of the sample's explosive performance to be made.

Thereafter, a whole range of chemical analyses can be performed, including wet chemical techniques such as spot-tests, TLC or solvent extraction and gravimetric analysis, and instrumental methods such as FTIR, GC, IC, CE, HPLC, SEM-EDAX, GC-MS and LC-MS.

9.4.3.2 Traces of Explosives. Recovery and sampling. It is necessary to recover samples for detection of invisible traces of explosives from a wide variety of types of evidence, and sampling techniques need to be chosen accordingly. Because of the enormous range of possible explosives which may be encountered it is necessary to devise an analytical approach for a given sample which covers the range of most likely candidate explosives whilst providing timely and reliable results. To do this the scientist will need to make use of any available information that may provide guidance about the nature of substances likely to have been used.

The following sampling methods are often used:

1. Swabbing, either dry or with solvent. A cotton wool swab is moistened with the selected solvent and gently rubbed over the test surface. When the swab becomes dirty or dry a further fresh swab is moistened with the solvent and the process continued in this fashion until the entire area has been sampled. The resulting swabs from any one test area are combined for later analysis.
2. Solvent washing. Small items are placed in a beaker, covered with the chosen solvent, and then the whole agitated in an ultrasonic bath to aid recovery of soluble materials.

3. Contact heater for adsorbed and absorbed volatile explosives. This is a specially designed device with an electrically heated, ventilated, metal platen which is placed on the surface to be sampled. Air is drawn in across the test surface, out through the centre of the platen and passed through a glass tube containing an adsorbent such as Tenax-GC (R) which traps any vapours evolved from the sampled surface.

4. Vacuum sampling to trap particulates. A glass syringe barrel is fitted with a filter support and a microporous filter, and then a length of disposable tubing connected between the needle fitting and a vacuum pump to apply suction. The open end of the syringe barrel is rubbed over the sample surface, so that any mobile particles are trapped on the filter for subsequent examination.

5. Adhesive lifts for particulates. The sticky surface of a clean adhesive tape is applied to the sample and any recovered particles or fibres examined.

Solvents for swabbing or washing need to be chosen to suit the type of explosive being sought and the analytical method which is to be used. Table 9.3 gives some commonly used solvents.

Table 9.4 gives some examples of the use of these different methods in the collection of samples for trace explosives analysis from a motor vehicle.

Table 9.3 *Solvents used for recovering traces of explosives*

Solvent	Type of Explosive	Comments
Water	Inorganics, *e.g.* nitrates, chlorates, perchlorates	Slow to evaporate
Methanol	Organic, *e.g.* nitrate esters, nitroaromatics, nitramines	Toxic, poor volatility
Ethanol	Organic *e.g.* nitrate esters, nitroaromatics, nitramines	Poor volatility
Acetone	Good solvent for organics, especially nitrocellulose	Good volatility, but leaves aqueous residue unless dried thoroughly
Isopropanol	Organics	Moderate volatility
Ethyl acetate	Good solvent for a wide range of explosives	Available in high purity
Methyl *t*-butyl ether	Organics	Good volatility
Diethyl ether	Organics	Good volatility, but serious fire hazard

Table 9.4 *Sampling of the surfaces and materials in a motor vehicle*

Substrate	Method
Glass and painted metal	Solvent swabbing
Impermeable plastic surfaces	Solvent swabbing or contact heater
Porous surfaces, woven materials	Vacuum sampling, contact heater, adhesive lifts
Small loose items, *e.g.* keys, coins	Solvent washing

Controls. With any of the methods outlined above, no results can be considered to have forensic significance unless proper environmental control measures are taken before the work of trace analysis starts. These need to be designed to take account of the sensitivity of the analytical method used and the degree of exposure of the exhibits and sample solutions during the whole process of sampling and analysis. As more sensitive detection methods are used, more stringent environmental control measures are needed. When quantities of a few hundred nanograms or less are being sought, operators should, as a minimum, wear fresh disposable overgarments and gloves to preclude transfer of suspect materials either between casework items or between samples and the general laboratory environment or the operator. Control swabs should always be taken from the actual work surface and also the operator's gloves and oversuit, and these control swabs should be prepared and analysed in the same way and at the same time as the actual trace analysis samples from the case under examination.

Sampling kits. A number of forensic laboratories prepare kits of materials ready for use in sampling at scenes of crime. Kits for the collection of explosives traces should contain items such as swabs and solvent which, by means of a rigorous quality control process, have been shown to be free of explosives traces. The official kit issued to police in Great Britain for recovery of explosives traces is shown in Figure 9.8. A similar but smaller kit, with fewer components, is used for swabbing suspects' hands.

Sample pretreatment and concentration. This will depend again on both the likely candidate explosives and the analytical method chosen. In favourable cases some information about microscopic samples of particulate matter can be obtained by SEM examination of adhesive tape lifts, yielding data on morphology and elemental composition. More generally trace chemical analysis methods have to be employed.

For example, particles of sugar and sodium chlorate mixture recovered by vacuum sampling can be dissolved in water and analysed

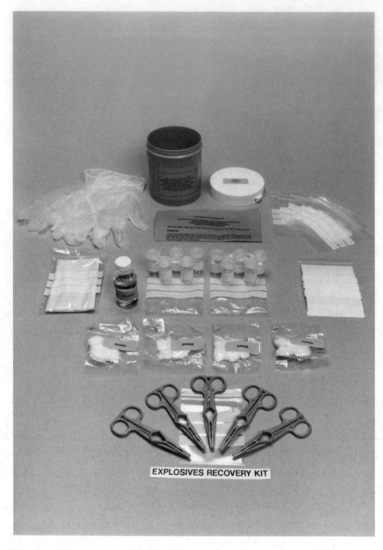

Figure 9.8 *Explosives traces recovery kit*

by ion chromatography or capillary zone electrophoresis to identify anions, cations and sugars, and by atomic absorption spectroscopy to confirm the presence of metal cations. The same basic technique can be applied to aqueous extracts from vacuum sampling tubes, water wetted swabs or washings. The extracts can be concentrated by partial evaporation to increase analytical sensitivity.

Liquid chromatography and gas chromatography are used in conjunction with different selective detectors for determination of organic explosives traces. For organic explosives of moderate to high volatility, such as nitroglycerine or RDX, capillary gas chromatography with a selective chemiluminescence detector or a mass spectrometer is suitable. For involatile explosives, liquid chromatography provides an alternative approach. Although these techniques are extremely sensitive, being able to detect picogram amounts of some explosives in pure solutions, they are generally liable to be seriously affected by the large amounts of non-explosive material, such as oils, fats, soot and soil, present in real samples. A good technique is to elute the initial sample extract through a simple solid phase chromatographic column which adsorbs the explosives selectively for subsequent elution with a different solvent.

Interpretation of explosives trace evidence. A surface such as the seat of a car or a person's hand may acquire traces of explosive either by direct contact with explosive, or by secondary transfer from another item contaminated with explosive. The quantity of explosives traces found will depend on a whole range of factors – the amount of explosive present on the donor, the duration and intimacy of contact between the donor and recipient, the nature of the respective donor and recipient surfaces, the elapsed time between the contact and sampling and whether the recipient surface has been cleaned since the contact. Experiments done on transfer when a person directly handled a stick of gelignite showed that they had micrograms to milligrams of nitroglycerine on their hands shortly afterwards. People who shook hands directly with them acquired much smaller traces, with transfer efficiencies of a few percent. When a third person shook hands with the second, higher transfer efficiencies of up to 35% between the second and third person were observed for short time intervals. It was suggested that this enhanced transfer efficiency was a function of the distribution of the explosives traces on the second person's hand. Some very limited experiments with the transfer of PETN between people suggested that it behaved in a broadly similar fashion to nitroglycerine. Other experiments have suggested decay rates for traces of nitroglycerine and RDX on hands of three to five orders of magnitude in the first day for normal hand-washing regimes. Much lower decay rates might be expected for traces on inanimate surfaces subject to infrequent cleaning and for solid explosives, such as RDX or PETN, having a lower vapour pressure than the relatively volatile nitroglycerine. Consideration must also be given to the likelihood that an explosives trace might have been acquired in the course of legitimate activity, for example quarry workers and military personnel may come into contact with explosives in the course of their normal duties. Nonetheless, the finding of traces of the organic high explosives is quite significant. Studies of casework samples received in our laboratory show that only a few percent of samples taken from suspects' hands, clothing or premises exhibit detectable explosives traces, with a somewhat higher proportion of suspects' vehicles yielding positive results. In addition a survey of traces of organic high explosives in the general public environment in England showed that such traces were rare. Higher levels of inorganic species of possible explosives significance, such as ammonium, nitrate, or chlorate ions, are likely to be encountered and so it is equally necessary to take into account available information on likely environmental levels of inorganic species when interpreting the significance of inorganic trace findings.

9.5 PHOTOGRAPHY

Because of the hazardous nature of explosives and bombs, it is not possible to take the majority of explosives exhibits to court. As a consequence it is essential to produce a high quality photographic record of exhibits, particularly when devices are being taken apart, in order to show all the relevant details to the court. A larger negative gives greater detail in the final print, and in practice 120 or 5 × 4 formats are more suitable than 35 mm film for production of courtroom quality photographs.

9.6 LINKS WITH OTHER FORENSIC DISCIPLINES

Explosives examinations generally take precedence over other forensic tests for two fundamental reasons: (i) safety and (ii) because unless there are items of explosives significance then it is unlikely that a case can be established under explosives-related legislation.

As in all complex forensic cases, discussion of requirements between scientists from different disciplines and the investigating officer is essential if the best evidence is to be protected and realised. Practical compromises between the requirements of explosives trace analysis, fingerprints, fibres and toolmarks need to be agreed and the work planned accordingly since each aspect can make a vital contribution in many cases.

9.7 A CASE STUDY

As in all forensic science, correct interpretation of the results is crucial. A delicate balance has to be struck ensuring that all the relevant facts are made known, that the appropriate weight is given to each aspect of the evidence and that the work is recorded in a manner which allows meaningful review by others skilled in the art so that, for example, other possible interpretations of the results can be explored by different experts assisting the various parties in a case. That said, how are the various scientific and technical threads pulled together? The following hypothetical case serves as an illustration.

9.7.1 The Scenario

A violent sharp report is heard just after 3 o'clock in the morning, apparently from the grounds of a large country house, whose alarmed inhabitants summon the local police.

Upon arrival the police make a search of the grounds and find that the base of one of the large (and valuable) statues in the ornamental gardens has been damaged; they decide to secure the scene and wait until daylight before making a detailed search.

Meanwhile, other officers have been called to an accident just a few miles away involving a motor car and a small lorry. The behaviour of the lorry driver and his passenger causes them to make further inquiries and their suspicions are increased when the passenger is reported to have been convicted of burglary some years before.

Further inquiries are made and a link to the attempted theft of the statue suspected. By now daylight has broken and examination of the scene around the statue has yielded a bootprint from a patch of soft ground and has revealed a pattern of damage on the stonework of the statue's base. Further searching results in the discovery of a pair of long thin wires, various fragments of broken stonework and an assortment of general litter and debris of uncertain relevance; in addition a tyre print is found in mud alongside the driveway. Scene photographs are taken, the recovered items carefully bagged and identified, and the damaged area of the statue's base is swabbed using an explosives trace recovery kit; the whole is then submitted to the laboratory for examination.

Chemical analysis of the swabs rapidly indicates the presence of PETN residues and this, coupled with examination of the damage to the statue, suggests an unsuccessful attempt to cut the statue from its base using detonating cord. After discussion with the scientist the police decide to have the suspects' lorry, their clothing and their hands tested for explosives traces. It is agreed that in this instance the scientist will give priority to the trace explosives analysis and leave examination of the various items from the scene of the explosion until later; this simplifies matters since it neatly allows clearly separate examination of the 'clean' and 'dirty' exhibits and avoids any possible suggestion of contamination between them.

Most of the results from the trace analysis are negative. However positive results for PETN are found for a pair of rubber gloves found in the cab of the lorry, from a penknife found in the pocket of the passenger's jacket, from the area of the pocket which held the penknife and from the knees of the passenger's trousers. The swab samples taken from both men's hands are negative.

A search of both men's homes reveals nothing relevant, but a subsequent search of a lockup garage rented by the passenger yields a bag containing what initially seems to be a length of plastic clothesline; however, it has the words 'Danger – Explosive' printed along its length. This is therefore examined by a bomb disposal operator who packages it safely and advises forensic examination to confirm if it is in fact detonating cord. The rubbish bin from outside the garage is also searched; some discarded packaging appears significant and is also submitted to the laboratory.

These new items are identified at the laboratory as a particular brand of commercial detonating cord containing PETN, and the remnants of a box which once contained a particular type of commercial electric detonator widely used in quarrying.

Meanwhile, the inhabitants of the house where the crime occurred have found another pair of rubber gloves discarded in their garden at some distance from the statue; unfortunately there have been several nights of heavy rain before the gloves were found. Examination of the gloves for explosives traces does indeed prove negative.

Searching of the debris from the explosion scene reveals a small burnt and deformed rubber plug, which the scientist is nonetheless able to identify as similar to the plug from detonators made by a particular company; there are also the wires: these prove to be a long length of a standard twin core cable attached to two short lengths of plastic insulated single strand wire whose colour and diameter match that from a particular range of commercial detonators. In fact, the packaging from the lockup garage also matches, and the three items: wires, plug, and packaging all belong to the same product range from the same company.

Further examination of the lorry reveals a partially discharged battery lying in a corner of the loading area and a torn packet matching the brand of rubber gloves found both in the lorry's cab and at the scene. One of the tyres on the lorry is found to match the tyre print in the garden.

9.7.2 The Prosecution Case

It was suggested that the facts were consistent with a bungled attempt to steal a valuable statue: detonating cord had been wrapped around the base in an effort to cut the statue loose so that it could be removed. In fact all that had resulted was noise and damage. The frustrated thieves had then fled leaving behind the rubber gloves worn by one of them; the second pair of gloves was found in the cab. No fingerprints were found at the scene and the hands of the accused were free of explosive traces because they had worn gloves. However, the penknife that was found in the passenger's jacket pocket had been used to cut the detonating cord and hence had contaminated the pocket with traces of PETN. Likewise the gloves found in the cab had been worn when handling the detonating cord, hence the traces of PETN on them. The absence of traces on the gloves found later in the garden could be due to their having been washed away by the heavy rain.

The detonating cord found in the garage contained PETN; the packaging found in the rubbish bin at the garage matched the detonator plug and wires found at the scene.

9.7.3 The Passenger's Defence

The lockup garage had been rented by a group of five men, including himself, each of whom had a key. The bag found in the garage belonged to one of the others.

Whilst working in the garage on the previous day he had needed some rope to secure a bundle of wood; when he looked in the bag he saw what seemed to be a coil of clothesline and so he cut a length off with his penknife. However a white powder then fell out of the line. He didn't know what it was, but realised it wasn't suitable for his purpose so he put it back in the bag, wiped the knife blade on one of his gloves and put the knife and glove back in his pocket. He did all this while kneeling on the garage floor.

The reason he was in the lorry was that he had been out poaching and had been walking home when his friend drove by in the lorry and gave him a lift.

9.7.4 The Lorry Driver's Defence

He had been employed recently to remove heavy rubbish from the gardens of the house and had legitimately been near the statue in the course of his work, hence the bootprint. Likewise, he had parked his lorry in the driveway, and this must be the explanation for the tyreprint.

9.7.5   What Really Happened?

The passenger was telling the truth.

The lorry driver was in league with the owner of the bag, who was one of the group renting the garage. The bag-owner had left the crime scene on foot when he realised that the noise of the explosion had alerted the inhabitants of the house. In his haste he had discarded the gloves found later in the garden. The lorry driver had had no choice but to leave *via* the driveway unless he wished to abandon his vehicle at the scene.

This case study illustrates the fact that there are often a number of explanations for a given set of circumstances. The forensic scientist's task is to conduct impartial examinations and formulate unbiased conclusions within the remit of their specific discipline, in order to help the court reach an informed judgement. In the majority of cases the scientific findings will only form a part of the jigsaw, the other pieces of which may radically affect the interpretation of the science.

Acknowledgement

We wish to thank our colleagues in the Forensic Explosives Laboratory for their generous assistance and constructive suggestions and the Defence Science & Technology Laboratory for permission to publish the photographs.

9.8   BIBLIOGRAPHY

Guide to Explosives Acts 1875 and 1923, HMSO, London, 1992.

J. Akhavan, *The Chemistry of Explosives*, The Royal Society of Chemistry, Cambridge, 1998.

A. Bailey and S.G. Murray, *Explosives, Propellants and Pyrotechnics*, Brassey, London, 1989.

A. Beveridge (Editor), *Forensic Investigation of Explosives*, Taylor & Francis, London, 1998.

P.W. Cooper and S.R. Kurowski, *Introduction to the Technology of Explosives*, John Wiley & Sons Inc, New York, 1996.

Tenney L. Davis, *The Chemistry of Powder and Explosives*, Angriff Press, Hollywood, 1943.

S. Fordham, *High Explosives and Propellants*, 2nd edn, Pergamon Press, Oxford, 1980.

J. Kohler and R. Meyer, *Explosives*, 4th edn, VCH, Weinheim, 1993.

T. Urbanski, *Chemistry and Technology of Explosives*, (3 volumes), Pergamon Press, Oxford, 1964.

H.J. Yallop, *Explosion Investigation*, The Forensic Science Society, Harrogate, 1980.

CHAPTER 10

Firearms

JAMES WALLACE and VICTOR BEAVIS

10.1 INTRODUCTION

Firearms are primarily designed to discharge projectiles from a distance with the intent to kill or at least incapacitate. Some will disagree with this statement, but nevertheless a designer of firearms will attempt to produce a device that will be effective for the selected quarry. For example, a garden gun is designed to kill a few pigeons and frighten the others that are destroying the apple harvest, whereas a much larger projectile, and therefore a different firearm, is necessary in big game hunting. The legal definition of a firearm according to the UK Firearm Act 1968 is: '. . . a lethal barrelled weapon of any description from which any shot, bullet or other missile can be discharged. . .' Put simply the job of a firearm is to:

1. Discharge a projectile with sufficient energy to kill.
2. Ensure that the projectile travels in the required direction.
3. Ensure that a bullet, as distinct from shotgun pellets, arrives at its target nose first.

The firearm is the means of aiming and discharging the projectile and imparting stability to it.

The firearm may, of course, have other features such as the ability to carry more than one cartridge, to load and unload automatically and to fire repeatedly on a single pressing of the trigger, but these are all refinements on the basic design. The means of discharge of firearms is gas pressure, ranging from the rapid release of stored compressed air or carbon dioxide, as in compressed air or gas guns, to the rapid burning of a propellant as in conventional firearms. No matter in whatever way this gas pressure is generated, the effect is the same in that the projectile is forced down and out of the barrel of the firearm, giving it speed, direction and stability in flight.

A complete study of firearms would deal with the firearm itself, its history and development, its ammunition, its discharge and a discussion of the materials resulting from the discharge, *i.e.* the spent bullets, spent cartridge cases (sometimes referred to as shells) and discharge residues. All this is not possible in this chapter, but it is nevertheless necessary to introduce each of the topics and to refer the interested reader to more comprehensive literature for further study.

Many terms, often contradictory, are used to describe firearms and ammunition, and some of the more common ones will be used and defined in this chapter. It is probably best at this early stage to define a cartridge, which, using the Oxford dictionary, is 'a case containing a charge of propellant explosive for firearms or blasting, with bullet or shot if for small arms'. Other terms, such as ammunition, round of ammunition or just round are also used for cartridges and are equally acceptable. Bullet, however, is wrongly used in this context and should be reserved for the projectile only.

10.2   INTERIOR AND EXTERIOR BALLISTICS

The forensic scientist is mainly concerned with a firearm that has been discharged and the immediate consequences. When a firearm is discharged, a sequence of events occurs and it is usual to group these under the headings of 'Interior Ballistics' and 'Exterior Ballistics'.

Interior ballistics is the study of the events that occur in the very short time between the trigger being pulled and the bullet leaving the muzzle of the firearm. It has been estimated that this period is considerably less than 1/10 second, and may be split up as follows:

1. 1/20 second – the striker (firing pin or hammer) impacts on the cartridge.
2. 1/500 second – the primer explodes and ignites the powder.
3. 1/100 second – the bullet travels along and leaves the barrel.

Exterior ballistics is the study of the bullet's flight from the moment it leaves the muzzle of the firearm until it comes to rest, and this period will vary from a fraction of a second to possibly several seconds, depending on the type of firearm and the range of fire. This area of study is not normally of relevance to the forensic examiner except in a very qualitative way and will not be covered in this chapter.

10.3 THE FIREARM

To understand fully and to evaluate the events that occur in the discharge of a firearm, it is essential to consider the components of the firearm and its ammunition and to discover the contribution of each to the complete picture. The obvious place to start is with the firearm itself. Only the briefest, non-technical, description is possible in this chapter and terms will be used which may not be familiar to the general reader. It is therefore convenient to define some of these here:

1. *Single shot.* Each cartridge must be separately and manually loaded and unloaded after firing.
2. *Bolt action.* After each shot is fired a bolt is manually drawn back and pushed onward again, thus ejecting the spent cartridge case and loading another cartridge from the magazine, if present.
3. *Manual.* This is very similar to the bolt action, but is mechanically more sophisticated. Examples are lever and pump actions.
4. *Self-loading.* After firing, the spent cartridge case is ejected from the firearm, and the next cartridge from the magazine loaded into the chamber. The firearm can be fired again by pulling the trigger.
5. *Automatic.* When the trigger is pulled the firearm will continue firing until the magazine is empty or the pressure on the trigger is released.

After manufacture and before being placed on the market for sale each firearm must be declared safe, as far as is practicable, to the user. This is accomplished by putting it through a series of tests, called proof firing, and, if satisfactory, the firearm is stamped with a proof mark. Proving of firearms may be done by the manufacturer, or at a government proof house.

10.3.1 Classes of Firearm

Firearms commonly encountered in crime fall into one of four main classes, *viz.* shotguns, rifles, pistols and submachine guns.

The interior of the barrel of rifles, pistols and submachine guns is 'cut' with a number of spiral grooves, called the rifling, which causes the bullet to spin as it travels along the barrel, thus giving it stability in flight. The number of spiral grooves and their dimensions can vary between different types and models of firearms and the grooves can be inclined to either the right or the left, *i.e.* clockwise and anti-clockwise, as viewed from the chamber end of the barrel. The raised area between two grooves is called a land.

10.3.1.1 Shotguns.   A shotgun is a smooth barrelled firearm, intended
to be fired from the shoulder and designed to discharge pellets, unlike
the other classes which discharge bullets. The pellets, called 'shot', are
contained within the cartridge case and are for shooting game on the
ground or in flight, where the spread of the large number of shot after
discharge increases the chance of striking the target. There are of course
many different types of shotgun, but they are normally either single
barrel or double barrel. Single barrel shotguns may be single shot,
manual action or self-loading and may come in several calibres,
normally referred to as bore, or gauge, although the most common is 12
bore (see Section 10.4). Double barrel shotguns are the most common,
12 bore being the most popular. Frequently, the barrels of shotguns
have a slight tapered restriction at the muzzle, causing the shot to stay
closer together as it leaves the muzzle and travels towards the target.
This restriction is termed the choke and the reduction in the internal
diameter of the barrel due to the choke is measured in thousandths of
an inch, and is referred to as degrees of choke. Thus a barrel with a
choke of 40 degrees, called Full Choke, will measure 0.04 inches less at
the muzzle than at any other position along the barrel.

10.3.1.2 Rifles.   A rifle is a firearm with a long rifled barrel and is
designed to be fired from the shoulder. Such firearms are manufactured
in many different types and calibres and are primarily designed
for game hunting or warfare. They can be single shot, bolt action, self-
loading, manual action or automatic, and are generally intended for
aimed shots at fairly long range (a few hundred metres). Most are self
loading and are equipped with a magazine holding up to 30 cartridges,
although single shot rifles are still fairly common. A variation of the
rifle is the assault rifle, which has a shorter barrel and is capable of fully
automatic fire besides single aimed shots. Modern military rifles are
now of the assault rifle variety.

10.3.1.3 Pistols.   A pistol is any firearm designed to be discharged
from the hand as distinct from the shoulder and has a much shorter
barrel than a rifle. As such it is less accurate and so is intended for
shorter range firing (up to about 50 metres). The barrel is rifled and
pistols are capable of firing either in single shot or self loading mode.
They are normally equipped with magazines holding up to about ten
cartridges. A revolver is a type of pistol, but has a revolving magazine
(called the cylinder, usually holding six cartridges), so that after every
shot the next cartridge is brought into line with the barrel as the cylinder revolves. It should be noted that with revolvers the spent cartridge
case remains in the cylinder unless manually ejected, whereas in most
other pistols it is automatically ejected after firing.

10.3.1.4 Submachine Guns.   A submachine gun (SMG) is essentially a pistol designed to be fired using both hands, either in single shot or fully automatic mode, but having a longer rifled barrel and a larger capacity magazine. Its main function is for sustained fire at reasonably close range so it is normally fired in the automatic mode.

10.3.2 Loading Mechanisms

Firearms that are designed to be fired without reloading after each shot require a mechanism to unload each spent cartridge case after firing and to chamber another round from a receptacle (magazine) for the ammunition.

10.3.2.1 Revolving Cylinder.   Pressure on the trigger of a revolver rotates the cylinder to bring a cartridge into the firing position. Such firearms, where the breech does not move on firing, are said to have a standing breech. The spent cartridge case remains in the cylinder.

10.3.2.2 Bolt Action.   After discharge the bolt is manually drawn back to eject the spent cartridge case. When it is pushed forward again it picks up another cartridge from the magazine, positioning it in the chamber and cocking the trigger ready for the next shot.

10.3.2.3 Manual Action.   Unloading and reloading are performed by the manipulation of a mechanical device on the firearm, such as a lever in the lever action, or a slide in the pump action.

10.3.2.4 Blow Back (Recoil Operated).   The pressure generated by the 'explosion' of the propellant, as well as driving the bullet down the barrel also drives back the bolt, thus 'automatically' having the same effect as the manually operated bolt action. Should the bolt open too quickly, however, the enormous pressures generated at the moment of discharge may result in the cartridge exploding in the chamber, or open breech, causing damage to the gun and injury to the firer. This hazard is greatly reduced by using a heavy bolt (or breech block) and small calibre bullets. Such mechanisms are common in pistols up to calibre 9 mm Short and in submachine guns. For larger calibre pistols, some submachine guns and high velocity rifles, it is essential to ensure that the bolt is locked to the barrel at the moment of discharge and remains locked until the high pressure has subsided. Many ingenious methods have been designed to delay the opening of the bolt to enable the pressure to drop to a 'safe level' and such firearms are described as having a locked breech.

10.3.2.5 Gas Operated.   This action is generally one in which at the moment of discharge, the bolt is locked to the barrel and cannot be opened until the bullet has travelled down the barrel beyond a small hole, called a gas port. Some of the gas escapes into this port and acts on a piston, above or below the barrel, to unlock the bolt from the barrel and to drive the bolt rearwards, thus extracting and ejecting the spent cartridge case and reloading another cartridge.

10.3.3   Forensic Significance

All this may seem rather cumbersome, but in a modern self-loading rifle from the moment the trigger is pulled, until the rifle is ready for firing again considerably less than 1/10 of a second has elapsed. The firearm can therefore be fired repeatedly as quickly as the trigger can be pulled, or in fully automatic mode, in the region of ten rounds every second. Another reason for explaining the mechanism of discharge is from the forensic examiner's viewpoint. As has already been stated elsewhere in this book, one of the tenets of forensic science is that 'every contact leaves a trace'. In the discharge of a firearm, the contact between the metal of a bullet and cartridge case with the metal of the firearm will leave unique marks on the softer metals, that is on the bullet and cartridge case. This means that the transfer of a cartridge into a firearm, the closing of the bolt, the discharge of the bullet and the extraction and ejection of the spent cartridge case are all potential sources of marks. Each of these, and their relative positions on the spent cartridge case, may indicate the type of firearm involved in the incident and can also be used to link, often uniquely, a spent cartridge case or bullet to the firearm used.

10.4   CARTRIDGES AND CALIBRE

Before considering the examination of spent cartridge cases and spent bullets, some explanations on the 'calibre' may be appropriate here. Calibre is often a very much misunderstood and confusing term, not helped at all by the many ways of attributing dimensions to it. In general, it gives an indication of the diameter of the bore of the firearm, but frequently it gives other information about the cartridge or the firearm in which it is designed to be fired, and is best defined as 'an alphanumeric term used to specify a cartridge that is designed to be discharged from a compatible firearm or group of firearms'.

This may seem confusing but, in general, a cartridge is originally designed so that when discharged from a suitable firearm, the projectile will strike the target, nose first, at the range intended, with sufficient energy to kill or at least incapacitate the selected quarry. The firearm is designed to permit this to happen with complete safety to the firer. So generally the cartridge is designed first and is given a name to distinguish it from other similar cartridges, the name bearing some similarity to the bore of the firearm. Calibre is usually given in inches or millimetres.

The calibre, or gauge, of shotguns is defined differently and originates from the early muzzle loading days, when the gauge of a musket was the number of lead balls of the diameter of the bore that made a pound weight. So a 12 bore shotgun means that 12 solid lead balls, each the diameter of the bore, weigh a total of 1lb. From the density of lead the diameter of the bore can be calculated and for 12 bore the internal diameter is 0.729 inches. There are several types of cartridge used in the many different firearms and an illustration of the common types is given in Figure 10.1.

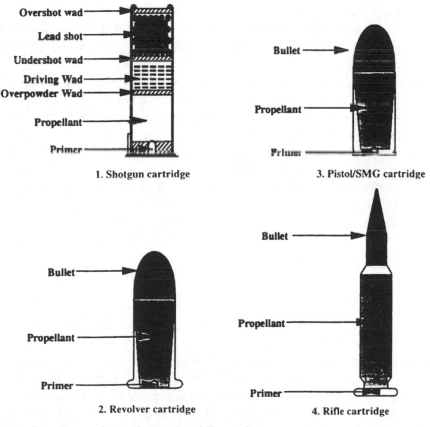

Figure 10.1   *Types of cartridge used in different firearms*

The following few examples of calibre will illustrate the confusing nomenclature that often exists:

1. 22L and .22LR – the same cartridge case is used in each of these, but the bullet weights are different. The bore diameter is 0.215 inches.
2. .380 Rev, .38 S & W, .38 Special and .357 Magnum – in each of these the diameter of the bore is about 0.35 inches, but the cartridge case dimensions, bullet weights and propellant charges are different, and each is designed for a different firearm, although some interchange is possible.
3. 9 mm P – this cartridge was introduced with the German Luger self-loading pistol, the 'P' standing for Parabellum. It is probably the most widely used pistol and submachine cartridge in the world today.
4. .30-06 – the diameter of the bore is 0.30 inches. The second figure, -06, refers to the year of adoptation, 1906, by the US army for the 1903 Springfield rifle.
5. .30-30 Winchester – the diameter of the bore is 0.30 inches, but the second number is the weight in grains of smokeless powder in the propellant charge.
6. 7.62 × 39 – this is the official Russian military cartridge since the end of the Second World War. The first number is the diameter of the bullet in mm and the second number is the length of the cartridge case, also in mm.
7. .308 Winchester – this was originally designed as a sporting cartridge in the USA, but is also known as 7.62 × 51 NATO, adopted by NATO as the official military cartridge. The case length is 51 mm.
8. .45 ACP and .45 Auto-Rim – the same bullet is used in each of these, but the design of the cartridge case differs. They are not inter changeable.

10.5   THE DISCHARGE

When a cartridge is fired in a firearm a sequence of events occurs and each has significance for the forensic firearms examiner. The process of discharging a round of ammunition in a firearm involves a mechanical device that causes a hammer to fly forward and deliver a blow to the firing pin when the trigger is pulled. In some firearms the hammer and firing pin are made in one piece. The firing pin goes through a small hole in the breech face and strikes the primer cup (also called primer cap, percussion cap or just cap). The primer cup contains a mixture of chemicals that are sensitive to percussion. Consequently, the composition burns rapidly, producing a flame and a shower of hot particles that penetrates and ignites the propellant.

The burning of the propellant very quickly produces a large volume of gases in the confined space of the cartridge case, accompanied by a substantial rise in temperature and pressure. This gas pressure is exerted in all directions, rearwards on the face of the bolt, sideways on the walls of the chamber and forwards on the base of the bullet, thus causing the bullet to move down the barrel and the bolt to open unless otherwise locked. The bullet, normally of lead or copper-jacketed lead, both comparatively soft metals, moves along the barrel and tends to 'fill' into the grooves of the rifled barrel, thus providing maximum seal for the gas pressure (this is referred to as obturation) and causing the bullet to spin. After discharge, in self-loading firearms, the bolt moves to the rear and the spent cartridge case is dragged out of the chamber by means of an extractor claw acting over the top of the rim of the cartridge case. On the rearward travel of the cartridge case it impinges on a projection in the open breech thus causing it to exit the firearm through the ejection port. This projection is called an ejector. As the breech opens, and following the exit of the bullet from the muzzle of the firearm, there will be an issue of discharge residues that will contain ingredients from the primer, from the propellant, from the bullet and possibly from the cartridge case. All, or most, of this material may be ejected from the firearm in the form of smoke, gases and particles, and is sometimes termed cartridge discharge residue (CDR), firearms discharge residue (FDR) or gunshot residue (GSR). This is demonstrated in Figure 10.2. These residues are significant to the firearms examiner as they may be deposited on any surface nearby, including the hands, face, head hair and clothing of the firer.

10.6 SCENE EXAMINATION

10.6.1 Examination of the Scene

The importance of a full understanding of the process of discharge is in its significance to a firearms examiner investigating the scene of a shooting incident and, as with every other type of examination, it is important that the scene is examined thoroughly by an experienced officer who is aware of, and capable of dealing with, any non-firearms related forensic evidence which may be present at the scene. Among the many things looked for and recovered will be:

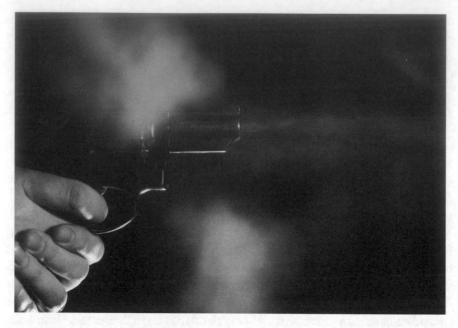

Figure 10.2 *High speed photograph of a firearm being fired showing ejection of discharge residues*

1. The firearm – its state noted, *i.e.* loaded, cocked, safety on or off *etc.*, and its exact location.
2. Spent bullets or spent cartridge cases – the location of each recorded and each individually labelled.
3. Bullet damage to any property – photographed if immovable, and fully described.
4. The possible line of flight of any bullet as determined by two or more sites of damage.
5. The possibility of discharge residues on surfaces that may have been close to the point of discharge.
6. Any evidence from an injured party, although this would normally be obtained by a trained medical medical practitioner or pathologist.

As will be seen later, the quantities of discharge residues normally found at the scene, of on the clothing of an injured party or suspect, is extremely small and their significance, when found, can only be interpreted relative to background levels. Therefore those examining the scene, or indeed present at the scene, must themselves be free from such residues.

To assist the subsequent investigation it may be possible to give the following information:

1. The number of shots fired, the direction of firing, and the point of discharge.
2. The type of firearm used and the means of discharge, which is particularly important in suspected suicides.
3. Whether any holes in windows or other materials are bullet entry or exit holes.

Frequently the examiner is requested to attend a post-mortem examination, when useful advice may be given to the pathologist regarding the scene of the shooting and additional information may be gained from the pathologist's examination. For example, swabs from the body may be required, an accurate measurement or details of the extent of the damage may be obtained, or, in the case of multiple injuries, information on which shot would have proved fatal.

At a later stage it may be necessary to examine a person suspected of the shooting and obviously the closer in time to the incident the better. Residues on the hands and face are lost very quickly, although they may remain on clothing for several days. The dangers of cross contamination are obvious and precautions must be taken to prevent this.

All the recovered material will be further examined at the firearms laboratory. Ideally, a well-equipped laboratory will have facilities for the examination of discharge residues, microscope facilities for the examination of the spent bullets and spent cartridge cases, firing range facilities to examine and test-fire the firearm recovered from the scene or submitted later, an area for clothing examination and a reference collection containing a variety of firearms and ammunition. Again the dangers of contamination are obvious and precautions must be strictly enforced.

10.7 EXAMINATION OF FIREARMS

Before examining any firearm it must be seen clearly by the examiner to be safe and unloaded, both in the breech and in the magazine. There have been occasions when a carelessly handled firearm at a scene has been discharged accidentally by inexperienced examiners, not only embarrassing the examiner but certainly presenting a danger and possibly destroying essential evidence.

An examination of the firearm should start with a detailed record of its condition as received. Is it loaded, or cocked, what is the position of the safety catch, how many cartridges are in the magazine or cylinder, what is the position of each spent and live cartridge in the cylinder and what is its general condition? The nature and location of any foreign material adhering to the firearm should be noted and samples taken for subsequent detailed examination. The same applies to the examination of spent bullets. It is good practice to complete an examination proforma sheet, as the examiner thus addresses all features and not just the ones that seem relevant at the time. An accurate measurement of the trigger pressure should always be carried out, especially in cases of suspected suicide or accidental discharge. The state of the bore of the barrel should be recorded and the presence or absence of discharge fouling, or the presence of any other foreign material, should be noted. A blocked or partially blocked barrel may result in excessive pressures when the firearm is test-fired and may cause serious injury should the firearm 'explode'.

After a full examination, an estimation of its suitability for firing can be made, and it may then be discharged to determine its effectiveness and to recover test bullets and cartridge cases. If in reasonably good condition it may be fired by hand, but if in poor or uncertain condition, it may be considered prudent to fire at least the first shot from it using a remote firing device. Only the correct calibre of ammunition for the firearm should be used. The spent bullets are usually recovered by firing into cotton waste, water or gelatine, the material chosen depending on availability or preference. It is important always to recover each bullet immediately after firing and to examine each to ensure it is undamaged and that satisfactory rifling marks are present.

Sometimes, when test-firing high velocity firearms, it may prove difficult to recover undamaged bullets by conventional methods, as the means of stopping the bullets in the available space causes too much friction between the bullet and the material. In such instances the quantity of propellant charge may be reduced by first pulling the bullet from the cartridge case and removing some of the propellant and reinserting the bullet. Such a practice is inherently dangerous as the burning characteristics of the propellant may be altered, resulting in too fast burning leading to excessive pressure, causing the breech to burst. This practice must only be carried out by an experienced examiner and even then a remote firing device should be used. The same calibre of cartridge may be manufactured by many different companies, each using differing hardnesses of lead or copper and as this may affect the marks left on the bullet or cartridge case it is desirable, though not essential, to fire the same type of ammunition as used in the shooting.

The spent cartridge cases are also recovered from the test-firing. With self-loading firearms, the ejection pattern, that is the direction of ejection of the spent cases – to the left or right, forward of the firer or to the rear – and the distance travelled should be noted. It is good practice to clean and lightly oil the firearm after firing, to ensure that it remains in the same condition as received. The recovered test-fired bullets and spent cartridge cases should be clearly labelled.

Occasionally, especially with air or gas guns, it may be necessary to determine the kinetic energy of the fired projectile. Originally this was done by firing the projectile into a freely suspended block of heavy absorbing material, such as lead. From the resulting swing and using the principle of conservation of momentum, the velocity of the projectile could be determined. Modern methods use a device known as a chronograph, which very accurately times the passage of the projectile over a known distance and produces a printout of its velocity. From the velocity and the weight of the projectile measured in grains, which is always used for bullet weights (7000 grains to the lb), the kinetic energy, in foot pounds (ft lb) can be calculated using the simple formula:

$$KE \text{ (ft lb)} = \frac{0.5MV^2}{700\,g}$$

where: M is the mass of the projectile in grains, V is the velocity in ft s^{-1} and g is the acceleration due to gravity, usually taken as 32ft s^{-2}.

For example, a .22 LR bullet, weight 37 grains, and fired with a measured muzzle velocity of 1205 ft s^{-1}, will have a muzzle energy of 120 ft lb.

The muzzle energy of the projectile is of interest to a forensic firearms examiner, particularly concerning low velocity projectiles such as those discharged from air guns, in order to give an indication of wounding ability. Table 10.1 gives typical KE figures for a few cartridges when fired in suitable firearms.

Table 10.1 *Kinetic energy values for some cartridges*

Firearm	Kinetic Energy (ft lb)
.177 air gun	6
.22 air gun	11
.22 LR rifle	120
.320 revolver	54
.38 S&W revolver	173
9 mm P pistol	408
.45 ACP pistol	450
.223 rifle	1530
.303 rifle	2240

10.7.1   Examination of Spent Bullets and Spent Cartridge Cases

The examination of spent bullets, spent cartridge cases, shot and wads recovered from the scene of a shooting incident can provide important information as illustrated below.

10.7.1.1   Spent Bullets.   If the bullet is undamaged, the calibre of the original cartridge can usually be readily determined by the experienced examiner from its overall appearance; otherwise its physical dimensions and weight will indicate its calibre. On some occasions it may be necessary to consult specialist books that give dimensions, weights and diagrams of all commercial and military cartridges. The rifling marks will give some indication of the type of firearm from which it had been discharged, using such features as the number of rifling grooves, their dimensions and whether they are inclined to the right or left. The exact width of the grooves and the angle of inclination is sometimes measured, but these parameters are not frequently used.

If the bullet is damaged, the amount of information gained will, of course, depend on the extent of the damage, but usually sufficient features are present to enable a good indication of the calibre to be made. From the calibre and the firing characteristics the examiner may be able to suggest the type of firearm used in the incident.

10.7.1.2   Spent Cartridge Cases.   The calibre of the spent cartridge case is usually easily determined purely by visual examination, although specialist books are available for unusual cartridge cases. From the action of firing, as described earlier, the following marks are looked for on spent cartridge cases:

1. The firing pin on the primer cup.
2. The ejector mark on the base of the case.
3. The extractor mark inside the rim.
4. The breech face marks on the base.
5. Any significant marks on the case wall made during chambering, discharge, or extraction.

Each of these marks, whether present or absent, and the shape and position of the extractor and ejector marks relative to each other has significance, and will help to determine the type of firearm involved, since various models of firearms will have differing designs of firing pin, extractor and ejector.

10.7.1.3   Shot and Wads.   The size and weight of each undamaged shot will indicate the shot size loaded in the cartridge. The number of shot particles recovered, or the extent of the damage by the discharge, may indicate the calibre (the bore) of the shotgun.

The diameter of any recovered wads will indicate the bore of the firearm used and the design and nature of the wads may indicate the manufacturer. Plastic wads are fairly common now and are fairly hard, compared to the traditional 'felt' wad. As they pass down a barrel they may on rare occasions pick up marks that can be used to link the wad to a particular barrel. By close examination of the wads it may be possible to see the marks left by the shot and so estimate the shot size.

10.8 COMPARATIVE MICROSCOPY

The aim of a microscopic comparison is to determine whether the objects have a common origin, that is if the bullet or wads were fired through the same barrel or if the cartridge cases were discharged in the same firearm.

In the manufacture of any object, the tool used in the process will leave marks on the object (see Chapter 4). So when the rifling grooves are being 'cut' in the barrel of a firearm, the 'cutting' tool will leave marks, called striations, on the lands and grooves. Since the metal of the barrel is extremely hard to withstand the high pressures involved, the cutting tool will itself change as it passes down the barrel so the marks will continually change. Not only will each barrel be different, but each part of the barrel will be different. This fact has been demonstrated by comparing bullets fired from barrels manufactured consecutively, with significant differences being found. It is said that the striations on a fired bullet are made by the last few centimetres of the barrel and this fact has also been demonstrated by comparing bullets fired from the same barrel, which was shortened between shots. Again the fired bullets had significant differences. So the theory, substantiated by experience, is that each barrel will leave unique marks on bullets fired through it. The striations in a barrel may, of course, change with time especially if not treated with care or if treated with abuse such as the use of an abrasive tool.

A comparison microscope is used for comparing evidence. Any two objects for examination are positioned one on each side of the microscope and by suitable arrangements of the stage of the microscope, and with suitable lighting, the images are presented side by side at the objective lens and are viewed through the eyepiece. By careful manipulation of each object on the stage, similar features of each object can be viewed together and compared. It is necessary to stress that only similar features should be examined in this way. For example, when comparing bullets the rifling marks on each must be similar and, if one bullet is rifled with six grooves left twist and another one is rifled six grooves right twist there is no purpose at all in carrying out a comparative microscopic examination. Also, with spent cartridge cases, the marks being examined must have similarity in shape and relative position.

Usually the initial examination is carried out at about ×5 magnification and each mark, groove or land on one object is compared to the equivalent mark, groove or land on the other. (Note that the groove on a fired bullet is caused by the land of the barrel and vice versa). The examination consists of examining each mark in the metal that has been made by the firearm during loading, firing or unloading.

Having found areas of comparison, the magnification may be raised to about ×20 but seldom much more than this. If an exact match between the striations is obtained the examiner will conclude that such marks have been caused by a common source, that is, both bullets have been fired from, or both cartridge cases were discharged in, the same firearm. However, before such a firm conclusion can be drawn, it must be established that the marks examined are consistently made by the firearm in question. Test-fired bullets or cartridge cases should similarly be examined and compared to confirm the consistency of the marks. Examples of comparisons of two bullets and two cartridge cases are given in Figure 10.3.

10.9 COMPOSITION OF CARTRIDGES

Before dealing with the chemical investigations relating to firearms casework, it is useful to understand something of the composition of the various cartridge components, namely cartridge cases, primer cups, primers, propellants and projectiles.

10.9.1 Cartridge Cases

The cartridge case houses all the components of the cartridge – the primer, the propellant and the projectile. The vast majority are made of brass (an alloy of copper and zinc) but other materials, including steel and aluminium, are also used. Shotgun cartridges are usually plastic with a brass or coated steel base and the projectiles are contained within the cartridge case.

10.9.2 Primer Cups

The primer composition for centrefire ammunition is contained in a small metal cup, recessed in the centre of the base of the cartridge case. In rimfire ammunition there is no primer cup and the primer composition is contained inside the hollow perimeter (the rim) of the base of the

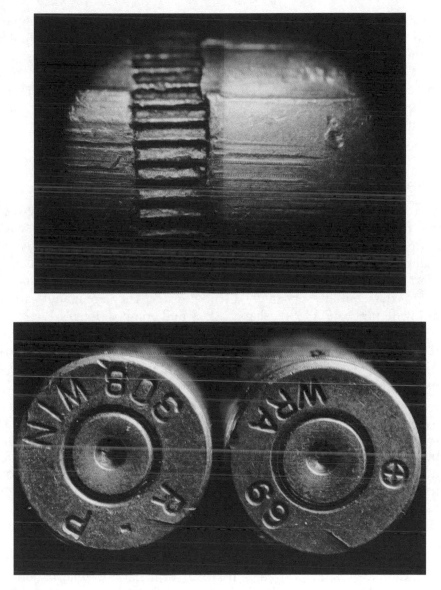

Figure 10.3 *Comparison microscopy of (a) two bullets and (b) two cartridge cases*

cartridge case. Primer cups are usually made of brass, although other metals and alloys are also used.

10.9.3 Primer Compositions

Primer compositions normally consist of an explosive, an oxidizer, a fuel and a frictionator, with other materials acting as sensitizers and binders. In 'modern' type primers the explosive is usually lead styphnate (lead trinitroresorcinate), the oxidizer is barium nitrate and the fuel is antimony sulphide. As a general rule, modern type primers are used in the USA, the UK and most European countries, whereas 'old' type primers, based on mercury fulminate as the explosive, are currently manufactured by some of the Eastern Bloc countries, although they also manufacture modern type primers. Other primer compositions may also be encountered in forensic casework.

10.9.4 Propellants

A propellant consists of a mixture of substances and may be defined as: 'Explosive materials formulated, designed and ignited in such a way as to permit the generation of large volumes of hot gases, at highly controlled predetermined rates'. These explosives are frequently referred to as gunpowder, the powder charge or simply as 'the charge' or 'powder', but are very rarely a true powder. They are manufactured in a wide range of colours, shapes, sizes and compositions.

Black powder was the first propellant used and consisted of a mixture of charcoal, saltpetre (potassium nitrate) and sulphur in the typical proportions of 15:75:10 respectively. It has now almost completely been replaced by smokeless powders, although it is still used as a propellant in some specialised cartridges and ammunition assemblies.

Modern smokeless propellants almost exclusively contain nitrocellulose as the major oxidising ingredient. Single base propellants contain nitrocellulose and double base propellants contain nitrocellulose and nitroglycerine. Other ingredients may be added to increase performance, or chemical stability, to facilitate processing and handling, or to minimise muzzle flash.

10.9.5 Projectiles

In most instances the projectiles are made of metal but other materials, such as rubber or plastic, are occasionally encountered. The metal used for bullets is normally lead, which is sometimes hardened by other metals, such as antimony or tin. Other metals, including iron, aluminium and brass are also used to manufacture bullets. The bullet may or may not be enclosed, or partly enclosed, in a jacket of copper or another metal. Lead bullets enclosed in a copper alloy jacket are by far the most common.

Shotgun cartridges are normally loaded with round metal balls, called pellets or shot, which are available in various sizes, degrees of hardness and materials, depending on their intended use. Occasionally a single projectile of the same diameter as the bore is used. The shot is contained within the cartridge case and sometimes a plastic buffer material is mixed with it to help prevent distortion of the shot during discharge. Lead is by far the most common shot material and can be hardened with antimony. The lead shot is sometimes plated with copper or nickel.

Up to four or five wads can be present in a shotgun cartridge, although two or three are more common. In close range shotgun shootings, wads and plastic buffer material may be recovered from the target. The wads can provide useful information as they can be chemically examined, which may reveal information about the primer and propellant, and physical examination can usually identify the calibre, approximate shot size and may indicate the manufacture of the cartridge.

There are many specialised types of projectiles, such as tracer, incendiary, armour piercing, exploding, saboted sub-calibre, flare, frangible, tear gas, multiple loads, baton rounds and fletchette cartridges. Although less likely to be involved in forensic casework, they may be encountered occasionally. Such projectiles may have distinctive chemical compounds or elements present, which can be identified and consequently aid the investigation of the incident.

10.10 FIREARMS DISCHARGE RESIDUES (FDR)

The gases, vapours and particulate matter formed by the discharge of a cartridge in a firearm are collectively known as firearms discharge residues (FDR). Anything present in the cartridge, and even the firearm itself, may contribute to the residues, which therefore contain both organic and inorganic components. The main sources of the inorganic components are the bullet and the primer composition, and the main source of the organic components is the propellant. FDR are therefore a complex heterogeneous mixture. The detection, identification and quantification of FDR can provide significant evidence in several areas associated with incidents involving the use of firearms.

10.10.1 Formation of FDR

Gases arise from the burning of the propellant and the primer. Organic particulate matter arises from unburnt or partially burnt propellant. Inorganic particulate matter is formed by the hot gases at high pressure acting on the base of the bullet and by strong frictional heating as the bullet passes along the barrel. The metal vapours from the bullet mix to some extent with the vapours from the inorganic compounds of the primer and condense as particles on cooling. This gives the inorganic particles a characteristic appearance, a sort of three-dimensional round-ness, typical of something condensed from a vapour or melt. The rifling of the barrel also produces inorganic particulate matter by stripping metal from the surface of the bullet. As already stated FDR may be deposited on the hands, face, head hair and clothing of the firer.

10.10.2 Collection of FDR

Some form of kit containing sampling materials is normally prepared, whether commercially, by police forces or by forensic laboratories, to sample suspects for FDR. Hand sampling methods include:

1. Swabbing – using cloth, cotton wool or filter paper, moistened with dilute acid.
2. Washing – using very dilute acid.
3. Film lifts – using a paraffin 'glove', or film-forming polymers.
4. Adhesive lifts – using adhesive tape or adhesive stubs.

The subsequent analytical technique to be used strongly influences the type of sampling procedure. Whatever recovery method is used, the lifting efficiency depends to a large extent on the care taken by the sampler. A suspect's clothing is usually sampled in the laboratory using either swabbing, adhesive lifts or suction sampling techniques.

10.10.3 Detection of FDR

When a firearm is discharged, the residues are emitted mainly from the muzzle, but also from cylinder gaps, ejection ports and other vents. The residues exiting from the muzzle can be deposited on the area surrounding the bullet entrance hole in close range shootings (less than 1 m). The density, shape and extent of the residue pattern can be used to estimate the muzzle-to-target distance and angle of fire. Unburnt or partially consumed propellant recovered from items examined, can be compared, using physical characteristics plus chemical analysis, with propellant associated with a suspect. Such examinations are rarely necessary, but have been important in several instances. In the case of shootings involving shotguns an additional range of fire indicator is the spread of the shot pattern. This is illustrated in Figure 10.4.

The detection of such residues on a suspect implies recent involvement with firearms and for many years forensic scientists have sought

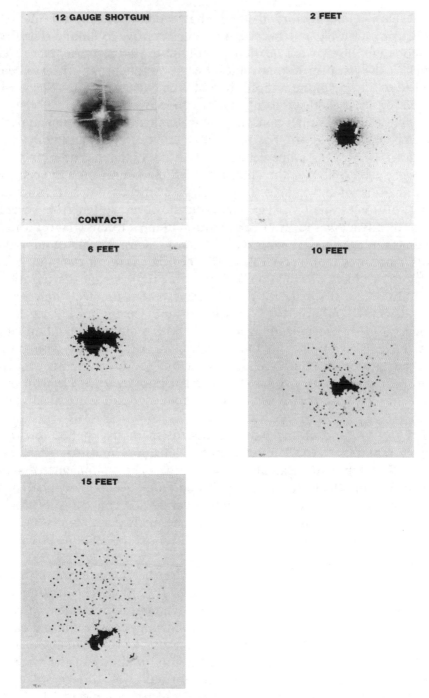

Figure 10.4 *Spread of shot pattern after firing a shotgun at different distances from a cloth target. Muzzle (a) in contact with cloth, (b) 2 feet, (c) 6 feet, (d) 10 feet and (e) 15 feet from the cloth*

to develop a satisfactory method that would conclusively identify FDR using a simple, reliable, rapid and inexpensive technique. The first methods employed were qualitative colour tests, the most famous of which is the paraffin test, so called because molten paraffin wax was applied to the skin of the suspect's hand to remove the residues. Such tests lacked sensitivity and specificity and are no longer used.

A major breakthrough came in 1964 when neutron activation analysis (NAA) was applied to the quantitative detection of antimony and barium in FDR. This is an excellent analytical tool that has been, and still is, used successfully for the detection of these metals on suspects, the identification of bullet holes in a variety of target materials, and for 'range of fire' estimations. The technique, however, suffers from several major disadvantages for routine operation in most forensic laboratories. These include the need for access to a nuclear reactor, the high equipment costs, the need for specialist staff and the slow throughput time of samples, due to irradiation, cooling and radiochemical separation. Possibly the most significant disadvantage is that this technique has a poor detection limit for lead, a very important element in FDR work.

Several other instrumental techniques, both quantitative and qualitative, have been applied to the detection of FDR, some with more success than others. Such techniques include flameless atomic absorption spectrophotometry (FAAS), molecular luminescence, electron spin resonance (ESR) spectrometry, X-ray analysis and electroanalytical methods.

FAAS was one of the more successful methods applied and has largely replaced NAA. It offers similar detection limits as NAA, but above all it can detect lead at very low levels. The equipment costs are reasonable and the instrumentation is commonplace in many analytical laboratories. Many metallic elements, over a wide concentration range extending down to ultra trace level, can be analysed, thus making the technique versatile and useful for other forensic applications as well as FDR detection.

Apart from the relatively low cost, the other main advantages are simplicity, speed of analysis, and 'in house' operation. This technique is the most popular for the quantitative determination of elements associated with FDR, such as lead, antimony, barium, copper and mercury.

All of these methods are referred to as 'bulk elemental analysis methods', that is, they detect and analyse the total quantity of the material present. They suffer, however, from the serious disadvantage of lack of specificity, in that the elements detected are not unique to FDR, but also occur from occupational and environmental sources. The best that can ever be stated is that their presence on a suspect is consistent with the discharge of a firearm, but cannot be taken as conclusive proof of the presence of FDR.

A more definitive method was sought, resulting in the particle analysis method, which can identify FDR. This method employs a scanning electron microscope, equipped with elemental analysis capability (SEM/EDX) and gives details of the morphology and elemental composition of individual FDR particles as small as one micron in size (1 micron = 10^{-6} m). A particle classification scheme was developed, based on the morphology and elemental composition of individual FDR particles, and this is used to classify particles as either (a) of non-firearm origin, (b) consistent with originating from the discharge of a firearm, or (c) definitely from the discharge of a firearm. The ability to identify FDR particles uniquely and to distinguish them from environmental sources of lead, antimony and barium eliminates the problem inherent in bulk elemental analysis. Such particles have also been observed to originate from blank cartridges and consequently are best referred to as cartridge discharge residues (CDR), rather than FDR, since some uses of blank cartridges are not firearms related.

Particle analysis is the most informative method to date for the identification of FDR particles. It does however, suffer from several major disadvantages, including high cost of instrumentation and lengthy and tedious procedures requiring specialised staff. Since its introduction, serious attempts have been made to solve the time problem, resulting in major improvements, but despite this the technique remains lengthy and costly. These disadvantages have recently renewed interest in the possibility of detecting the organic components of FDR either as a primary method or as a screening technique.

Many of the organic constituents of FDR are explosive or explosive-related compounds and their residue is usually detected using chromatographic techniques. To date, the majority of forensic laboratories have concentrated on the inorganic components of FDR and it is only relatively recently that some laboratories have started routinely to look for the presence of certain organic constituents in addition to inorganics. Whilst the particle analysis method has been well tried and tested in both casework and court and is a very satisfactory method, the presence of organic constituents of FDR would substantially raise the significance level of inorganic particles in the 'consistent with' category. At the present time the detection of organics is unsuitable as a screening technique. It is much more likely to be a complementary method rather than a primary one and has already been used to complement both bulk elemental analysis and particle analysis.

10.11 CONCLUSION

It has been possible to give only a brief outline of some aspects of forensic firearms examination in this chapter. However, forensic firearms examination can be seen to encompass both the physical and chemical examination of FDR, firearms and related items, with these being recovered from the scene, the injured party and the suspect. Therefore, the results of these examinations can provide very useful information about a sequence of events and assist substantially in the investigation of an incident.

10.12 BIBLIOGRAPHY

Gunshot Residue Detection, *Aerospace Report ATR-75 (7915)-1,* The Aerospace Corporation, California 1975.

W. Lichtenberg, Methods for the Determination of Shooting Distance, *Forensic Sci. Rev.*, 1990, **2**, 37–44.

R. Saferstein, *Criminalistics – An Introduction to Forensic Science*, Prentice Hall, New York, 1981.

J.S. Wallace, Chemical Aspects of Firearms Ammunition, *A.F.T.E. Journal*, 1990, **22**, 364–389.

F. Wilkison, *Firearms*, Camden House, 1977.

CHAPTER 11

Drugs of Abuse

MICHAEL COLE

11.1 INTRODUCTION

The problem of drug abuse continues to increase and the drug analyst is always being presented with new and interesting challenges. No two cases are ever the same, each presenting unique problems to the analyst. The legislation in the United Kingdom prescribes different charges and sentences for different drugs and for differing amounts of those drugs and hence, in addition to the identification of any controlled substance which might be present, the analyst may also be required to quantify any drugs of abuse in the sample. Depending upon the drugs thought to be involved, some techniques are more appropriate than others for these tasks and it is a knowledge of the drugs involved and which techniques to apply, and when, that form the major part of a drug analyst's expertise and activity. There is also increasing interest, both nationally and internationally, in the comparison of drug samples with each other. This provides information concerning the manufacture, distribution and supply networks and movement of drugs at the street (user) level. These comparisons require different methodologies again and understanding them is an important part of the drug chemist's work.

11.2 DRUG CONTROL LEGISLATION IN THE UNITED KINGDOM

Two principle pieces of legislation are employed to control the drugs of abuse, namely the Misuse of Drugs Act, 1971 and the Misuse of Drugs Regulations, 2001. The 1971 Act lists the drugs in one of three classes as part of Schedule 2 to the Act. A number of classes and individual drugs have subsequently been added and deleted, as detailed in the 17 subsequent Amendment Orders, the Misuse of Drugs Regulations 1985 (which controlled benzodiazepines and barbiturates for the first time) and the Misuse of Drugs Regulations, 2001.

Other pieces of legislation, for example the Traffic Offences Act, 1986, makes provision for the confiscation of the proceeds in trafficking in drugs of abuse. The legislative materials also make provision for the control of any salt, ester or stereoisomer of the compounds listed in the legislation. A controlled substance is therefore defined as one which falls order the umbrella of being controlled by one or other, or both, of these pieces of legislation. In short, The Misuse of Drugs Act prescribes that which cannot be done and the regulations prescribe what may be done, with the correct authority to do so.

11.3   DRUGS OF ABUSE AND THEIR SOURCES

During the course of a drug analyst's career, he or she is likely to encounter a wide variety of drugs. What follows is a description of the principle classes of drugs likely to be encountered.

11.3.1   Cannabis and Its Products

The products of *Cannabis sativa* L. form the major part of the drug materials currently submitted to forensic science laboratories. The materials come in three major forms: herbal material, resin and oil. All contain the active constituent, Δ^9-tetrahydrocannabinol (Δ^9-THC), sometimes with a small amount of its isomer, Δ^8-tetrahydrocannabinol. The precursor

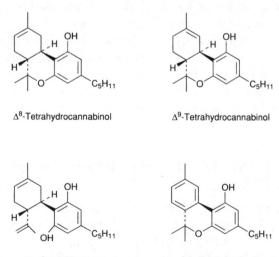

Δ^8-Tetrahydrocannabinol Δ^9-Tetrahydrocannabinol

Cannabinol   Cannabinol

Figure 11.1   *Chemical structures of some cannabinoids present in* Cannabis sativa *L*

of this is cannabidiol and in older samples in which decomposition of the sample has started, cannabinol will also be observed.

Herbal material of *C. sativa* comes in a variety of forms, from whole plants to small amounts of leaves, stems and flowering tops, and can also be encountered fresh or, more commonly, dried. The fresh material has a very characteristic smell, of spearmint, although not exactly the same. The leaves are palmate, with serrated edges. The stems are square and four-cornered. The flowering tops are recognised by the presence of buds, flowers and sometimes seeds. Three principle morphological characteristics allow the recognition of *Cannabis* plant material. The first is single-celled trichomes, which are colourless and, unusually, point unidirectionally up the stem. The second feature is the cystolithic trichomes, with crystalline materials at their base. The third feature is the presence of glandular trichomes, which consist of a glandular structure on a thin stalk of cells. It is within the glandular trichomes that the resin is stored. If all three of these morphological features are observed, the plant is almost certainly *C. sativa*. Even the closest relative of *C. sativa*, the hop (*Humulus lupulus*) does not have these morphological features, although it does have glandular trichomes at the base of the leaves which may, by the inexperienced eye, be confused with the trichomes of *C. sativa*. The herbal materials may be packaged in a number of different ways and contain differing amounts of leaf, stem, flowering tops and seed material. The form of the packaging and its contents may be used to identify the country or region of origin of the plant materials.

Resin comprises the material collected when the glandular trichomes are collected, by a variety of means. These include passing large hessian nets through the crop, onto which the resin becomes stuck, as is performed in North Africa, and rubbing of the plant materials between the palms of the hands, as occurs in Pakistan. Either way, the resin is compressed into blocks and shipped. The resin may contain other cellular debris and it is frequently possible to observe parts of both the glandular hairs and the cystolithic trichomes under microscopic examination.

The oil, commonly referred to as 'hash oil', represents the material that is extracted using solvent extraction of the plant materials. The concentration of the active constituents and other cannabinoids is much higher (as a percentage by weight) in this material than in the resin, which in turn contains a higher percentage by weight than the plant materials.

11.3.2 Heroin

Heroin is derived from the opiate alkaloids produced by the field poppy, *Papaver somniferum* L. It is not a pure compound, but a mixture of the impurities, reaction intermediates, breakdown products manufactured during the synthesis of diamorphine and excipients and adulterants used to dilute, or cut, the drug. Diamorphine is derived from morphine, one of the alkaloids from opium resin. The seed pods of the plant are lacerated and latex exudes which then dries. From the dried latex, morphine is extracted using a series of solvent/solvent extractions and precipitations. The morphine is then acetylated, producing the mono-acetylmorphines, and diamorphine which are substances controlled under the Misuse of Drugs Act 1971. Since codeine is also extracted from the opium resin using these methods, acetylcodeine will be present in the final product. The chemical structures of alkaloids commonly found in heroin are shown in Figure 11.2.

Thebaine is also present because it is extracted from the opium resin in the same way. Two other alkaloids, of different classes, are also co-extracted, namely noscapine and papaverine. The amounts of all of these alkaloids present in the final products will depend upon a number of factors, including geographical origin of the drug, quality of the extraction chemistry, quality of the acetylation and quality of the purification processes. Since heroin is produced in a batchwise manner,

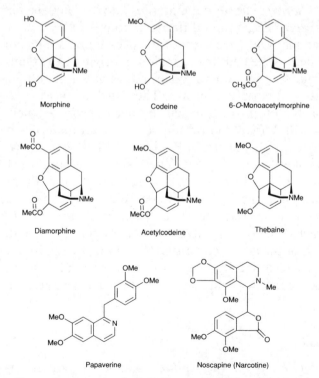

Figure 11.2 *Chemical structures of alkaloids commonly found in Heroin*

no two samples will ever be the same. However, heroin samples from different geographical regions may exhibit certain chemical characteristics, outlined in Table 11.1.

It is due to this variation in appearance and chemical nature that the origins and relationships of drugs can be identified. This is examined in greater depth later in this chapter.

Cutting agents are used to dilute the drug and hide its impure nature and/or lack of active ingredient. Examples of common cutting agents include sugars, caffeine, methaqualone and barbiturates. These too can be used to compare drug samples.

11.3.3 Cocaine

Cocaine is produced from the tropane alkaloids produced by the trees *Erythroxylon coca* Lam, *E. novogranatense* (Morris) Hieron and *E. novogranatense* var. *truxillense* (Rusby) Plowman.

The species involved depends upon the region in South America from which the sample originates. As with heroin, the tropane alkaloids are

Table 11.1 *Examples of how heroin samples can vary both chemically and in appearance, within and between regional sources*

Origin	Drugs Present			
	Acetyl codeine	6-O-monoacetyl morphine	Noscapine	Papaverine
S.W. Asia				
type I Beige – brown, powder with small aggregates. 60% diamorphine, free base.	5%	3%	10%	4%
type II White – off white 80% diamorphine, hydrochloride salt.	3%	2%	—	—
Middle East				
type I Beige – light brown powder, aggregates rare 50–70 % diamorphine, hydrochloride salt.	3%	2%	—	—
type II White – off white, fine powder 70–80 % diamorphine, hydrochloride salt.	2–3%	2%	Data not provided	

— Drug not detected.

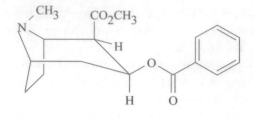

Cocaine

extracted from the leaves by a series of solvent/solvent extractions and precipitations. A hydrolysis step follows, so that all of the ecgonine-based tropane alkaloids are hydrolysed to ecgonine. From this ecgonine methyl ester is manufactured, from which the final product, cocaine, is synthesised. Since the material, like heroin, is produced in a batch process, no two samples are the same and contain impurities from the natural products in the leaves, including other tropane alkaloids, *e.g.* cinnamoylcocaines and truxillines, which are carried over into the final product.

Again, it is the composition of the product that allows samples to be compared and geographical origin to be determined. Two principle forms of cocaine may be encountered. The first is cocaine hydrochloride. This is a white powder, which has a fluffy, snow-like

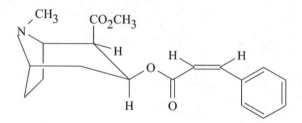

cis- Cinnamoylcocaine

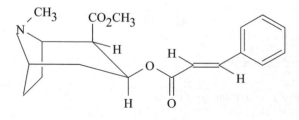

trans- Cinnamoylcocaine

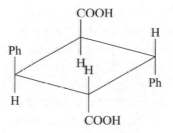

α-truxillic acid, on which many truxillines are base

appearance, hence one of its street names, 'snow'. If the free base is then re-prepared from the hydrochloride salt, hard, granular lumps of cocaine are formed. This form is known as 'crack'.

As with heroin, cocaine may also be cut. The cutting agent that is employed depends upon the region in which the drug is encountered. In the United States, the drug is commonly adulterated with sugars, for example, mannitol or sucrose, whilst in the United Kingdom, local anaesthetics (*e.g.* lignocaine, procaine) are frequently used.

11.3.4 Amphetamines

In the United Kingdom, under the Misuse of Drugs Regulations, 1985, Schedule 1, Part (c), any compound structurally derived from phenethylamine is controlled. These compounds include amphetamine, methylamphetamine (MA), methylenedioxy-amphetamine (MDA), methylenedioxymethylamphetamine (MDMA) and their chemical structures are shown in Figure 11.3.

These compounds can be derived *via* a number of different synthetic routes, with either natural products derived from plants (such as ephedrine, safrole and isosafrole) as starting materials, or chemicals available from chemical companies (for example benzenemethylketone,

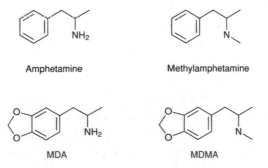

Amphetamine   Methylamphetamine

MDA   MDMA

Figure 11.3 *Chemical structures of some amphetamines*

BMK). It is unlikely that the natural products will ever come under control, but certain purified precursor chemicals are now only supplied to organisations who have the requisite licence. Examples include BMK and ephedrine, both precursors of amphetamine.

These compounds can be synthesised *via* a number of different routes, including, for amphetamine, the Leuckart synthesis, the nitrostyrene route and the reductive amination of BMK. The most popular routes depend upon the country in which the drug is being synthesised. For example, in Western Europe, the synthesis of amphetamine is principally *via* the Leuckart synthesis, whilst in the United States the drug is manufactured principally *via* the reductive amination of BMK, and in Japan ephedrine is currently favoured as the starting material. Since different synthetic routes are employed and the drugs are made in batches, the impurities and reaction by-products will also vary between batches. Some will be route specific, indicating which of the synthetic routes has been employed. Further, the overall composition of the drugs and impurities will vary between batches, allowing differentiation between samples from different batches and identifying those street samples that have come from the same parent batch. This is discussed in detail later in this chapter.

11.3.5   *Psilocybe* Mushrooms

Occasionally, material from mushrooms with hallucinogenic properties may be encountered by the forensic chemist. The hallucinogens in the fungi are based on an indole-amine nucleus. In the genus *Psilocybe*, of the approximately 140 species, 80 are known to contain hallucinogens. The three most important species are *Psilocybe semilanceata* (Fr.) Quel., *P. cubensis* (Earle) Singer and *P. mexicana* (Heim). These contain psilocybin (4-phosphoryloxy-*N*,*N*-dimethyltryptamine), its biochemical precursor, baeocystin (4-phosphoryloxy-*N*-methyltryptamine, norpsilocybin), and psilocin, the dephosphorylated compound, which is also a hallucinogenic metabolite. The chemical structures of the major constituents found in *Psilocybe semilanceata* are shown in Figure 11.4

The effects of baeocystin are not known. The materials may be encountered as either dried mushrooms or as powdered materials. Identification of these species (as with any mushrooms) is a difficult area in which to work, and ideally, a qualified mycologist should be engaged to positively identify the species of mushroom found in a criminal case.

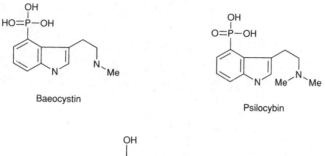

Baeocystin

Psilocybin

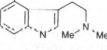

Psilocin

Figure 11.4 *Chemical structures of major constituents found in* Psilocybe semilanceata

11.3.6 Mescal Buttons

The Peyote cactus, *Lophophora williamsii* (Lem. ex Salm-Dyck) Coult. is known to produce the hallucinogen, mescaline.

The material is usually encountered as disc-like slices of the body of the cactus which have been dried. Occasionally (particularly in the United States), seizures are encountered where the active constituent has been synthesised from 3,4,5-trimethoxybenzaldehyde.

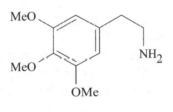

Mescaline

11.3.7 Lysergic Acid Diethylamide

Lysergic acid diethylamide (Lysergide, LSD) is one of the most potent hallucinogens known to man and is usually encountered in four forms: blotter acids, gelatin blocks, small tablets and microdots. It can be synthesised following the isolation of lysergamide from Morning Glory or from Hawaiian Baby Woodrose seeds, or from ergot alkaloids from *Claviceps purpurea* or *Aspergillus clavatus.* No single route dominates over the others.

LSD can be encountered in gelatin blocks, into which the LSD has been incorporated whilst the gelatin has been a liquid. It can also be

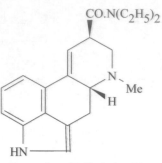

Lysergic acid diethylamide

encountered as small, highly coloured tablets, known as microdots. Both of these carrier media present problems in terms of homogeneity through the dosage units, one unit may contain a very much larger or smaller dose than another. It is for this reason that the most frequently encountered dosage form is the blotter acid. Adsorbent paper is passed through a solution of LSD and then the solvent allowed to dry, leaving the drug impregnated in the paper. The papers are often decorated with emblems, patterns or symbols, that may pass over the whole sheet, or individual dose units. The dosing on LSD blotter acids is far more uniform, giving rise to doses of 30–50 µg per dose unit.

11.3.8   Barbiturates and Benzodiazepines

In addition to synthetic drugs and drugs derived from plant products, some drugs are 'diverted' from licit sources. This is particularly true of the benzodiazepines and barbiturates. The principle dosage forms of both of these drug classes are tablets and capsules. However, a few of the drugs are prescribed as solutions for injection and the drug chemist ought to be aware of the possibility of encountering these drugs in this form.

11.4   IDENTIFICATION OF DRUGS OF ABUSE

There are a number of materials which may be encountered during the course of a drug analyst's career. These include drugs in bulk form (anything which can be seen), trace samples, wrapping materials and paraphernalia that may be associated with drugs of abuse.

The drugs are likely to come in a number of forms as shown in Table 11.2.

Whilst not perfect by any means for eliminating drug classes, the visual appearance of a sample may well indicate the class of drug which

Table 11.2 *Dosage forms of commonly encountered drugs of abuse*

Drug	Dosage form
Amphetamines	Powder Capsules Tablets
Barbiturates	Tablets Capsules Powder Solution for injection
Benzodiazepines	Tablets Capsules Powder Solution for injection
Cannabis	Leaf Resin Oil
Cocaine	Powder (snow and crack)
Lysergide	Blotter acids Microdots Gelatin blocks
Opium	Resin White powder
Heroin	Powder

may be expected in the sample. Trace samples should not be examined in the same room as bulk samples. Trace samples can be considered to be materials in or on which drugs are likely to be found but cannot be seen. Bulk samples are anything where the drug can be seen. The potential for contamination is enormous and great care should be taken to eliminate it.

In addition to drug samples themselves, there are a number of items of paraphernalia that may well be encountered on a regular basis by the drug chemist. These include glass surfaces (*e.g.* oven doors) on which drug samples may be cut or mixed; scales or balances on which drugs are weighed and measured; knives and cutting implements which have been used to divide and apportion the drug materials; and items which have been used to administer or take the drug. These include spoons, syringes and items for smoking cannabis or methamphetamine-based products. Further, drug debris may also be present. For example, there may be residues from heating of a heroin sample on a spoon or smoked material from cannabis.

Provided that there is sufficient sample, the analysis should proceed through a logical sequence of events leading to drug identification and if necessary, quantification. If only a tiny amount of sample is present, the non-specific non instrumental methods are usually omitted and the analysis proceeds straight to the instrumental methods that provide definitive information concerning the identity of the drug. The analysis proceeds as shown in Figure 11.5.

1.1.4.1 Sampling

Initially a full physical description of the item should be made, including a note of the integrity of the packaging. Following this, decisions should be made as to how the item(s) should be sampled. Each item that appears different should be considered to be a separate item. Powders are sampled after thorough homogenisation by, for example, placing the sample in a polythene bag or Waring blender to simply mix it up, or using the cone and square method. (The cone and square method is one where the sample is mixed thoroughly and divided into four equal parts. The process is repeated for two opposite quarters. This process is repeated until a sample suitable for analysis is obtained). Small samples are easily removed from liquid samples. Stains and marks can be sampled using a swab soaked in an appropriate solvent. Rinsing is often used to remove drug samples from some items of paraphernalia, for example, syringe barrels. Methanol is suitable for most drug classes, whereas ethanol is preferred for surfaces thought to have been in contact with cannabis products, since these are most stable in ethanol. Chloroform is the best solvent for heroin and cocaine samples, which are easily hydrolysed in methanol due to the presence of water. Whichever sampling technique is employed for a single item, sufficient should be left for a re-examination by any other interested party. This is particularly important in any adversarial legal system. If leaving a sample for re-analysis is not possible, the value of the data should be very carefully considered before the analysis is undertaken.

11.4.2 Presumptive Tests

If the sample is a trace sample, such as a swabbing, then there may only be sufficient to perform instrumental tests. If there is sufficient sample, however, the analyst should proceed to the next stage, the presumptive tests. These are cost effective, easy, simple tests, which provide accurate and reproducible colour changes on the addition of certain reagents to certain classes of drug. The principle importance of these tests is that

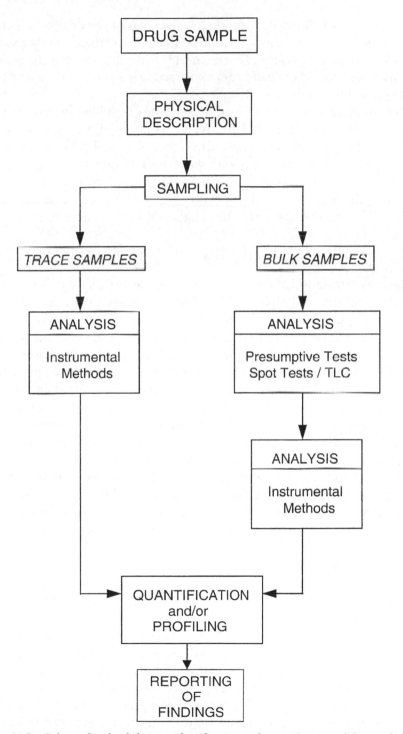

Figure 11.5 *Scheme for the definitive identification and quantification of drugs of abuse*

they provide for the analyst an idea of the class of the drug to which the seizure is thought to belong. Additionally, due to their simplicity, they are ideal for use by police officers and customs and excise officers, who are not necessarily trained laboratory scientists, at airports, sea-ports and other duty stations.

Presumptive tests have several disadvantages. These include the fact that they do not identify, definitively, which member of a particular class of drug is present. They also react with substances which are not under control, thus providing what are somewhat inaccurately described as "false positives". This term should be avoided if at all possible, the reaction is just as positive, only with a compound which is different to the one for which the analyst may be testing.

11.4.3 Thin Layer Chromatography

Following presumptive tests, the next stage in the analysis is the use of thin-layer chromatography (TLC). The stationary phase, mobile phase and developing reagent selected will have all been determined on the basis of the class of drug present, identified through the presumptive tests. Aliquots of the sample(s), together with both positive and negative controls, are spotted seperately onto the plate which is then developed with the mobile phase. On completion of this stage the plate is removed from the developing tank and dried. Next it is examined visually under ultraviolet light (long and short wavelengths) and any compounds observed are marked. The plate is then sprayed with a developing reagent. The distances moved by the drug relative to the distance moved by the mobile phase (R_f values) and colour reactions of any compounds observed are recorded in the laboratory notes. TLC is relatively rapid and separates many of the major components of different classes of drugs of abuse. Compounds are identified on the basis of two physico-chemical parameters: R_f value and colour reaction with a developing reagent.

The positive controls provide R_f and colour reaction data against which the sample can be compared, whilst the negative control demonstrates that the compounds observed are due to the sample applied and not to contaminants of the solvent. These controls are essential on every plate since the R_f values obtained will not be identical between operators or replicates of the same analysis, due to a number of physico-chemical properties of the analytical system.

Due to the nature of the chromatographic system, two principle limitations are inherent to the technique. Firstly, the separation between compounds (resolution) is limited, and secondly, even with these data generated there is no definitive proof of the identity of the drug. The power of the technique lies in the ability to determine which members of a particular class of drug are present, and on some occasions, which adulterants are present too, in a short time interval.

11.4.4 Instrumental Techniques

Following the identification of the members of the class of drug present, instrumental methods are employed to identify the drug. The techniques used fall into three categories: chromatographic, spectroscopic and hyphenated techniques.

11.4.4.1 Chromatographic Techniques. The techniques that find common application are performance high liquid chromatography (HPLC) and gas chromatography (GC). Other techniques, such as supercritical fluid chromatography, are not used routinely and hence not considered any further in this chapter although on occasion they do have certain advantages. HPLC separates analytes on the basis of chromatographic mobility in a stream of liquid flowing through a matrix of materials on which the compounds are retained to different extents and hence are separated. The time a compound is retained on the column (retention time) is recorded. The choice of stationary and mobile phase is again dictated by the drug class in question, illustrating once again the central importance of the correct interpretation of the presumptive tests. Gas chromatography separates the analytes in the gas phase, again providing a retention time. The principle means by which compounds are identified through these methods is the comparison of the retention time of the components of the sample with those of a standard compound, chromatographed on the same instrument under identical conditions.

HPLC requires that the compound exhibits good liquid chromatographic properties and is readily detected through, for example, the absorbance of ultraviolet light or by fluorescence. It represents an improvement over TLC because of its greater resolving power of components in a mixture. However, alone, it does not definitively prove the identity of the drug. Gas chromatography has similar advantages and disadvantages. The latter technique also presents problems, in that some drugs, for example certain benzodiazepines, are thermally labile (decompose when heated), or others, which are polar, acidic or basic, require modification (derivatisation) of their chemical structure prior to their analysis. This requires the analyst to ensure that each compound is fully derivatised prior to analysis.

11.4.4.2   Spectroscopic Techniques.   The principle spectroscopic technique that is employed is infrared spectroscopy (IR). The IR spectrum is obtained by measuring the interaction of IR radiation with a drug. The interpretation of the spectrum can be achieved in four ways. One method is to compare the spectrum of the sample with that obtained for a standard of the pure drug. Another is method is by comparison of the fingerprint region of the spectrum (frequency range of 450–500 cm^{-1}) of the sample to the standard or to literature values. The fingerprint region is of particular value because it represents the low energy vibrations of the molecule due to absorbance of the IR radiation and is unique to any one analyte, and therefore provides definitive identification of the drug. Alternatively, the principle peaks (strongest absorptions) can be listed and the compound identified through examination of the literature values. The final method is where the sample is compared with a library of data prepared in the laboratory, preferably by computer. This has a particular attraction because the spectra will have been prepared on the same instrument, eliminating the differences observed due to any inherent variation between instruments. This should also be borne in mind when comparing data to literature values – operator dependent and instrument-dependent differences will always be observed.

An HPLC system using a diode array multiwavelength UV-visible detector not only provides the retention time of a compound, but also its ultraviolet, or ultraviolet/visible absorption spectrum as it elutes from the end of the chromatographic column. This provides another means of compound identification. The spectrum should always be compared to a sample analysed in the same batch of solvent since variation in solvent strength and pH may lead to variation in the observed ultraviolet spectrum. HPLC coupled to diode array detection provides stronger evidence for the identity of the compound – the degree of proof that this represents is arguable and the discussion as to whether or not this technique provides definitive proof of the identity of a drug is beyond the scope of this chapter.

11.4.4.3   Hyphenated Techniques.   Techniques which can provide definitive identification of drugs of abuse include the so called hyphenated techniques, *i.e.* gas chromatography/mass spectroscopy (GC-MS) and gas chromatography/fourier transform infrared spectroscopy (GC-FTIR). The GC separates the analytes and the spectrometers provide unique signatures which can be used to identify a compound. The degree to which such techniques are successful depends upon a number of factors. The more concentrated the sample, the better the signature for any given compound. The better the instrument tuning, the more accurate the signature. Thus in these cases, it is the unique nature of the mass spectrum for GC-MS or IR spectrum for GC-FTIR, that provides data for identification of a compound. If the sample is very weak, or the detector dirty, then interferences in the spectra will be observed and it will be difficult, if not impossible, to call a match between samples analysed under identical conditions to the standards, or between the samples and the literature values. Literature values should always be treated with caution, since the spectra obtained by both MS and IR are dependent on the conditions under which they were obtained. The instrument and operator in the laboratory will be different, in most cases, to those from which the literature values were obtained and this should always be borne in mind when data interpretation is undertaken.

11.5 QUANTIFICATION OF DRUGS OF ABUSE

Apart from forensic scientists wishing to identify drugs, diluents and other components in samples, a knowledge of the quality and quantity of the drug present is often required. Under the Misuse of Drugs Act, 1971, there is also a requirement, for certain controlled drugs, to quantify the amount of drug in a sample, since the length of imprisonment or level of a fine are dependent upon the quantity involved in a case. Furthermore, by determining the quantity of drugs and other components, comparisons between samples can be made which can aid drug profiling and intelligence studies.

To quantify the amount of drugs present in a sample, a number of techniques, including UV, HPLC, GC and GC-MS, may be employed. All are based on a measure of a change in a physico-chemical property measured by the detector in response to different amounts of analyte passing through the detector.

A few points are pertinent to drug quantification. Whether single point, two point or full linear regression methods are used, linearity of response must be established prior to use of the method as a quantitative technique. This is essential when the instrument is installed, but also whenever a component of the instrument is changed and is vital if the data generated is to stand up in court. Further, data should be generated from the lowest to the highest expected concentration, thus reducing the risk of column priming and carry-over between samples. The samples should always be separated by blanks, *i.e.* the solution or solvent in which the sample has been prepared. This best demonstrates the cleanliness of the instrument, equipment and operator. With the instrumental chromatographic systems described, the output from these instruments is a graph (chromatogram) of time versus detector responses. Hence, peaks are detected for components as they elute from a column, but for quantification it is important that there is baseline resolution between the peaks of interest and any others, and the response for the compound being quantified is due solely to that compound. If these general rules are observed, then valid quantification data will be obtained.

11.6 PROFILING OF DRUGS OF ABUSE

In addition to quantification of drugs of abuse, as indicated earlier it may be necessary to compare two or more samples of drugs to determine whether or not they came from the same parent batch. The term 'common origin' should be avoided since it is plainly obvious, for example, that all cannabis resin has a common origin: the plant *Cannabis sativa*. What is really meant is 'do the resin samples form part of the same batch?'. The same question can be asked of other plant-derived drugs, such as heroin and cocaine, or synthetic drugs, such as the amphetamines. It is now recognised by the European Network of Forensic Science Institutes Working Group on Drugs that there are four levels of drug comparison which can be used to determine whether the drug samples came from more than one batch. The levels are: (i) drug identification, (ii) drug quantification, (iii) identification of the cutting agents and (iv) chemical impurity profiling. Each of these will yield different information concerning the drugs and their relationships. In addition it is possible to determine any relationship between wrapping materials – the drugs may be different but the wrapper might be linked.

The first three steps in the process, the identification and quantification of the drugs and examination of the cutting agents, are carried out using all of the principles described above. The methodologies and controls are identical and represent the first stage in comparison. However, the information that such analyses reveal about sample relationships is variable. If a drug is identified and found to have been mixed with a very uncommon cutting agent then a link may be inferred but it may also be that the same batch of drug was cut with two different agents and a match may be missed. It is vital, therefore, to examine the chemical impurity profile, the signature of impurities themselves, in the drug samples if meaningful data is to be obtained. This is because unless the cutting agent is itself one of these impurities, the proportions of these in the sample will remain unchanged on cutting the drug.

Some drugs, such as cannabis products and heroin, require only an extraction step prior to analysis. Others require a step involving the concentration of impurities, because the impurities are present at extremely low concentrations. Drugs requiring this approach include the amphetamines and cocaine. Either solid phase extraction or liquid/liquid extraction steps can be used. However, the essential point is that they must be rapid, cheap, quantitative and extract all of the compounds relevant to the profiling with virtually 100% efficiency, without decomposition or production of artifactual impurities. Alternative methods of comparison include examination of DNA profiles and such methods have been used recently for the comparison of herbal drugs, particularly cannabis products.

11.6.1 Profiling of Cannabis Products

Cannabis resin and oil (herbal material does not give meaningful chemical profiling results but is especially useful as a source of DNA) contain tetrahydrocannabinolic acids. These are thermally labile and hence GC is not a good means of drug profiling because the drugs decompose on heating in the injection block of the instrument. GC is used in some laboratories to quantify total Δ^9-tetrahydrocannabinol content, but HPLC is preferable since no decomposition occurs and both the acidic and neutral cannabinoids can be detected and quantified directly. It is also for these reasons that HPLC can be used for profiling these samples.

The profiling of *Cannabis* samples is performed by extracting the herbal material, resin or oil with ethanol, and then an aliquot of the centrifuged solution is analysed by HPLC. The chromatograms obtained are then either compared directly as shown in the example below in Figure 11.6, where three *Cannabis* samples have been analysed, or normalised (scaled with respect to a component in the mixture) and overlaid if computing facilities are available.

From these results the profiles for samples (a) and (b) cannot be discriminated, but are different to the profile for sample (c). Hence this example illustrates how profiling can be used to determine if materials are related to each other. Such data does, however, illustrate the difficulty in drug profiling when HPLC is used because of its lack of resolution when compared to GC.

In addition to chemical profiling, DNA profiling is now finding application in drug comparison, particularly for *Cannabis* comparisons. This is because many plant stocks are taken from vegetative cuttings, which will be genetically identical to the parent plant, but which may exhibit phenotypic plasticity within and between plants in terms of cannabinoid content. DNA techniques for identification and, particularly, comparison of *Cannabis* samples, may include the use of the following:

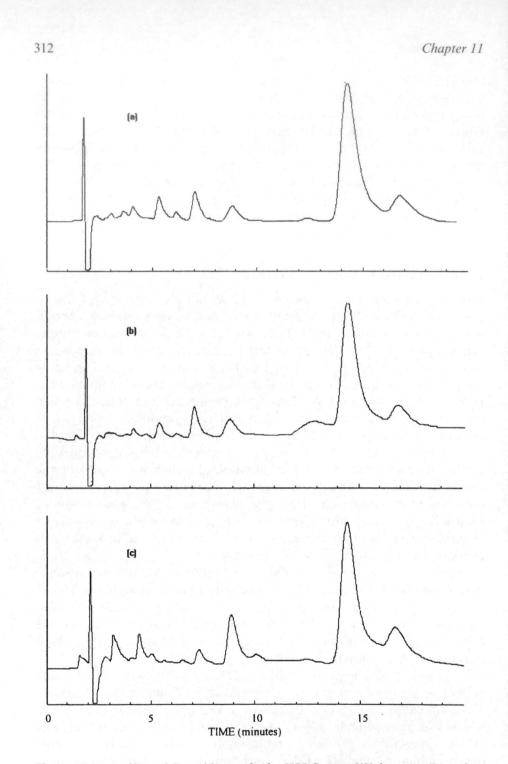

Figure 11.6   *Profiling of* Cannabis *samples by HPLC using UV detection. From these profiles it impossible to discriminate between samples (a) and (b) but these are different to sample (c)*

1. RAPD analysis.
2. ITS I and II.
3. Inter-Simple Sequence Repeat (ISSR) fingerprinting.

Of particular use are maternally inherited markers such as those found in the chloroplast since these will be passed on through seedstock and cuttings taken from plant material.

11.6.2 Profiling of Heroin

In recent times, with advances in instrumentation, both HPLC and GC have been applied to heroin profiling. The sample requires a simple pretreatment to extract the opiate drugs, noscapine and papaverine, which are then subjected to either HPLC or GC analysis. If GC is used, it is common to derivatise the sample prior to analysis to eliminate the problem of artefact formation. For example, it is known that transacetylation of diamorphine (3,6-diacetylated product of morphine) and morphine can occur if both are present in an injection block of a GC and certain solvents are used at the point of injection, with the production of both 3- and 6-monoacetylmorphine. Typical profiles obtained from two related heroin samples are shown below in Figure 11.7.

A great deal of work has been carried out on data processing for heroin comparisons. This ranges from comparison of paper copies of chromatograms and examination of the ratios of different compounds, through to cluster analysis and the use of Euclidean distances for retrospective comparisons of large data sets. Such methods have been used in, for example, determination of country of origin of heroin samples and also to examine the relationships of samples at the street level. However, there are still problems in terms of data exchange, comparison and the setup of large databases for heroin comparison. Problems with the methods include intra-and inter-laboratory variation and obtaining the same result from identical samples. It has been argued that, for heroin comparisons at least, there is some argument for using a centralised laboratory.

11.6.3 Profiling of Amphetamine

Whilst there have been various studies of on how to carry out amphetamine comparisons there are a number of problems associated with these methods because they have been formulated around amphetamine synthesised by the Leuckart route. A number of different synthetic routes are now used (including variations on the Leuckart, reductive

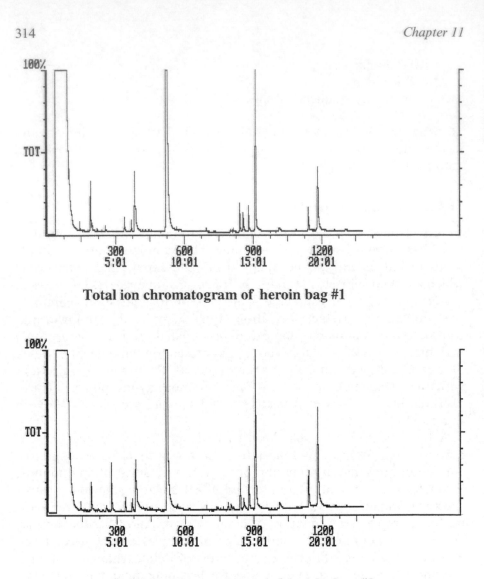

Total ion chromatogram of heroin bag #1

Total ion chromatogram of heroin bag #2

Figure 11.7 *Profiles of two related heroin samples following derivatisation with N,O- bistrimethylsilyl acetamide and analysis by GC – MS*

amination and nitrostyrene routes) and it is only now that they are being fully investigated. The situation is further complicated by the fact that the impurities occur at extremely low concentrations, require a complex extraction and recovery step, are relatively polar and reactive in nature and profiles often contain a very large number of compounds including stereo and regioisomers. An example of an impurity profile obtained from the GC-MS analysis of a synthetic amphetamine is shown in Figure 11.8.

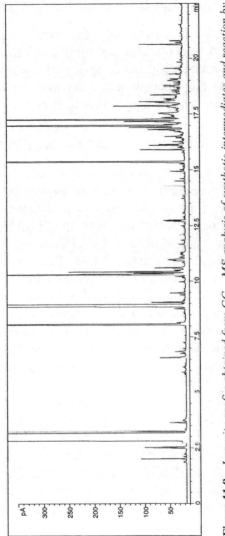

Figure 11.8 *Impurity profile obtained from GC – MS analysis of synthetic intermediates and reaction by products of amphetamine synthesised by the Leuckart route*

11.6.4 Profiling of Other Drugs

It is possible to profile a number of other drugs including cocaine, LSD, MDA, MDMA and the less common amphetamines and other synthetic drugs. However, with the exception of cocaine, there is still a great deal of work to be undertaken to fully understand the chemistry and data interpretation of the drug profiling processes.

11.7 QUALITY ASSURANCE IN DRUG ANALYSIS

In addition to simply analysing the drugs of abuse, it is important that quality assurance is considered. Every necessary control should be taken, and these have been considered above. Further, the instruments should be demonstrably functional, both qualitatively and quantitatively. This requires that standard test samples providing defined responses are analysed at the start and end of each analysis, and if the analysis is long, within the analytical sequence too.

Computerised data handling should be backed up by analogue output. In the United Kingdom this is an important requirement. This is because there is a myth amongst the lay population (and unfortunately amongst some scientists too) that if the answer comes out of a computer, it must be right. With numerical data handling and quantitative analyses, this is especially so.

Recent developments include the introduction of internationally accepted quality assurance practices *e.g.* ILAC 17025. These require all the above-mentioned precautions to be taken and in addition, good laboratory note-taking. Everything that is done, right down to changing a pair of protective gloves should be recorded in writing. Glove changes are necessary to prevent contamination between samples and it is unlikely that in a court room 6 or 12 months after the analysis was performed, you will remember whether or not you changed gloves between samples. If all of the above points are considered, the drug analysis will be valid and successful.

11.8 BIBLIOGRAPHY

J. Ballany, B. Caddy, M. Cole, Y. Finnon, L. Aalberg, K. Janhunen, E. Sippola, K. Andersson, C. Bertler, J. Dahlén, I. Kopp, L. Dujourdy, E. Lock, P. Margot, H. Huizer, A. Poortman, E. Kaa and A. Lopes, *Science and Justice*, 2001, **41**, 193–196.

A. Braithwaite and F.J. Smith, *Chromatographic Methods*, 4th edn, Chapman and Hall, London, 1985, 414.

M.D. Cole and B. Caddy, *The Analysis of Drugs of Abuse – An Instruction Manual*, Ellis Horwood, New York. T.A. Gough, 1995. *The Analysis of Drugs of Abuse* John Wiley, Chichester, 1991.

M. Cole, *The Analysis of Controlled Substances*, Wiley, 2004.

L.A. King, *The Misuse of Drugs Act – A Guide for Forensic Scientists*, Royal Society of Chemistry, Cambridge, 2003.

Rapid Testing Methods of Drugs of Abuse, *United Nations Drug Control Programme Handbook ST/NAR/13*, United Nations, Vienna, 1988.

D.H. Williams and I. Fleming, *Spectroscopic Methods in Organic Chemistry*, 4th edn, revised, McGraw Hill, London, 1989.

CHAPTER 12

Forensic Toxicology

ROBERT ANDERSON

12.1 INTRODUCTION

12.1.1 What is Toxicology?

Toxicology is the study of poisons, including their origins and properties, their effects on living organisms and the remedial measures which might be taken in poisoning incidents. The meaning of the term poison is dealt with later. A toxicologist is primarily concerned with analysing specimens such as blood and urine for the presence of a poison and subsequently with interpreting the significance of the results obtained.

Most readers of this book will by now have surmised that appending the word 'forensic' to the name 'toxicology' simply means the use of toxicological investigations in court. Any type of toxicological work, in its widest sense, may appear in criminal or civil court proceedings and a list of examples is given below. However, because most court cases currently are concerned with the effects of poisons on humans, 'Forensic Toxicology' is usually considered to be the study of the effects of drugs and poisons on human beings and the investigation of fatal intoxications for the purpose of some sort of medico-legal enquiry, the exact nature of which depends on the country involved. This is the context with which this chapter will be primarily concerned. Typically, such deaths are caused by suicidal or accidental overdose of drugs, or else the death may have occurred under the influence of a drug or poison but have been the direct consequence of another agent such as a car accident or falling from a height. Very occasionally homicidal poisonings occur involving drugs, such as insulin, but also other substances such as carbon monoxide gas, cyanide or pesticides. In these investigations, the toxicologist works in close collaboration with the forensic pathologist, who carries out the medico-legal autopsy and furnishes samples of blood, urine and other tissues for analysis in the laboratory. The toxicologist is rarely required to attend the locus of a death or even the autopsy room, and specimens for analysis are normally delivered to the laboratory by an appropriate method, as described in Chapter 2, to preserve their integrity (preventing any tampering) and legal validity (preserving a chain of custody).

The forensic toxicologist is also often employed in other types of case which do not involve fatalities, notably traffic safety (drivers of motor vehicles allegedly driving under the influence of alcohol and or drugs), drug abuse screening (the clients of drug abuse clinics are regularly tested to determine whether they are using controlled drugs) and employment screening (when the employment contract prohibits employees from using or working under the influence of a proscribed panel of substances). In some European countries the courts use the forensic toxicologist to ascertain if a suspect is a drug-user, as a user found in possession of controlled drugs may attract a lighter penalty than a dealer.

Other types of toxicological investigation that may appear in court proceedings are summarised as follows:

1. *Clinical Toxicology* – The investigation of poisoning in a supervised medical setting, usually a hospital.
2. *Environmental Toxicology* – The investigation of deleterious effects of poisonous substances in the environment.
3. *Regulatory/Animal Toxicology* – The investigation of the toxic properties of substances to fulfil the legal requirements for the licensing of commercial products.

12.1.2 Origins and Development of Forensic Toxicology in the United Kingdom

Systematic toxicological investigation of human organs for the presence of poisons is widely considered to have been initiated by M.P. Orfila, a Paris physician whose treatise on toxicology was the first substantial work on the subject. Several students of Orfila returned to the United Kingdom in the 19th century and established the subject here in their home universities within the wider discipline of forensic medicine.

Toxicology services have therefore been provided traditionally by university Forensic Medicine Departments and this is still true in some cases today. However, other laboratories now largely provide these services including the Forensic Science Service in England, the Local Authority 'Police' Laboratories in Scotland and the Northern Ireland

Forensic Science Laboratory in Belfast. In other countries, laboratories are operated by law enforcement agencies, by the governments, by the medical services and by commercial organisations. Unfortunately, defence counsel may not have access to many of these laboratories, either because they work only for the law enforcement agencies, or else, having been engaged on a case by the prosecution, are unable to work simultaneously for the defence.

Qualifications accepted by the courts for forensic toxicologists are degrees in medicine or science in an appropriate subject (chemistry, biochemistry or pharmacology) plus experience in a routine forensic toxicology laboratory. In common with other fields of forensic science, most toxicologists are trained in the laboratory in which they are employed, although as yet there is no National Vocational Qualification in toxicology. However, toxicologists in the UK can now be accredited through the CRFP (Section 1.6.1). In other countries, notably the USA, systems have been created for certification of individual scientists and for accreditation of laboratories.

12.2 POISONS

12.2.1 Definition

A poison is 'any substance which, taken into or formed in the body, destroys life or impairs health'. This definition does not restrict poisonous substances to particular classes in any way. Almost all substances are capable of causing harm if given at a high enough dose, including such essentials for life as salt, water and even oxygen. As examples, consider the use of salt and water as an emetic to induce vomiting when a poison has been swallowed. If care is not taken to restrict the quantity of salt to a spoonful or two, a dangerously high amount of salt can be ingested and this can have its own toxicological consequences. Also, at least one teenager, recognising the dangers of dehydration when using ecstasy at a 'rave' discotheque, has almost died by drinking too much water (about six litres) too quickly. A number of factors are involved in determining the dose at which toxic effects are produced and these are considered below.

12.2.2 Factors Affecting the Toxic Dose of a Substance

It will already be apparent that most substances can act as poisons at a high enough dose and that the crucial question of whether or not a substance is a poison relates to the dose administered rather than to the substance involved. Figure 12.1 may be useful in visualising the

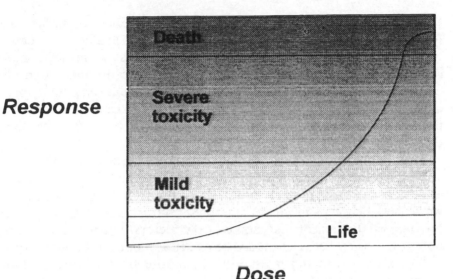

Figure 12.1 *Relationship between the dose of a substance administered and the effects or response on the organism*

relationship between the size of the dose administered and the effects on the subject.

The graph indicates that low doses of a substance are consistent with life; for example, if a drug is administered at the normal dose (the therapeutic dose), then life will continue and the organism or person may benefit from the substance. As the dose increases, however, mild toxicity may occur, shown by the appearance of adverse or side-effects. At this stage, the beneficial effects of a drug may be outweighed by the appearance of these undesirabe effects. As the dose continues to rise, the adverse effects become more severe and may start to become life-threatening. Severe toxicity occurs at too high a dose and is seen, for example, as depression of respiration: the breathing becomes more shallow and the rate slows down until it finally ceases altogether. At this point the threshold for fatal poisoning is passed. Note that there is no indication of timescale in the diagram. The dose may be administered at one time or in small amounts over a period of time and similarly the response may appear in the short term (acute poisoning) or long term (chronic poisoning).

It should also be noted that, for some substances, the lower end of the dose curve starts to rise again as the dose approaches zero, giving a U-shaped curve. Such substances are often trace elements and vitamins which must be present in the diet. Both too much and too little is harmful, for example copper, iron and zinc are essential trace metals. If the dietary intake is too low, the term used is deficiency rather than toxicity. In the context of forensic toxicology, an analogous situation can arise in which a patient may take too low a dose of medication and suffer as a result. Common examples are under-dosing, on insulin (resulting in a diabetic coma), anti-epileptic drugs (resulting in an epileptic fit, which can be fatal) or antipsychotic drugs (resulting in a psychotic episode in which the individual can be harmed or can hurt those in the vicinity).

What has been described qualitatively above is a dose-response curve and in an ideal situation, the curve is predictable and the thresholds for each transition stage are clear-cut. This does not happen in the real world and hence problems of interpretation arise, and these are the province of the toxicologist. Some of the factors that upset the simple dose-response curve are examined below. It is important to note here, as well as later, that there is no specific 'fatal dose' or 'fatal threshold' for a drug or other substance.

12.2.2.1   Carcinogenic and Mutagenic Substances.   These include, for example, asbestos and ionising radiation, which cause cancer or mutations, respectively. There is no simple dose-response relationship for them because usually there is no safe exposure level. Any exposure can cause a catastrophic injury to the living organism and once initiated it is essentially irreversible. In the case of ionising radiation this can cause damage to genetic material in living cells *i.e.* a gene mutation – an alternation to the normal DNA sequence. This mutation, however, may not be apparent in the generation exposed to the radiation, but several generations later.

12.2.2.2   Age and Size.   The doses of a substance required to cause mild toxicity, severe toxicity and death vary with the age and size of an individual. Children, for example are often given a lower dose of a drug than adults because their body weight is less. Doses are normally calculated as quantity of substance per unit of body weight, *e.g.* milligrams per kilo, to compensate for different body sizes. Children also differ from adults in the metabolism and disposition of substances which they ingest. The adult liver, for example, may behave differently from that of a child because it has already encountered substances during its lifetime. As individuals reach old age, their ability to cope with exogenous (foreign) substances is diminished and the threshold for toxicity may be expected to become lower.

12.2.2.3   State of Health.   Ill-health can also affect the ability of the body to cope with a toxic insult due to a poison. This is obvious if the liver, the main metabolic organ of the body, is diseased, but many other organs can be involved, including the kidneys which excrete poisonous substances into the urine, and the heart, which may fail at a lower dose of substance than normal if there is already significant heart disease.

12.2.2.4 History of Exposure.   The response of an individual to a particular dose of a substance is often affected by the previous exposure of the individual to the substance and can take two forms. The most common, and better known, effect is the development of tolerance. This means that the individual becomes used to the effects of the drug such that a higher dose or more frequent administration of the same dose is necessary to obtain the original effects. For example, regular heroin users (*i.e.* injecting heroin several times per day) develop tolerance to the effects of the drug over a period of several weeks and may end up using doses which would have caused death if they had taken them at the beginning of their habit. Equally important in the forensic context is the loss of tolerance. Again considering the case of a heroin abuser who has acquired tolerance to the drug over a period of weeks or months, this tolerance will be lost fairly rapidly during a period of abstinence from the drug, for example, during a custodial prison sentence. If this individual returns to the same heroin habit after release from prison there is a severe risk of accidental heroin overdose and death.

The second way in which previous exposure to a substance can affect the dose-response relationship is through the development of sensitisation. In effect, this means the development of an especially sensitive, idiosyncratic response at doses that may be lower than those normally required to elicit a toxic effect. One example of this type is volatile substance abuse (sometimes called 'glue-sniffing' or 'solvent abuse'). Regular sniffing of butane (the gas used in cigarette lighters) or aerosol propellants containing halogenated hydrocarbons (ozone-damaging propellants such as freons) can cause sensitisation of the heart such that further exposure can cause cardiac arrythmia (irregular heart beat) and death.

12.2.2.5 Paradoxical Reactions.   For some substances there exists the possibility of a paradoxical reaction by a small number of individuals. The dose-response curve described above does not apply to these individuals as the effects produced are quite different from the normal ones experienced by the majority of the population. An interesting example is a paradoxical excitement caused by some of the benzodiazepine tranquillisers. These drugs normally cause sedation, relaxation and loss of anxiety, but in a very small number of cases the drugs can cause excitation and aggression. Although this effect is very rare, its occurrence is unpredictable and it is sometimes invoked by the defence to explain bizarre or uncharacteristic behaviour in a client.

12.2.3   Types and Examples of Poisons

Poisons may be grouped in different classes depending on their chemical composition and properties, and their effects on the human body. Some basic subdivisions can be made according to the mode of action.

12.2.3.1   Corrosive Poisons.   This group includes acids, alkalis and other substances which cause physical corrosion of body tissues through direct contact. Corrosive poisons such as vitriol (sulphuric acid) and Lysol (a mixture of substances similar to carbolic acid) were more commonly encountered earlier this century, when there were fewer drugs to hand. This type of substance is usually taken orally and causes severe and very visible damage to the soft tissues of the mouth, the oesophagus, stomach and intestine.

12.2.3.2   Irritant Poisons.   These include inorganic and metallic poisons such as lead, mercury and arsenic compounds. They are most often ingested orally and cause severe gastro-intestinal irritation and upset, with nausea and vomiting, colic and diarrhoea or constipation depending on the substance involved. They are absorbed from the digestive system and are transported in the blood circulation to the organs of the body. These poisons are compounds of chemical elements which cannot be broken down or metabolised by the body to inactive substances: the body can only excrete them or deposit them where they will do least damage, for example, in the bones of the skeleton. The passage of metals such as lead through the kidneys into the urine causes severe tissue irritation and kidney damage is a common result.

12.2.3.3   Systemic Poisons.   This is the largest group of poisons and it includes those substances that act directly on biochemical processes within cells. The poisons must reach their site of action *via* the circulatory system after they are taken into the body by any one or more of a variety of routes, considered in more detail below. Within this group are the well-known alkaloid poisons such as strychnine, inorganic poisons such as cyanide and carbon monoxide, pesticides and drugs. The remainder of this chapter will be concerned primarily with poisons of this type.

12.2.3.4   Toxins.   Naturally-occurring substances produced in living organisms are usually referred to as toxins rather than poisons. The best known examples are venoms from poisonous snakes and insects, but the group also includes substances produced by bacteria such as salmonella and algae such as the blue-green algae which has recently caused problems in both inland water reservoirs and in the sea in the neighbourhood of shellfish beds.

12.2.4   Routes of Administration and Excretion

Administration and excretion are topics of particular importance in the interpretation of analytical results contained within toxicology reports. They are frequently the basis of questions asked in court. Drugs and other substances can be taken into the body by a number of different routes, which affect:

1. The speed with which they act on the body.
2. The fraction of the dose which is absorbed and is able to reach the site of action of the substance (the bioavailability).
3. The ultimate concentration of the substance that is achieved in a particular body tissue.
4. The effects experienced by the person receiving the substance.

Similarly, substances are cleared from the body by different routes, which affect:

1. The length of time over which the substance remains active in the body.
2. The methods by which the body inactivates or detoxifies biologically-active materials.
3. The selection of samples for analysis.
4. The length of time over which the substance can be detected.

A generalised diagram illustrating the absorption and excretion of a substance is shown in Figure 12.2. This relates the level of the substance

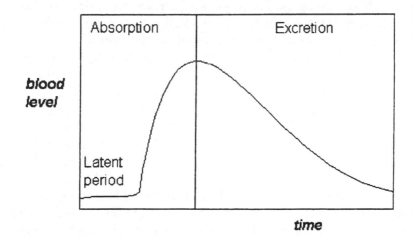

Figure 12.2   *Absorption and excretion profile for a substance entering and leaving the human body*

in blood to the time after administration. By and large, substances must enter the blood circulation before they can exert an effect on the body and often, but not always, the time when the strongest effects are observed can be related to the maximum blood concentration. The units of concentration currently favoured are milligrams per litre of blood (or urine) and milligrams per kilogram of tissue (such as liver).

12.2.4.1 Routes of Administration. Oral. Oral administration implies entry to the body through the mouth and digestive tract, whether taken deliberately, accidentally or inadvertently as a result of administration by someone else. To the toxicologist, entry to the body begins not when the substance is swallowed and reaches the stomach or digestive tract, but when it is absorbed from these organs into the bloodstream. In this respect the digestive tract from the mouth to the anus can be considered as a tunnel passing through the body: the space within it is not an integral part of the body itself. Some substances are not well absorbed when taken orally, including liquid mercury and some other toxic metals and their compounds. These mostly pass through the digestive system without entering the body tissues.

The most important questions usually raised in court concern, 'How long did it take for the substance to be absorbed into the body?' and 'How much of it was actually absorbed?'

The first question depends on the substance involved: as mentioned above some substances are not absorbed at all. However, most drugs are formulated to be taken orally and are usually absorbed within a period of 30 minutes to two hours. This also applies to many other poisons when taken orally. The pharmacologist can alter the rate of absorption by changing the physical form (formulation) of the substance. Slow-release capsules contain small granules with different rates of dissolution. Information on rate of absorption can be obtained from the medical literature for the drug in question, contained in pharmacopoeias and, for example, the Association of the British Pharmaceutical Industry (ABPI) Data Sheet Compendium.

A related question is, 'Where does the absorption take place?' Most substances are absorbed from the intestine rather than from the earlier part of the digestive system (mouth, oesophagus and stomach) or the end section (the large intestine and rectum). There are exceptions, for example, some substances are absorbed through the mucous membranes of the mouth. This is usually considered to be a different route of administration (buccal) and is similar to the intranasal route considered below. Similarly, drugs are absorbed through the rectum but this is also usually considered as a separate route when the substance is introduced as a suppository.

The second question, 'How much of it was actually absorbed?', refers to the pharmacological term, bioavailability, which is important in determining the dose required for a desired effect. Bioavailability simply gives the percentage of a dose that reaches the blood circulation and is thereby available to exert a biological effect. If this is small, the dose must be increased proportionately. There are several factors that affect the bioavailability, two of which will be mentioned here. Firstly, the physical and chemical form is important. Thus cannabis resin is poorly absorbed from the digestive tract because it is an oily resin. Some salts of lead are poorly absorbed because they do not dissolve very well in water.

Secondly, the effect of the liver must be taken into account. When drugs, in particular, are absorbed from the intestine, they enter the bloodstream through a network of blood vessels that surrounds the gut. The blood flows into the hepatic portal vein and thence to the liver, where the vein subdivides again into smaller blood vessels that pass through the liver (the portal blood vessels). From the liver, the blood is collected once more into the hepatic veins, which return blood to the heart *via* the inferior vena cava. In this way food absorbed from the gut is taken directly to the liver to be metabolised for energy. However, medicinal drugs are also carried in the same direction and are immediately confronted by the metabolic capabilities of the liver. This is called the first-pass metabolism and explains why many drugs are rapidly inactivated after they are swallowed and an alternative route of administration is used. Heroin is one example: if taken orally, there is almost complete first-pass metabolism and the products formed, morphine and other metabolites (breakdown products) do not give the same euphoric effect as when heroin is administered by injection.

Intravenous. The intravenous route became available at the end of the 19th century, when the hypodermic syringe was first introduced. This device was almost immediately recognised as an efficient means of introducing abusable drugs such as cocaine and heroin directly into the bloodstream. The bioavailability of substances administered intravenously is almost 100% (some is left in the syringe) and, more importantly, the time taken to reach the blood circulation is essentially zero. This route of administration can give rise to a very rapid rise in the blood concentration of a drug such as heroin or cocaine, causing potentially lethal side-effects on the body's physiological systems, notably on the heart rate and blood pressure. Drugs administered intravenously under medical care are usually diluted in a transfusion of glucose or saline (salt) solution or are pumped in slowly using a syringe pump.

Inhalation. Drugs and other substances can readily enter the blood circulation following absorption through the lungs, which are richly supplied with blood vessels. Common instances of this type of administration are the use of an inhalant (such as an anti-asthmatic device) and smoking (tobacco and other drugs). Smoke particles and gases enter the lungs and their constituents pass through the thin walls of the air sacs (alveoli) and the surrounding capillary blood vessels. Substances are absorbed rapidly by this route and effects are seen within seconds or minutes of administration. The intense 'high' obtained after smoking crack cocaine contributes greatly to its addictive properties. The bioavailability of inhaled drugs is about 40%.

Through the mucous membranes. The moist surfaces of mucous membranes are found in the nose, mouth, eyes, ears, vagina, rectum and lungs, amongst others, and all of these can and have been used to administer drugs. Perhaps the best known instances are the use of eye and ear drops, nitro-glycerine tablets taken sublingually (*i.e.* under the tongue), chewing tobacco and snuff, and also the snorting (a more formal name is nasal insufflation) of drugs such as cocaine and amphetamine. The bioavailability of drugs by snorting is similar to smoking, about 40%.

12.2.4.2   Routes of Excretion.   The excretion phase shown in Figure 12.2 consists of two main components, at least for many substances of interest here. The first consists of the metabolism or breakdown of substances such as drugs and the second involves voiding of the waste products from the body.

The role of the liver in metabolism has already been described. Most other tissues, including the blood itself, also have metabolic capabilities, which are designed to break down food and nutrients to provide energy and sustenance for the body. However, the biochemical reactions involved are not specific to foodstuffs but also act on most exogenous (foreign) substances. The metabolic process in this case often serves to inactivate or detoxify the exogenous materials prior to their excretion from the body. As part of the metabolic sequence (many biochemical reactions may take place one after the other), sugar molecules may be attached to the metabolites to render them more water-soluble and easier to discharge into urine. These end products are called conjugates and as illustrated in Figure 12.3, the metabolism of diamorphine (the controlled drug present in heroin) is *via* monoacetylmorphine, morphine and then finally ending up as the glucuronide conjugate of morphine.

The main routes for actual excretion of substances from the body are: (i) through the kidney into the urine, (ii) through the liver into bile and

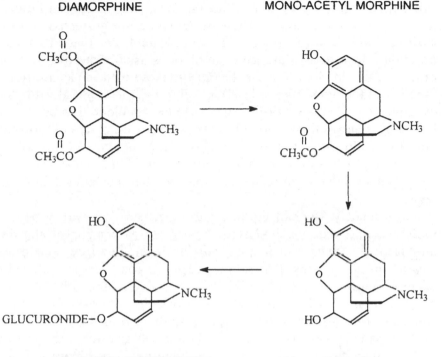

DIAMORPHINE   MONO-ACETYL MORPHINE

MORPHINE GLUCURONIDE   MORPHINE

Figure 12.3 *Metabolic sequence for the controlled drug diamorphine found in heroin. Mono-acetylmorphine and morphine are intermediate metabolites. The final product, morphine glucuronide, is a water-soluble product excreted in urine*

faeces, (iii) through the skin, and (iv) through the lungs, exhaled in breath. An interesting consequence of excretion *via* bile into the digestive tract is that substances can be reabsorbed into the bloodstream further along the gut, giving rise to a loop called the enterohepatic circulation, *i.e.* from the gut to the liver, from the liver to the bile, discharge of the bile into the gut, re-absorption from the gut and then back to the liver. The net effect of this circulation can be to retain substances in the body longer than might otherwise be expected and in turn, this can lead to a prolongation of the time over which they can be detected by the toxicologist. A common example of this is the retention of cannabis metabolites in the body and these can be detected for up to several weeks after a single administration, in favourable instances. In one exceptional case, cannabis metabolites were detected for four weeks after a single cannabis cigarette was smoked.

Court cases often raise the question of, 'How quickly was the substance cleared from the blood?' or, knowing the concentration of a substance in blood at the time of death, 'What was its concentration at a given time prior to death?' These questions can be answered if graphs such as those shown in Figure 12.2 are available for the substance in question. The rate of clearance from blood is usually specified in terms of the half-life in blood. This is the time it takes for the concentration of the drug in the blood (blood level) to fall to half of its initial value. For example, the average half-life of morphine in blood is about three hours. This means that a blood level of, say, one milligram of morphine per litre of blood will fall to 0.5 in about three hours, and this will fall to 0.25 after a further three hour period, and so on. Half-life data for drugs are available from reference tables in pharmacopoeias and other texts.

Unfortunately, the half-life of a substance can vary widely between individuals for reasons similar to those given earlier concerning dose and response. Data which are available from literature sources are usually given in terms of a range and mean (or average) value. For example, morphine has a half-life of one to five hours with a mean or average value of three hours. When the toxicologist is asked in court to calculate blood levels at a time prior to death, or prior to the time when a blood sample was taken, the calculation will give the average value and a range, based on data from the literature. For obvious reasons these calculations are referred to as back-calculations. The range can be very wide if a significant time interval is involved, in fact so wide as to be useless for the purposes of the court.

12.2.5 Patterns of Poisoning

12.2.5.1 Time and Place. As indicated earlier, poisoning incidents have tended to involve substances which are readily available in the locality or are familiar to those involved. These have changed during the course of history as the number and variety of chemicals available has increased but also, today, vary according to geographical location around the world. In Europe, USA and other developed industrial regions, the most common substances involved in fatal poisoning are drugs, both licit and illicit, whereas, in most developing countries, agricultural chemicals (mostly pesticides) cause the highest proportion of deaths. Interesting regional variations occur even in the types of illicit drugs used, for example, amphetamine is commonly abused in the UK and Europe but in Japan and South East Asia, methamphetamine has always been more popular. Similarly, in South and Central America and the Caribbean, cocaine is the major drug of abuse and heroin is uncommon whereas the pattern is reversed in South East Asia.

Nevertheless, these patterns are subject to change, sometimes suddenly. The former Warsaw Pact countries of Central and Eastern Europe had a very small problem with illicit drugs until recently, but these countries are now faced with a rapid rise in the abuse of illicit drugs and the attendant rise in drug-related deaths.

Finally, it should be noted that unexpected incidents can happen as illustrated by poisoning incidents in recent years. In Tokyo, the underground system was deliberately poisoned with the chemical war gas, sarin, in a mass homicide incident for political reasons. In Switzerland and France, mass suicides by ingestion of cyanide have been encountered in several locations among members of religious organisations.

12.2.5.2 Fatal Poisoning. Suicide. Most cases of fatal poisoning occur as a result of suicide or accident (37% and 17% of drug-related deaths, respectively) and most of these in the UK involve an overdose of medicinal drugs, consistent with the principle that people tend to use what is readily available. In the UK suicidal poisoning with other substances such as cyanide usually involves a laboratory or factory worker who has access to the substance in their place of work. The important medico-legal question in these cases concerns the intent of the deceased, *i.e.* did they intend to commit suicide or not? Circumstantial evidence from the locus of the death can be helpful in this respect, including the presence of a note written by the deceased stating intent, the presence of medicine containers such as empty pill bottles and the absence of any indication of violence or struggle. The medical history of the deceased is also relevant, especially if suicide had been attempted previously. Suicide by deliberate overdose is not necessarily easy to perform: the quantity taken can be too little to cause death or be so large as to cause vomiting of the substance swallowed.

Accident. Accidental overdose cases also occur regularly, although it can be difficult to distinguish between accidental and deliberate overdose on the basis of toxicology alone. The toxicologist is often asked to estimate the quantity of substance taken, to assess if it could have been taken inadvertently. Two points should be taken into consideration when making this estimation. Firstly, the dose cannot be calculated simply and accurately by multiplying the blood concentration of substance by the volume of blood in the average body. This is because the substance does not reside only in the blood but is distributed throughout the body tissues in varying amounts. An accurate estimation of the dose would require the analysis of each of the body organs, including the stomach contents and contents of the digestive tract. Secondly, the minimum dose taken can be estimated very approximately by comparing the concentration of the substance in autopsy blood with the levels obtained therapeutically when it is administered in a known amount. In many overdose cases, however, the blood level is ten or more times higher than the normal therapeutic level and the likelihood of this being accidental is lower.

It is important to note that the blood level obtained after a given dose of substance varies widely between individuals and that, as for half-life calculations, a range and average value should be calculated. Also, the estimation of dose taken must take into account the length of time the deceased survived before dying and therefore involves a back-calculation to the time when the substance was administered. As described above, these calculations have their own uncertainties which compound the problems of dose estimation.

Industrial. Deaths from poisoning in industry are unusual in this country. Deaths in industry resulting from an accident caused by intoxication with alcohol or drugs are more common. Poisoning cases inevitably involve exposure to substances present in the workplace, and Health and Safety legislation carefully controls this type of exposure. Deaths are therefore the result of an accidental or unregulated exposure, unless the premises involved are not conforming to legal requirements. Typical problems are exposure to carbon monoxide through faulty ventilation, cyanide through splashing with cyanide solutions (rapidly absorbed through the skin) and asphyxiation due to entering a closed space in which the oxygen has been displaced by an inert gas such as carbon dioxide (for example in a brewing tank). If industrial cases of non-fatal poisoning are encountered, relevant data for interpretation purposes can be obtained from tables published by the Health and Safety Executive, amongst others.

Iatrogenic. These deaths are the result of medical treatment or intervention. They are infrequent and are often the result of errors in calculating the dose of a drug to be administered. Deaths due to administration of a general anaesthetic are usually considered separately.

Homicide. Homicidal poisoning is unusual in the UK today, but occurs more frequently in other parts of the world, again often involving agricultural chemicals. Mass deaths due to cyanide poisoning which may have been at least partly homicidal have also occurred in different countries. Such cases attract media attention because of their relative rarity. Examples which have occurred in the UK over the last few years include poisoning of spouses with cyanide, paraquat, carbon monoxide and atropine. The use of carbon monoxide can involve car exhaust but in one case involved a cylinder of the gas, which was obtained by a technician from his laboratory.

12.3 THE WORK OF THE FORENSIC TOXICOLOGIST

12.3.1 The Role of the Forensic Toxicologist in Medico-Legal Investigations

The relationship of the toxicologist to the legal authorities in charge of the case is normally that of a consultant with established expertise in the field of forensic toxicology. In almost all contexts, specimens for analysis are taken at autopsy by a pathologist authorised by the legal system to do so. The main alternative is when specimens are taken by a police surgeon from an alleged offender, typically under the Road Traffic Acts (1988 and 1991), Sections 5 and 8. These specimens remain the property of the legal authorities and are delivered to the toxicologist who is instructed to undertake the analysis.

Decisions on whether or not to have a toxicological analysis carried out, and what types of analysis are to be requested, are usually taken by the pathologist or by the law enforcement personnel involved in the case of road traffic cases. The factors involved in these decisions are varied but include the facilities available in the toxicology laboratory, the time taken to obtain results and the cost of the analyses. It is recommended that the toxicologist is involved in making these decisions and that background information concerning the case is made available in order to direct the toxicological investigation. Relevant information is contained in the pathologist's report, the police report, the medical history and (if involved) the fire brigade report.

The purpose of involving a toxicologist in an investigation is to establish if drugs or other substances played a significant role in the case, for example a death, and access to relevant case background can be of great assistance in indicating what substances to look for. The work can then proceed effectively and produce results in the shortest time and at the minimum cost. In most cases, once the immediate objective of the work has been obtained, for example, a cause of death or a cause of impairment has been established, work can cease unless there is a specific need for further information. This might concern the presence of other drugs or substances available to the deceased or the accused. There is certainly no need for blanket coverage of all possible substances which the laboratory is capable of detecting and measuring, as this would take too long and cost too much.

12.3.2 The Forensic Toxicological Investigation

Different scenarios are possible for the toxicology investigation, depending on the nature of the case. The possibilities are as follows:

1. *Specific substance.* The simplest situation is when a sample is submitted with a request for analysis of one or more specific substances. Typical examples are requests for analysis of alcohol in blood (almost all cases) or for analysis of alcohol and carbon monoxide in a fire death.
2. *Partial unknown.* In this situation, the toxicologist is informed of a number of possible substances which were available to the deceased or the accused, all or none of which might be present in the samples. In this case, each of the possible substances is targeted specifically to determine whether or not it is present.
3. *Complete unknown.* Usually, there is no information available concerning what was available to the deceased or accused. The pathologist may not have a cause of death but may suspect that drugs are involved. Alternatively, there may be an established cause of death or impairment and the intention is to rule out the involvement of drugs and other substances in the case. In this situation, the toxicological investigation consists of a systematic application of all of the analytical methods available, beginning with the analysis of drugs and substances commonly found in the locality. The profile of the deceased or accused may be used as a guide: for example, pensioners may be unlikely to abuse illicit drugs. Also, as indicated earlier, if a significant level of a drug or other substance is detected, the toxicologist may stop further work after consultation with the pathologist or police.

Because cases often involve a fatality, there is not the same degree of urgency as in other contexts, with the exception that a homicide involving a poison may require the investigation to be carried out as rapidly as possible. Depending on the complexity of the investigation, the time required to produce a report may vary from less than 24 hours (*e.g.* a simple blood alcohol measurement) to several weeks (for a full toxicological investigation of all types of drugs and other possible substances).

12.3.3 General Analytical Approach

By virtue of the nature of the cases handled by the forensic toxicologist, most involve some type of legal proceedings. In this context, two different analyses are required to provide a high standard of proof that the results are reliable, beyond reasonable doubt. The first test carried out is normally a screening test, which is used to establish if a drug or substance is present. Ideally, this type of test should be quick and inexpensive, but also reliable. If the first test is positive, a second analysis, called a confirmatory test, is carried out. The primary characteristic of this test is that it should be unambiguous and give unequivocal proof of the presence (and also the amount) of a substance.

12.3.4   Different Types of Specimen

The different types of specimen and their value to the toxicologist are summarised in Table 12.1.

12.3.5   Tools of the Trade – Methods of Analysis

A generalised approach to the methods of analysis is outlined in Figure 12.4.

Table 12.1   *Types of specimen and their value to the toxicologist*

Specimen	Attributes
Blood	– Available in almost all cases and can be obtained at autopsy without opening the cadaver if it is known to present a risk of infection; – blood levels of substances can be interpreted using literature data;note, however, that autopsy blood is usually partially clotted and haemolysed (i.e. the red blood cells have burst open) so it is not exactly equivalent to blood taken by a clinician from the living patient.
Urine	– A "non-invasive" specimen available from living patients; – available from only a percentage of autopsy cases; – contains most substances, including drugs and their metabolites, which have been taken into the body; the concentrations of drugs and metabolites in urine are often higher than in blood; – drugs and their metabolites are often detectable in urine for longer periods after administration than in blood;-lower infection risk than blood;the major disadvantage of urine is that concentrations of substances in urine cannot be correlated with response in the same way as blood levels and therefore cannot be used to establish a cause of death or a cause of impairment, but only indicate a possible cause.
Liver	– Available at autopsy from almost all cases; – when available, the liver is a large organ and there is no problem in obtaining an adequate specimen for analysis; – drug concentrations are usually higher than in blood; – interpretation of drug levels in liver can be difficult due to lack of reference data.
Bile	– Bile is obtained from the gall bladder in some autopsy cases; – drug concentrations can be higher than in blood; – often used for opiates such as morphine;-bile contains many constituents which can cause interference in the analytical methods; – interpretation of drug levels in bile can be difficult due to lack of reference data.

Table 12.1   *Continued*

Specimen	Attributes
Vitreous humor	– Can be obtained from the eyes in most autopsy cases but may be ethically unacceptable;-few interfering substances present; – volume available is small; – difficult to interpret levels of drugs and other substances because of lack of reference data; – sometimes used for drowning cases to measure potassium and sodium concentrations.
Cerebrospinal fluid	– Can be obtained from the brain in most autopsy cases; – difficult to avoid contamination with blood and brain tissue; – difficult to interpret levels of drugs and other substances because of lack of reference data.
Brain	– Available in most autopsy cases; – difficult to analyse because of the presence of fatty material; – difficult to interpret levels of drugs and other substances because of lack of reference data; – useful for detection of volatile substance abuse as volatiles are retained in the fatty tissue of the brain for longer periods than in blood.
Lung	– Used for detection of gases and volatile substances; – only qualitative results are obtained as concentrations are not meaningful.
Hair	– Useful for detecting both drugs and other substances such as metals an arsenic; – contains a "profile" of the sample donor's exposure during the lifetime of the hair; – only qualitative results are obtained as concentrations are not meaningful.
Nail clippings	– Used mainly for assessment of chronic exposure to environmental materials

12.3.5.1   Sample Pre-Treatment.   This involves physical and chemical manipulation of specimens to render them suitable for further work, for example, filtration of urine, centrifugation of blood, homogenisation of tissue, adjustment of pH (degree of acidity).

12.3.5.2   Extraction.   Drugs and other substances are often present at very low concentrations in specimens, typically comprising less than one part per million of specimen (*i.e.* one milligram per litre of blood or per kilo of tissue). To detect and quantify these substances, the bulk of the matrix containing the analyte (substance being analysed) must be removed and discarded. This entails a process of extraction and concentration of the analyte(s) using either an organic solvent such as ether and chloroform or a solid absorbent, currently often based on silica.

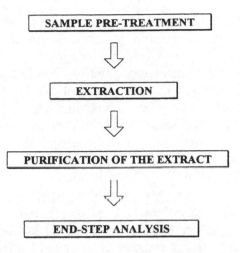

Figure 12.4 *Generalised analytical sequence used in most toxicological investigations*

12.3.5.3 Purification of the Extract. Also called the 'clean-up' stage, this purifies the initial crude extract to remove as many interfering substances as possible.

12.3.5.4 End-Step Analytical Procedure. The purified extract is analysed by one or more techniques based on spectrometry, chromatography or immunoassay.

Spectrometric techniques regularly used in toxicology include ultra-violet spectrophotometry and mass spectrometry. These depend on the isolation of a pure substance from the specimen and the information obtained from the technique is a spectrum. This is valuable information for the identification of unknown substances, and/or confirmation of the identity of a substance thought to be present. Spectrometric methods become more powerful when combined with another technique, especially a chromatographic separation method.

Chromatographic methods in this context should be considered simply as methods of separating out the constituents of extracts of biological specimens. The type of information obtained is a chromatogram or chart, characterised by the presence of peaks, each representing one or more constituents, along a time-scale generated by the recording device. The chromatogram gives two pieces of data: the time taken for each constituent to pass through the chromatographic system (the retention time) and the size (height or area) of the peak on the chromatogram. The former is characteristic of the substance concerned and the latter is proportional to the amount. A chromatographic method, therefore, can be used for identification (based on the retention time) by using a reference table of retention times of drugs and other substances, and also to quantify the amount of substance present (based on peak size). Taken alone, a single chromatographic method cannot give unequivocal identification, as more than one substance may have the same retention time. The reliability of identification can be improved by using more than one system or as indicated above by combining the chromatographic method with a spectrometric method. The most common examples of the latter are gas chromatography-mass spectrometry (GC-MS) and high performance liquid chromatography-diode array spectrometry (HPLC-DAD). GC-MS is often considered to be the most reliable method for the identification of substances. However, it is not applicable to all materials of interest and alternatives must also be used.

Immunoassay tests are based on the use of antibodies, the protein molecules formed in living organisms to control infection or invasion by foreign organisms such as viruses. Each different type of foreign material (antigen) entering the blood causes the formation of an antibody which will 'recognise' and bind to that material forming an antibody-antigen complex. Using a variety of methods, antibodies can be obtained which recognise drug molecules. These can be used to test for the presence of drugs in a specimen if there is some way of detecting the formation of an antibody-drug complex. Many ways of detecting the formation of these complexes have now been developed and these form the basis of different immunoassays, including radioimmunoassay (RIA), enzyme multiplied immunoassay technique (EMIT) and fluorescence polarisation immunoassay (FPIA, often known by the commercial trade name TDx). These tests are used as screening methods as they are relatively simple to perform and large numbers of specimens can be processed. They are, however, expensive to buy as commercial kits and are not available for all types of drugs – the most commonly available kits are for drugs of abuse. The nature of immunoassays is such that they are usually not specific for a single substance, but only for a group of similar substances. This can be advantageous, as a single kit can be used to screen for the presence of several substances, for example for the opiates. The disadvantage is that the lack of specificity renders the immunoassay test unsuitable for unequivocal identification of a particular substance or for quantification of an analyte in a specimen.

12.3.6   Chemical Classification of Drugs

An area of confusion that sometimes arises, especially during an examination of a toxicology report in court, concerns the difference between the pharmacological and chemical classification of drugs. The former is familiar to most people. Drugs are classified according to their effects on the body, for example, narcotic analgesics, stimulants, hallucinogens and anti-depressants. The chemical classification, however, deals only with the chemical structures and the resulting chemical properties of the drugs – in particular, whether a drug is acidic, basic or neutral. Acidic drugs contain, for example, carboxylic acid groups and include anti-inflammatory drugs (such as ibuprofen and salicylic acid) and the barbiturates. Basic drugs contain an amine structure and constitute the largest group of drugs. Included are the narcotic analgesics such as morphine, stimulants such as amphetamine and cocaine, major tranquillisers such as chlorpromazine, antidepressants such as amitriptyline and most other prescribed drugs. Neutral drugs are fewer in number and this group contains, for example, anabolic steroids. To the analyst, the chemical properties of drugs can be used to subdivide them into three fractions (acidic, basic and neutral drugs) and toxicology reports will normally refer to these chemical classes as well as to particular drugs (*e.g.* alcohol, paracetamol) and groups of drugs (*e.g.* benzodiazepines).

12.3.7 The Toxicology Report

A toxicology report is issued only by a qualified analyst after the results of the analyses have been reviewed and checked and, preferably, following consultation with the requester to ensure no further work is required. In current practice, the analytical work is carried out entirely by the reporting scientist or is carried out under the supervision of the scientist with assistance from technicians, or by a combination of these two. In practice, of course, much of the work is performed by machinery under the supervision of the analyst and results are printed by computer. The degree of personal involvement is steadily diminishing with time. In some contexts, the laboratory can be fully automated and all that is required of the personnel is to load the specimens in a rack. The reliability of the results produced must be safeguarded by the instigation of a quality system in the laboratory. More than just quality control, a quality system specifies the *modus operandi* of all aspects of the work of the laboratory.

At its simplest, a toxicology report contains the following information:

1. Date and reference number of the report.
2. The case name (the deceased or specimen donor).
3. The instructing authority *e.g.* the Procurator Fiscal or coroner.

4. The name of the pathologist or police officer from whom the specimen(s) was received.
5. The date on which the specimen(s) was received.
6. The tests that were carried out.
7. The results of the tests.
8. The signature of the analyst.

In addition, according to the practice of the laboratory and legal system concerned, the report can also contain the written opinion of the toxicologist who signed the report.

Presentation of a toxicology report in court will normally consist of attesting that the report and its contents are an accurate record of what was done and of the results obtained. This much is the province of the 'professional witness'. If, in addition, the toxicologist is asked to give an opinion on the results, expert testimony is involved.

It seems probable that, in future, the toxicologist will be subject to more searching cross-examination. In the fields of serology and haemogenetics, the use of statistics is well established for estimating the probability that a match is correct. Similarly, the process of method validation in toxicology will in future give an indication of the probability that an analytical result is true.

12.4 INTERPRETATION

As indicated previously, the results of toxicology are primarily either qualitative, indicating whether a substance is present or absent, or quantitative, indicating the concentration of a substance in a specimen. Bearing in mind the purpose of the investigation, the interpretation of results concerns their significance with respect to a cause of death or a cause of intoxication.

12.4.1 Qualitative Results

These are derived from either an analytical method that does not produce quantitative results, including immunoassays, thin layer chromatography and colour spot tests, or from analyses carried out on specimens in which the concentration is not of diagnostic significance. The most important examples of the latter are tests carried out on stomach contents or on urine. It is perhaps obvious that the concentration of material in the stomach is less important than what is present or perhaps the total quantity present.

12.4.2 Quantitative Results

These are normally concentrations of substances in blood, urine and other tissues. Their interpretation is intended to relate the concentrations to effects on the sample donors. However, concentrations of substances in urine are more open to misinterpretation and misunderstanding. The problem arises because urinary concentrations vary widely, even more than in blood, and this in turn is because the volume of urine excreted changes according to the state of hydration of the body. The volume is also affected by foodstuffs such as coffee, and drugs, such as diuretics, which influence the working of the kidneys. Analytical methods used for urine specimens nevertheless often yield quantitative results and the toxicologist is often pressed to use and interpret these concentrations with respect to incapacitation in, for example, road traffic samples.

While some experts are willing to provide such an opinion, it is generally accepted and recommended that such interpretations should not be made except under exceptional circumstances – usually where additional information is available. In these cases, the concentration of drugs and other materials in urine do not prove incapacitation, and it is better to base the case on an examination of the accused by a police surgeon or other physician at the time of the offence. Toxicology can then be used to support this examination.

One final point concerning urine concentrations. In the USA, threshold values (also called 'cut-off' values) have been specified for drugs of abuse in urine and if a drug is detected but its concentration is below the cut-off, then the sample is considered to be negative. These cut-off values were established to deal with problems of inadvertent exposure and analytical methodology, but have not been set in the UK and are in fact, not recommended by many European laboratories.

For reasons of variability the medical literature records in many cases that no concentration-effect relationship exists. This means that, in practice, the interpretation process is limited and may be, at best, an estimate based on the information available. Perhaps the best-known example of this is the wide variation observed in the effects of alcohol on individuals: some people may function quite well at blood alcohol levels that would cause others to be extremely unwell. As a guide, the toxicologist can indicate what would happen to the average person at a given level and point out the possibility of individual variation.

In all branches of forensic science, interpretation is based on a comparative approach. Toxicology results are compared with a database of concentrations found in previous cases. This reference information comes from a variety of sources:

1. The records of the toxicologist's own department.
2. Compilations of records from other departments, for example, those published by the International Association of Forensic Toxicologists (TIAFT).
3. Reference textbooks and pharmacopoeias.
4. The medical literature.

The information in this database is segregated according to the type of case, including normal-therapy cases, non-fatal poisoning cases and fatal poisonings. The toxicologist can then decide if the results are consistent with therapeutic administration, an overdose or a fatal overdose. What it is often not possible to do is give an opinion with certainty that a measured level of a substance would have produced a particular effect.

12.4.3 Specific Problems of Interpretation

Care must be taken in the interpretation of some results since there could be alternative explanations. Several instances have been recorded in the literature in which drugs were found in test specimens (particularly urine samples), but these were derived from legitimate sources rather than from illicit materials. For example, poppy seeds used for baking have sometimes been found to contain traces of opium acquired from the poppy capsule. If enough of these seeds are consumed, morphine can be detected in urine. Similarly, some cases were encountered in the USA in which tea was mixed with coca leaves ('Health Inca Tea'), giving rise to positive cocaine tests in urine. Thirdly, many drugs, which can be prescribed or even purchased, are metabolised in the body to controlled substances. A wide range of medicines give either amphetamine or methamphetamine, and codeine is metabolised to morphine. Finally, the analyst must take care in testing specimens for amphetamine to avoid false positives due to the presence of similar substances such as ephedrine, a constituent of cold remedies and ginseng tea.

The effects of passive smoking cannot be neglected. Cannabis and crack cocaine are usually smoked, and those present in the same room or car as the smoker can have an intake of the drug due to passive smoking. Recent work has also indicated that rooms used for smoking crack become contaminated by cocaine 'fall-out' – minute crystals of the drug which settle on the furnishings of the room. People coming in contact with these can give false positive results in a drug screening test. For this reason, cut-off values for urine tests make allowance for the maximum level which could possibly be obtained by passive exposure.

The fact that some drugs are very rapidly metabolised in the body, and cannot be detected a short time after dosage, must not be overlooked. The best known example is diamorphine, which is present in heroin samples and is converted rapidly to morphine. Heroin abuse is usually inferred from the presence of morphine in the blood and from other circumstantial evidence.

In addition to the problems of interpretation already mentioned, it is also known that drug concentrations can vary according to the source of the blood specimen (*e.g.* from the heart or from a major blood vessel) and can change after death due to passive physical processes of diffusion in the organs. It is generally not possible to predict or allow for these post-mortem changes and redistribution when reporting drug levels, and the part of the body from which blood samples are taken at an autopsy should be chosen with care, to minimise these effects.

12.5 SPECIFIC AREAS OF INTEREST AND CASE STUDIES

The purpose of this section is to provide some detailed examples of how toxicology cases are handled and how the results can be used in subsequent legal proceedings.

12.5.1 Fires

In the investigation of a fatal fire, the role of the toxicologist is often to establish the presence of carbon monoxide in the blood of the fatality. This has an important medico-legal significance as, taken in conjunction with the pathologist's observations concerning the presence or absence of soot in the air passages, it establishes whether or not the deceased was alive at the start of the fire. If death occurred before the fire, the possibility of homicide must be considered.

Carbon monoxide is a colourless, tasteless, odourless gas produced by partial combustion of carbon-containing materials. It is produced almost immediately during the course of a fire and those present in the fire locus are exposed essentially from the beginning of the fire. The gas is inhaled and absorbed through the lungs into the blood circulation system. The significance of carbon monoxide in fires is that casualties can be overcome by the effects of the gas while asleep or at rest, perhaps unaware that a fire has started. If they subsequently become aware of the fire and start to exert themselves, the sudden demand for more oxygen can cause fainting and collapse, resulting in failure to escape.

In addition to carbon monoxide, the toxicologist may be requested to measure other toxic materials in the blood which have originated in the fire, particularly hydrogen cyanide, which is produced by the effects of heat on nitrogenous materials such as polyurethane, polyacrylonitrile, wool and silk. Cyanide levels are usually measured in whole blood specimens, even if plasma or serum is available. Cyanide is present at low levels (up to 0.25 mg l^{-1} of blood) in the normal population and at slightly higher levels (up to 0.5 mg l^{-1}) in smokers, as cyanide is present in cigarette smoke. The toxic effects of cyanide result from its action within body tissues, where it blocks cellular respiration at the cytochrome level. Symptoms of cyanide poisoning include dizziness, headache, chest pain, confusion, staggering, slowing of the heart and breathing rates, unconsciousness, coma and death. The significance of cyanide in fires lies in its rapid incapacitating effect, which can render casualties unconscious in a very short time. This is in contrast to carbon monoxide, which may build up in blood more slowly.

Most fire deaths in the UK and elsewhere occur in dwellings and research has shown that more than 80% of these have been exposed to hydrogen cyanide in the fire. However, the toxicologist may be asked to analyse blood specimens for the presence of alcohol and drugs, because these might have affected the ability of the deceased to escape from the fire scene or even have contributed to the events leading to the initiation of the fire.

12.5.1.1   Case Study: A Fire in a Leisure Centre.   A fire occurred in a leisure centre in which six people died. The fire was small and did not spread outside a single room. The casualties occurred due to the spread of smoke, heat and toxic gases along a corridor, blocking the escape of the occupants who had been in side-rooms. The results of the toxicological analysis of blood samples are given in Table 12.2 and show that the casualties had breathed in significant amounts of both carbon monoxide and cyanide, leading to their collapse and death.

Table 12.2   *Carbon monoxide and cyanide levels in fatalities resulting from a fire in a leisure centre*

Casualty	%Combined Haemoglobin/Carbon Monoxide in Blood (%HbCO)	Cyanide Level (mg/l)
1	48	0.9
2	48	0.8
3	41	3.1
4	34	0.5
5	29	2.0
6	38	2.3

12.5.2 Explosions

The toxicologist can often contribute to the investigation of deaths resulting from explosions, as many of these fatalities are actually caused by a fire accompanying the explosion. Also, analysis of blood from fatalities in a gas explosion may assist the explosion investigator in identifying the nature of the gas involved. Usually this is either consumed in the explosion or dissipated after the event, and cannot be detected by the investigator at the locus. However, it may still be detected in the blood of the fatalities who were alive before the explosion and inhaled the explosive mixture of gas and air involved.

12.5.2.1 Case Study: A Gas Explosion in a Dwelling House. Early one winter morning, a gas explosion partially demolished a building containing four flats in a city suburb. Five people were killed, including the four members of one family in a ground-floor flat (fatalities 1–4) and the resident of an upper flat (fatality 5). The causes of death were established at autopsy as blast injury, crush asphyxia (due to collapse of the building) and inhalation of smoke and fire gases from the subsequent fire.

The initial investigation produced three possible sources of flammable gas: the mains gas supply (methane), old coal mine workings under the flats (firedamp, a mixture of methane and ethane), and bottled gas (butane). Subsequent investigation revealed that the actual source was the mains gas supply. The type of gas was confirmed by the toxicological analysis of the blood specimens taken at autopsy, which showed the presence of methane only, see Table 12.3. Also, the levels of methane were in inverse relationship to the levels of carbon monoxide, indicating which of the fatalities had survived longest after the explosion. The methane in the building had been vented by the explosion and survivors would have breathed out any gas in their blood, but simultaneously would have inhaled carbon monoxide from the fire which was ignited by the explosion.

Table 12.3 *Carbon monoxide and methane levels in fatalities resulting from an explosion and fire in a dwelling house*

Fatality	Combined Haemoglobin / Carbon Monoxide in Blood (%HbCO)	Methane (ml/litre)
1	2	1.14
2	8	Trace
3	11	0.28
4	32	0.15
5	71	0.04

The father of the family appeared to have initiated the explosion with his activities in the kitchen, and was found to have suffered blast injuries but not to have inhaled any carbon monoxide. By contrast, the upstairs neighbour (fatality 5) did not have any methane in her blood sample but had the highest level of carboxyhaemoglobin.

12.5.3   Drug Overdose Cases

Recent deaths in the UK due to the illicit use of drugs have often involved diamorphine (heroin) and amphetamine derivatives such as methylenedioxymethamphetamine (MDMA, 'ecstasy'). In these cases, the role of the toxicologist is central to the investigation of the case but often problems arise in court in providing the clear-cut interpretations needed by the legal process. The pharmacology of the drugs involved in the following case studies can be obtained from other sources.

12.5.3.1   Case Study: A Heroin-Related Death.   In this case, a known intravenous drug-user was found dead at home with a syringe beside him. Some heroin powder was also found nearby. Interviews with his friends suggested that he had not injected for some time. At autopsy, death was ascribed to the inhalation of gastric contents, due to drug intoxication (confirmed by subsequent toxicology). A blood sample taken at the autopsy was found to contain 0.6 milligrams of morphine per litre of blood. The heroin was analysed and found to contain 40% diamorphine by weight (the normal diamorphine content of street heroin in the UK is 30–50%, but in the locality of the case, the average purity was 10–15%).

In this investigation, there is a clear indication that drugs of abuse might be involved and the toxicologist was specifically directed to analyse the blood sample for the common street drugs. Heroin is rapidly metabolised in blood to morphine and this is the usual target substance for the analyst. The sample was screened using an immunoassay method, which gave a positive result for opiates but not for any other drugs. This test did not specifically indicate which opiate drug was present since it lacked specificity. As a result, attention was focused on determining which type of opiate was present, and the sample was analysed by GC-MS following the preparation of a basic drug extract. This analysis produced data for the positive identification of morphine and for the quantity present.

Interpretation of the result requires a knowledge of morphine levels found in clinical patients treated with morphine or diamorphine (the mean and range for normal therapeutic levels) and also morphine levels found in overdose cases. Consultation of a pharmacology text, pharma-copoeia or specialised textbook, indicates that blood levels of morphine found in normal therapy are usually less than 0.1 mg l^{-1}, although can be very high in cancer patients treated long-term with morphine, due to development of tolerance. Blood levels found in heroin-related deaths are usually above 0.2 mg l^{-1} although the range is from zero to several milligrams. The latter requires some comment. Many published studies of heroin-related deaths report cases in which no morphine was detected in blood. The reasons are not known but may include failure of the heroin to be uniformly distributed in the circulation, due to death occurring very suddenly. On the other hand, very high levels have been found in regular injectors who are tolerant to the drug.

In the present case, the level of 0.6 mg l^{-1} was clearly above the normal therapeutic range and provided a plausible cause of death. Regular users of heroin might live at such a level but the deceased in this case had not injected for some time and had presumably lost his acquired tolerance. He might also have overdosed because the heroin used was of higher purity than normal in his area.

The cause of death was attributed to inhalation of gastric contents: this is a common finding in drug-related deaths, as the normal swallow-ing reflex is lost due to depression of the nervous system by the drug. Also, nausea is a side-effect of heroin use, especially in new users.

During examination of this case in court, the toxicologist concerned was asked if the morphine level found in blood was a fatal level. It should be clear that a simple yes/no answer is inappropriate and that some qualifications need to be added. However, the toxicologist was able to confirm that the morphine level was in the range found in previ-ous heroin-related deaths and could provide a cause of death, when taken into account with the other information available.

12.5.3.2 Case Study: A Paracetamol Overdose. A middle-aged woman with a history of suicide attempts was admitted to hospital at 9.30 pm after taking an overdose of paracetamol and an antidepressant (prochlorperazine) earlier that evening, at 6 pm. A sample of blood was taken for analysis in the hospital laboratory while normal supportive treatment (gastric lavage) was carried out in the ward. At 1 am, the lab result indicated a level of 270 milligrams of paracetamol per litre of blood. Medical intervention was indicated by this result and the drug acetylcysteine was administered. Unfortunately, a ten-fold overdose of this drug was administered despite the dose being queried by the pharmacist and duty nurse. A booster dose of the drug was subsequently administered, also in a ten-fold overdose, at 4 am. The patient died at 6 am of a cardiac arrest.

The toxicological investigation of the case covered a full screen for drugs, including paracetamol and basic drugs (which include antidepressants). The results were: 102 milligrams of paracetamol per litre of blood and 0.07 milligrams of prochlorperazine per litre of blood. All other analyses were negative.

Interpretation of these results must first deal with the drugs individually. Paracetamol causes liver failure and death (after two to three days) if taken in a high dosage. There are clear guidelines available to physicians which relate the blood level of paracetamol to the time which has elapsed since the drug was taken and to the probability of liver failure. In the present case, a blood level of 270 mg l^{-1}, four hours after the drug was ingested, indicated a high probability of fatal liver damage. The treatment to be followed is to administer a drug such as acetylcysteine which protects the liver from such severe damage. The level of paracetamol measured in the autopsy blood specimen was 102 mg l^{-1}, indicating a drop of 170 mg l^{-1} in eight hours approximately. If the deceased's liver had been functioning normally, the level should have fallen to below 70 mg l^{-1} (the half-life of paracetamol is four hours). Incipient liver damage is indicated by this result.

The inadvertent overdose of the antidote, acetylcysteine, resulted from poor labelling of the drug formulation and from subsequent miscalculation of the dose. Several similar cases have been published in the medical literature, which led the manufacturer to design new packaging. There was no analytical procedure available for this drug and it was not measured in the autopsy blood specimen. The small amount of toxicity information available related to tests on dogs, although this indicated that the drug was safe even at high doses.

Prochlorperazine taken on prescription produces a therapeutic range of 0.01–0.04 mg l^{-1} blood. In the present case, the deceased had survived about twelve hours after taking the drug. The half-life is approximately six hours so the drug at its highest level might have been two to four times higher (*i.e.* up to 0.28 mg l^{-1}), which places it in the range associated with toxicity.

Conclusions reached on the basis of toxicology alone are unclear, especially concerning the role of the drug acetylcysteine in the death, in the absence of toxicity data. Paracetamol and prochlorperazine were at concentrations likely to cause toxicity. A death certificate was issued ascribing the death to drug overdose and in the subsequent inquiry, the court accepted this as the cause of death.

12.6 BIBLIOGRAPHY

R.E. Ferner, *Forensic Pharmacology – Medicines, Mayhem and Malpractice*, Oxford Medical Publications, Oxford, 1996.
The Pharmacological Basis of Therapeutics. 9th edn, Goodmans and Gilman's, McGraw-Hill, New York, 1996.
T.A. Gough (ed), *The Analysis of Drugs of Abuse*, Wiley, Chichester, 1991.
S.B. Karch, *The Pathology of Drug Abuse*, 2nd edn, CRC Press, Boca Raton, 1996.
B. Levine, *Principles of Forensic Toxicology* (2nd Edition), AACC Press, 2003.

CHAPTER 13

Alcohol Analysis

VIVIAN EMERSON

13.1 INTRODUCTION

The term alcohol is generic for a large class of chemical compounds which possess a common feature in their chemical structures. For the purpose of this text the word alcohol will be used, as in common parlance, to mean ethyl alcohol or ethanol.

The production of alcoholic drinks has been known by man for many thousands of years and the effect of the consumption of such drinks on the imbiber is equally well known. It is probably man's oldest and most widely used drug and is produced in vast quantities every day. Somebody once calculated that the annual global production was sufficient to provide every man, woman and child with 10 bottles of wine and 40 pints of beer each year.

The details of the mechanisms and conditions for the production of alcohol are clearly the subject for further reading but it is worth recording the fact that alcoholic liquors can be divided into three groups: (i) those made by fermentation only, such as beers and table wines; (ii) those where the alcoholic concentration has been increased by distillation, such as spirits and liqueurs; and (iii) fortified wines such as port, sherry and vermouth where the concentration is increased beyond the original fermentation level by the addition of distilled spirit such as brandy.

It is perhaps relevant at this point to draw attention to the strengths of such liquors and the confusion which can arise if care is not taken with the units used. Percent volume/volume is now widely used to express alcoholic strengths but percent weight/volume and weight/ weight have sometimes been used. To add to the possible confusion the term 'percentage proof' has also been used, which refers to an alcohol/water mixture known as 'proof spirit' which has an alcoholic strength of 49% by weight or 57% by volume.

It is also relevant to point out that the size of measure used to dispense drinks in a public house is in relation to their strengths. Therefore in general terms, a half pint of beer is equivalent to a glass of wine, a glass of sherry or a single whisky.

In consuming an alcoholic drink, many people are not aware that they are taking a drug. Ethanol is a depressant and its effect is similar to that of an anaesthetic. On consuming alcohol the drinker will become less aware of depression or fatigue and at moderate levels of consumption they will experience a loss of inhibition and self-restraint. Judgement in manual tasks becomes impaired, although the drinker with slurred speech, reduced visual awareness and, possibly, dizziness, may feel confident and think they have the ability to perform them. At still higher alcohol levels, loss of consciousness and death from respiratory paralysis may occur due to acute ethanol intoxication.

With drink-driving cases or sudden deaths, the courts often require the help of the forensic scientist to determine the amount of alcohol a person may have consumed. To be able to accomplish this, forensic scientists have to understand how alcohol is absorbed, distributed in the body and eliminated. They must also have knowledge of the legislation and technical defences for drink-driving cases. All of these aspects are discussed in this chapter and the methods used for the analysis of alcohol in body fluid and breath samples.

13.2 ABSORPTION, DISTRIBUTION AND ELIMINATION OF ALCOHOL

13.2.1 Absorption

Alcohol once ingested, passes through the oesophagus to the stomach and thence past the pyloric sphincter into the small intestine. Only about 20% of the ingested alcohol diffuses through the gastric mucosa, owing to the comparatively small surface area of the stomach and its relatively short period of retention. The alcohol then passes into the small intestine where the remaining 80% is absorbed.

Alcohol is absorbed into the body without undergoing any chemical changes and theoretically its transfer across biological membranes is by a process of simple diffusion which follow Fick's Law. It is extremely difficult to prove this conclusively due to the many and varied factors that influence any attempts at its measurement. However, in general, it is possible to show that individual factors such as alcohol concentration, regional blood flow and the nature of the absorbing surface affect the rate of alcohol absorption in a manner consistent with the process of simple diffusion.

Immediately after absorption, alcohol is distributed in the blood to the various organs and body fluids throughout the body by the cardio-vascular system. Differences in alcohol concentration can be seen in venous and arterial blood due to distribution lags that occur during the absorption process. When the alcohol in the body fluids reaches equilibrium, the concentration is in proportion to their water content.

As soon as any alcohol is present in the bloodstream, then various mechanisms for its removal start to act. The major one of these is destructive oxidation which takes place in the liver at a constant rate. This rate of removal of alcohol from the system is much slower than the rate of absorption and hence, depending on a number of factors which will be dealt with a little later, the blood alcohol concentration (BAC) in the body will rise. If a graph is plotted of BAC *versus* time a curve similar to that shown in Figure 13.1 might be expected.

At this point it is worth considering the various factors that affect the shape of this curve. The rate of absorption will be governed by a number of conditions such as the concentration of the drink consumed, whether food is present in the stomach and the type of food, and

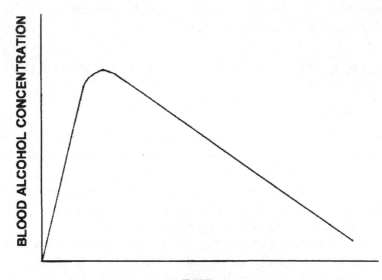

Figure 13.1 *The blood-alcohol curve. The blood alcohol concentration and time values are dependent upon several factors including quantity of alcohol consumed, period of consumption and body weight of an individual*

whether the drink consumed is carbonated. In general, on an empty stomach maximum absorption occurs when the drink contains 20% by volume. Hence the effects of a preprandial sherry can be felt very quickly. More concentrated drinks such as neat spirit irritate the gastro-intenstinal tissue and cause the absorptive surface to be coated with a protective mucus slowing the absorption rate. Dilute drinks such as beer produce a peak blood alcohol concentration later and lower than that produced from the same amount of alcohol at optimum concentration. If drinks are carbonated as in gin and tonic or champagne then this hastens the gastric emptying and hence also hastens the absorption.

Food present in the stomach will slow up the absorption of alcohol and there are many tales about secret remedies that reduce the absorption rate and hence delay the onset of the marked effects of alcohol consumption. Whilst each person may have their own view and special food in this respect, in general terms both fats and carbohydrates seem to show the same effect in reducing the rate of absorption. It should be noted that this is really only in relation to wines and spirits because, as mentioned earlier, the rate of absorption from beer is already much slower and therefore not particularly affected by the presence of food.

Hence the initial phase or the rising part of the curve can vary in steepness depending on the factors already mentioned. This will also affect the maximum level attained because elimination will be taking place simultaneously with absorption and if the rate of absorption is reduced the maximum level reached will thus be lower for the same quantity of alcohol consumed.

13.2.2 Distribution

There are two other factors that have a significant effect on the maximum level achieved. The body mass of the individual will clearly be relevant because the larger a person the greater the quantity of water present in the body and hence the more dilute the alcohol consumed will be when fully absorbed. Also fat does not absorb alcohol to any significant extent, so the extra body weight of an obese person does not automatically absorb more alcohol. Women's bodies contain a higher proportion of fatty tissue than men's and hence, weight for weight, require less alcohol to attain the same blood alcohol concentration.

The subject of what level can be reached from the consumption of a quantity of alcohol was extensively studied by E. Widmark in Sweden and a simple formula produced:

$$a = c\,p\,r$$

where a is the amount of alcohol taken, c is the peak concentration in the blood, p is the weight of the drinker and r is a factor known as the 'Widmark factor', which is derived from the proportion of the total body mass over which the alcohol can be distributed, probably better described as volume of distribution. The mean values of r found experimentally by Widmark were 0.68 for men and 0.55 for women.

Thus the maximum possible peak of blood alcohol concentration for the consumption of a given quantity of alcohol can easily be calculated provided that:

1. It is assumed that all alcohol is absorbed before any significant degree of elimination has taken place. This does not happen in practice as previously described.
2. It is assumed that the alcohol is uniformly distributed throughout all the water in the body.
3. The total quantity of water in the body is known.

For example, if an average man of say 70 kg consumes two double whiskies (a single contains approximately 7.5 g of alcohol) and the average Widmark factor of 0.68 is used, then:

$$c = \frac{a}{p \times r}$$
$$= \frac{30}{70 \times 0.068}$$
$$= 0.63 \text{ g l}^{-1}$$
$$= 63 \text{ mg}/100 \text{ ml of blood}$$

Although there are a number of important assumptions made in this calculation, it provides a good approximation of the peak concentration.

Warning. It has already been stressed that there are many factors which will have an effect on the level that can be reached from the consumption of a certain quantity of alcohol. Even though the Widmark calculation will give a very good indication of the maximum level that can be reached it should never be used to calculate the quantity of alcohol that can be consumed and still be below the legal limit to drive a motor vehicle. Even small amounts of alcohol can seriously affect a driver's judgement and reaction to an emergency. Therefore even if a person is below the statutory limit it does not necessarily mean they are safe to drive.

13.2.3 Elimination

Returning to the blood alcohol *versus* time curve, having dealt with the rising curve and peak level achieved we pass on to the falling curve or elimination rate.

The whole subject of the rate of elimination of alcohol in humans and the factors which influence the rate have been extensively studied. It is generally accepted that between 2 and 10% of alcohol absorbed is excreted unchanged mainly *via* the kidneys and the lungs. The remaining 90% plus is metabolised, largely by the liver the stages being:

Alcohol → Acetaldehyde → Acetic Acid → Carbon Dioxide + Water

It is often thought, quite erroneously, that significant quantities of alcohol can be eliminated from the body by violent exercise or by copious urination. One only has to pause for a moment to consider the amount of sweat and/or urine that can be produced compared with the total volume of water in the body to realise that these routes are not realistic. Because exercise will use up energy, and alcohol as a carbohydrate is a food which on metabolism will produce energy, then it is possible to understand the logic of such an argument but in practice the effect on the removal of alcohol from the system is at best marginal.

The rate at which alcohol is metabolised in the body by oxidation in the liver is remarkably constant, although there are differences between individuals from a variety of causes. It is generally accepted that the most probable rate of fall of blood alcohol is 15 mg/100 ml/hour although it can drop as low as 10 and rise as high as 25 in a few extreme cases. In consumption terms this relates to a loss of one unit of alcohol (one single whisky) per hour for a standard 70 kg man.

This is an over-simplification and there is some evidence that the rate of loss is not truly linear throughout the whole range. Greater rates have been identified at the higher levels and lower rates at the very low levels. However it is a good rule of thumb that can be used if a person knows the real amount they have consumed.

There is the popular misconception that a good night on the town can be swept away during a good night's sleep, this clearly is not the case. Many motorists who have not used their car to go to a party, where they have overindulged, get up the next morning to drive to work and are surprised to learn that they are still over the legal limit.

13.2.4 Concentration of Alcohol in Urine and Breath in Relation to Blood Alcohol Concentration

Whilst dealing with the distribution and elimination of alcohol in blood, it is perhaps relevant to deal with the concentration of it in urine and breath, although the actual legislation that governs such concentration in drivers and the analytical methods will be dealt with later.

Although, nowadays, urine is rarely used as the body fluid of choice for analytical purposes, there was a time when it was regularly taken for analysis in relation to road traffic offences. The analysis of urine as a method of determining blood alcohol concentration has never been ideal, its only redeeming virtue was that it was comparatively easy to collect and did not require the presence of a doctor. The major difficulty is that it is not possible to collect a sample of urine and relate it directly to the blood alcohol level at the same instant because it will only relate to the whole period over which the urine was excreted. It is important to remember that a drinking subject's BAC is varying all the time either up or down. The urine however is continually excreted by the kidneys from this changing blood concentration, is stored in the bladder and the excreted in batches. The urine in the bladder at time of passing will reflect the average blood alcohol concentration over the period during which the urine was excreted.

There is also another factor which has to be borne in mind, namely that is that urine to all intents and purposes comprises just water whereas blood contains 15–20% suspended solids. Therefore a straight comparison between the water content of urine and of blood levels is a ratio of 1.2:1. However, practical work over a period of time, and involving many subjects, has led in practice to a ratio of 1.27:1, although a ratio of 1.33:1 is more generally accepted and this has been used in Road Traffic Legislation.

The other substance in which the concentration of alcohol is relevant to the contents of this chapter is breath. We are all aware of the smell of alcohol that can be detected on a person's breath some time after drinking has ceased. Consider the anatomy of the respiratory tract: it starts at the nose and mouth and passes into the trachea, at the end of which it divides into two bronchi, one for each lung. The bronchi then subdivide into twenty smaller channels, which in turn, lead to numerous blind pouches called alveolar sacs. These sacs are partitioned by interalveolar septa into about 300 million alveoli, which are surrounded by a fine network of capillary blood vessels. The heart pumps the blood through the pulmonary artery and these capillaries. It is returned *via* the pulmonary vein and is then distributed throughout the body. Exchange of gases and fluids takes place through the alveolar and capillary walls.

Therefore with the surface area of all the alveoli being about 70 m^2 which is in contact *via* the capillaries with the pulmonary blood flow of about 6 l min^{-1} and with a daily respiratory volume of about 10^4 l it might be expected that breath should faithfully reflect the blood alcohol concentration.

The physical changes which take place during respiration are governed by the various gas laws. The most important of these with regard to breath analysis is Henry's law which states that the quantity of gas dissolved in a liquid at standard temperature and pressure is directly proportional to the partial pressure of that gas in the gas phase. At equilibrium the partial pressure of the gas in the gas phase is proportional to the concentration in the liquid phase.

Therefore, to use breath as a means of determining the BAC it is essential to know the blood:breath ratio. To a first approximation this certainly lies between 2000 and 2500:1. This figure is derived from the number of volumes of breath containing the same weight of alcohol as one volume of blood. Although much of the earlier work determined in practice that the ratio was 2100:1, much research by various workers produced other values. It is now generally accepted that the figure is 2300:1 which is now used in UK legislation discussed in more detail later.

13.2.5   The Effects of Alcohol

Before leaving this section it is perhaps pertinent to consider the effects which alcohol has on the imbiber. Alcohol is a drug that, contrary to popular belief, acts as a depressant on the central nervous system. Most people are aware of the effect that consumption of alcohol has on them and/or their friends without relating the behaviour to any particular blood alcohol level. Table 13.1 sets out what might be expected for the average occasional or moderate drinker. The effect in relation to blood alcohol levels are of course approximate.

There is some evidence that a given blood alcohol level produces more effect on the absorption phase, or rising part of the blood alcohol curve than during the falling section or elimination phase. This suggests that the drinker appears to adapt to the effects of alcohol to some extent while it is circulating in the body. This is particularly apparent in the alcoholic whose blood alcohol level rarely falls below about 100 mg/100 ml and whose body therefore adjusts to accept this as a baseline. This in no way suggests that they are not still under the influence of alcohol but rather indicates that the body's compensatory mechanism enables the alcoholic to behave comparatively normally at levels where the infrequent drinker would be seriously affected.

Table 13.1   *The Effect of Alcohol*

Blood Alcohol Concentration (mg/100ml)	Effects
<50	No obvious effect, except perhaps a tendency to a feeling of well being and to become more talkative
50–100	First obvious effects start to show. Some loss of co-ordination and sensory perception, possibly some slurred speech
100–150	More marked loss of co-ordination, poor sensory perception, possibly nausea and desire to lie down
150–200	Drunkenness and probably nausea
200–300	General inertia, inability to stand, vomiting, probably coma
>300	Approaching danger limit, coma and anaesthesia, impaired circulation and respiration, possible death
>450	Probable death due to respiratory paralysis

The other point to stress is that the same person will be affected differently on different occasions. The drink consumed and the food, if any, is taken with the drink will have a marked effect on the blood alcohol level reached and thus the behaviour of the drinker. Equally important is the drinker's diminished ability to evaluate their own performance. In fact it is often the case that drinkers will fail to remember and then underestimate the number of drinks they have consumed, believing that because the amount drunk is small they cannot be affected.

13.3   LEGISLATION

The legislation in relation to alcohol mainly concerns the effect it has on the individual who imbibes, either to excess or when driving a motor vehicle. The offence of being 'drunk and incapable' was recognised in law in the 18th century but it was really the antisocial results of persistent and public drunkenness that was tackled by legislation in relation to those in charge of animals or machinery. In 1872 the Licensing Act gave statutory recognition to the offence of being drunk in charge of any carriage, horse, cattle or steam engine.

It was, however, the Criminal Justice Act of 1925 which can really be considered to be the first attempt in legislation to deal with the drinking driver by penalising a person drunk while in charge, on a highway or other public place, of any mechanically propelled vehicle. This proved to be less effective than originally planned because of the lack of the exact definition of the term 'drunk'. With the increase in motor traffic it became necessary to be more precise with the definition of an offence. The Road Traffic Act 1930 redefined the offence as follows:

> 'Any person who, when driving or attempting to drive or when in charge of a motor vehicle on a road or other public place is under the influence of drink or a drug to such an extent as to be incapable of having proper control of the vehicle . . . shall be guilty of an offence.'

Despite this improved definition it was still difficult to obtain a conviction in anything other than the severely affected cases, particularly if juries were involved. Therefore the statistics in relation to those found guilty compared with the road accident figures were unreliable and in any case seriously understated the problem. The British Medical Association (BMA), realising this, produced an improved medical examination regime and later, having reviewed the situation in other countries, recommended that UK legislation be reformed to fix a level of alcohol in blood above which an offence would be committed. It further recommended that the level should be 50 mg/100 ml.

The Road Traffic Act 1962 defined 'unfitness' in terms of 'ability to drive' being 'impaired'; this definition has continued in subsequent legislation. This Act also entitled the court to take note of any evidence derived from analysis of blood, urine or breath although breath was never used. Despite this it failed to bring widespread use and reliance upon such evidence. Most samples provided were of urine and the ability of the courts to interpret the results in relation to blood and in relation to the quantity of alcohol consumed depended upon the availability of expert evidence and thus the use of such scientific evidence was restricted.

Eventually the legislature recognised the wisdom of the BMA proposal and in 1967 introduced the Road Safety Act which introduced two significant things:

1. That a policeman could, subject to certain conditions, require a motorist to take a screening breath test and, if this proved positive, a blood or urine sample for laboratory analysis.
2. A prescribed alcohol limit of 80 mg/100 ml in blood and 107 mg/100 ml in urine.

This Act, and some later ones, resulted in a plethora of case law because the judicial system largely interpreted the letter of the law rather than the spirit behind it. However it did have the effect of raising the public awareness of the hazards of mixing alcohol with driving a motor vehicle. There is more further reading on the subject than there is time or space for here.

The next significant step in relation to alcohol and road traffic legislation was the Road Traffic Act 1981, which introduced the use of substantive breath alcohol analysis equipment and a statutory breath alcohol concentration (BrAC) of 35 μg/100 ml. This significantly reduced the number of samples of blood submitted for analysis but did not eliminate them altogether. This was because subjects with a BrAC above 35 μg/100 ml but below 50 μg/100 ml had the statutory right to take up the option of providing a blood or urine sample if they did not wish to accept the breath result. The first instruments under this legislation were introduced and used in 1983. These were in continual use in England, Wales and Scotland under this Act and the later consolidating Road Traffic Act 1988 up to and beyond the time when the next generation of instruments received approval in 1998. They were phased out by the end of 1999. Details of the various types of analytical equipment follow shortly.

13.4 ANALYSIS OF BODY FLUID SAMPLES FOR ALCOHOL

The legislation first set out prescribed levels necessitating the analysis of blood and urine and later, as detailed above, moved on to breath. Clearly it is the concentration of alcohol in the blood passing through the brain that will affect the behaviour of the person consuming the alcohol and, if that person is also the driver of a motor vehicle, their ability to drive. The analysis of blood is therefore ideally the most relevant and neither urine nor breath is the medium of first choice for analysis from a purist point of view, but they are of course much more convenient to obtain. That is why urine was most popular after the Road Traffic Act 1962, was then replaced in most instances by blood following the Road Safety Act 1967 and, finally, by breath in 1983 following the Road Traffic Act 1981.

13.4.1 History

Early work on the analysis of both blood and urine started just before the turn of the century and a number of people in various parts of the world developed methods, most of which were based on the quantitative reaction of alcohol with acid dichromate solution. The various methods developed, and some further adapted by analysts, did not all produce very accurate results. The courts, when they started receiving such results, were not very impressed, particularly if there were significant differences between the results presented on behalf of the defence and the prosecution.

The BMA set up a committee to look into the situation, for only if courts could rely on the results of analyses would it be possible to start to combat the problem of the drinking driver. The outcome of this was that for court work it was highly desirable for the analyst to use one of two officially recommended methods. These were both modifications of an original method and both relied on the oxidation of alcohol by acid dichromate. These were: (i) the Cavett Method which used only 0.1 ml of blood or urine and relied on the passive evaporation of alcohol from the solution incubated at 37 °C for 4 hours; and (ii) the Kozelka and Hine Method which used 1 or 2 ml of blood or urine which were then steam distilled through alkaline mercuric chloride solution to remove aldehydes, ketones and volatile acids. The alcohol was then condensed and acid dichromate solution added, followed by potassium iodide. The liberated iodine was estimated with sodium thiosulphate solution using a starch glycolate indicator. These methods were proved to be both precise and accurate, although not to today's high standards, and were widely used for many years. However if new legislation was to be introduced to encourage the provision of a blood sample a move away from venous blood to capillary blood was necessary. Thus it was also necessary to move to analytical methods which only required small volumes of sample and were capable of high throughput.

Two such methods had been developed which came to the fore around 1965. These were the ADH method which depended on the enzymatic oxidation of alcohol in the presence of NAD (nicotinamide adenine dinucleotide), to produce acetaldehyde and NADH (the reduced form of NAD). NADH adsorbs radiation at 340 nm while NAD does not. Thus concentration of NADH and hence that of alcohol in the original sample could be determined using a spectrophotometer. The other method was that of gas chromatography, using liquid injection of a small quantity of the original sample diluted with water containing an internal standard.

The Road Safety Act 1967 completely changed the situation of blood/urine alcohol analysis for two reasons. Firstly, the use of roadside screening devices gave the police officer the power to require a sample of blood or urine, usually the former. Secondly, it introduced statutory levels for the concentration of alcohol in blood and urine above which an offence was committed. These, coupled with the use, initially, of capillary blood samples giving the analyst about 0.1 ml for duplicate analyses necessitated the move to a new method. For it to be accepted by the courts and be practical for those responsible for analysing the vastly increased numbers of samples submitted under the Act, the method had to meet the following requirements:

1. It should be sensitive and capable of providing results to the appropriate level of accuracy and precision when performed by a competent analyst.
2. It should be specific for alcohol (as ethanol).
3. It should be suitable for routine analysis of large numbers of samples.
4. It should be capable of multiple analysis from a small sample.
5. It should be as inexpensive as possible.

The only method which fulfilled all these criteria was gas chromatography, which was thus adopted widely by all forensic science laboratories.

13.4.2 Gas Chromatography

The original method involved the liquid injection of a diluted sample on to a gas chromatograph (GC). This practice was replaced by headspace analysis, which also requires the dilution of the original sample with internal standard, but this is then placed in a septum-topped phial which is heated to between 40 °C and 60 °C. The vapour above the liquid (headspace) is then analysed by the GC. The introduction of headspace analysis allowed automation of the analytical procedure, including dilution and analysis of the samples.

One of the first problems identified when using GC was that using a single column for separating alcohol from other components in a sample did not provide the required specificity for alcohol. To overcome the problem of any other substance present in the sample eluting at the same time as either the alcohol or the internal standard, a second column containing a different packing material was used. At the same time as this aspect of the method was developed, the practice of using a second person to perform a second independent analysis of the sample also developed. This duplication of analysis and equipment became standard practice, and has the advantage of not only overcoming the problems of handling errors, but also highlights malfunctions which may arise in the performance of any of the component parts of the system. These include the diluter, the chromatograph and the integrator, malfunctions that would not be apparent if only one set of equipment was used.

To achieve the highest degree of precision in reporting an alcohol level, it is essential to minimise any variation which may occur in the sampling, dilution and injection procedures. Whilst headspace analysis provides a reproducible injection system, the inclusion of an internal standard in the sample improves the overall precision of the alcohol analysis procedure.

It is normal practice for the diluent to contain the internal standard which should be chromatographically resolved from ethanol but have similar properties to those of ethanol. It should also be separable from all other compounds likely to be found in the sample under test. Propanol and isopropanol possess these properties and are the internal standards of choice. It is most convenient to use the same internal standard for each column, although some analysts use a separate one for each column. It is important to note that because isopropanol is used in the preparation of some 'mixers' it is possible to detect traces of it in blood samples on some occasions.

Sodium bisulphite is usually added to the diluent as an antioxidant and to remove any volatile aldehydes. Oxidation of ethanol and propanol may take place over long periods of time at elevated temperatures, conditions typically experienced by diluted samples awaiting analysis in an automated injection system. The products of oxidation are the corresponding aldehydes. This oxidation does not occur if bisulphite is added to the diluent.

The choice of a primary standard is up to the analyst but traceabilty is important to establish its authenticity. The Laboratory of the Government Chemist has long been the commercial source for a whole range of alcohol standards, which are available to anyone involved in alcohol analysis. The concentrations available include 80 and 107 mg/ml and go through several steps up to 200 mg/100 ml. These standards are prepared by dilution of a 20% w/v standard, the concentration of which is determined by specific gravity measurement.

If a standard is purchased and used to calibrate and check the linearity of the analytical system, then it is good practice for the analyst to prepare their own check standard to satisfy themselves of the validity of their results. The dilution of absolute alcohol could of course be considered by the analyst, but weighing out samples of ethanol can prove problematical due to it being both volatile and hygroscopic. An alternative novel method is available which relies on the alkaline hydrolysis of ethylaminoacetate hydrochloride (glycine ethyl ester hydrochloride). One gram of this compound produces 330.04 mg of alcohol. The advantage of this method is that the compound can be oven-dried to constant weight and then weighed out accurately to achieve an alcohol solution to the required accuracy.

One last check which any analyst should do is to ensure that the analytical regime does in fact separate potential interferents from ethanol and the internal standard. This may be achieved by preparing a mixture of all these substances and putting them through the system to ensure that all are separated by either of the columns used.

13.4.3 Accuracy and Precision

Mention was made earlier of precision and accuracy. It is important that these should be distinguished and clearly understood. The accuracy of a method is determined by how closely its results approach to the true ones. Hence the need for analysts to prepare their own standard to satisfy themselves that they can determine the true result is within the precision of the method. This brings us neatly on to precision, which is determined by the smallest difference that can be detected in the quantity measured. It is therefore, with a competent analyst, equivalent to the extent to which successive measurements of the same quantity will yield identical results, that is reproducibility. Precision or reproducibility is therefore a somewhat complex concept since it depends both on the smallest difference that the method used is capable of detecting, and on the skill of the operator. The high degree of automation introduced into the analytical system very largely removes the skill of the operator and therefore the precision rests with the reproducibility of the equipment used.

There is no such thing as absolute accuracy for any type of measurement. Any analysis is subject to error and the important thing is to recognise this and to know how large the error is likely to be. To evaluate this we must dip our toe into the ocean of statistics and tackle the concept of standard deviation. If a number of measurements of the same thing are made, and it is assumed there is no systematic source of error, then the errors in these measurements will be randomly distributed above and below the true value. The mean of these measurements should be very close to the true value. The standard deviation provides information about the extent to which the individual measurements differ from the mean or presumed true value. The standard deviation is defined as the square root of the mean of the squares of the deviations of the individual results from their mean. Because the standard deviation depends on the spread of a series of results around their mean it can be used as a measure of this spread. Therefore, if the mean and standard deviation are known for a large number of measurements then 68% of all the measurements will lie within one standard deviation either side of the mean, 95.5% within two standard deviations and

99.73% within three standard deviations. Therefore, if an analyst regularly finds that the results achieved for a known quantity show a certain standard deviation, it is perfectly acceptable to use that standard deviation to predict the accuracy of results of an unknown sample.

To put all this into context, the practice used in forensic science laboratories for the quantification of alcohol is as follows. A standard deviation of 2 mg/100 ml is assumed, although it is often smaller than this, and results from some samples of a standard must agree to within ±1.5% of the recorded mean value. Each body fluid specimen is analysed by two independent analysts using entirely separate sets of equipment, and their results, if within the critical region (*i.e.* 87 to 100 mg/ml inclusive for blood, or a comparable 115 to 133 mg/ml inclusive for urine) must agree to within ±2 mg/100 ml of the mean. Outside the critical region the criteria is ±3% of the mean. From the mean of these determinations 6% of its value (that is at least three standard deviations) is deducted if it is over 100 mg/100 ml or 6 mg/100 ml if it is less than 100 mg/100 ml. The alcohol content of the body fluid analysed is finally reported on the analysis certificate as 'not less than' the figure that is thus achieved. It can be shown that the chance of this figure being too high is several millions to one against.

13.5 ANALYSIS OF BREATH FOR ALCOHOL

Although the Road Traffic Act 1962 entitled the courts to take note of any evidence from analysis of blood, urine or breath, the latter was never utilised. As stated earlier, it was not until the Road Safety Act 1967 that the use of breath testing devices was introduced and then only for road-side screening. However, as screening devices they provided the police with a powerful weapon in the fight against the drink-driver. An officer could arrest, without warrant, a person who gave a positive screening test or who refused to take one. The person arrested had to be given the opportunity to take a second breath test at the police station before any requirement for a sample for laboratory analysis could be made.

13.5.1 Screening Devices

The only breath alcohol screening device that was available in 1967 was the ALCOTEST 80 which was manufactured by Drägewerk, in Germany, and was used for many years with considerable success. It was commonly called the 'Breathalyser', a name that crept in because of the use of the BREATHALYZER substantive breath alcohol test instrument used in the USA. Several years later a similar device the

ALCOLYSER, manufactured by Lion Laboratories in Cardiff was also approved for use under the Road Traffic legislation. Both of these devices work on the same principle. A glass tube, containing acid dichromate absorbed on crystals of silica gel, is sealed at both ends. To use the device the police officer snaps off the seal at each end of the tube and connects one end to a mouthpiece and the other to a non-elastic bag of one litre capacity. The subject is invited to blow through the tube to inflate the bag. Alcohol present on the breath reduces the acid dichromate and the yellow crystals turn green from the end nearest the mouthpiece. The length of the green stain is proportional to the concentration of alcohol in the breath of the subject and hence also to the blood alcohol concentration.

These devices have a number of disadvantages:

1. Only one litre of breath is sampled which comes from the top of the lung, thus it is not true alveolar air and it could not be expected to reflect truly the BAC.
2. They are notoriously inaccurate. About 20% of samples submitted to the laboratory as a result of these devices prove to be below the limit.
3. They are difficult to read, particularly at night under sodium street lighting.

If to these is added the problem of using glass tubes, the disposal of acid crystals, the limited shelf-life and the cost per test, it is amazing that they were used with such effectiveness at the commencement of the 1967 Act. However, the initial impact of the legislation in saving life on the roads gradually reduced and by 1974 great concern was again expressed about the efficiency of the legislation in its fight against the drinking driver.

A committee, set up under the chairmanship of Mr Frank (later His Honour Judge) Blennerhassett QC, made a number of recommendations resulting in the 1981 legislation. It also stated that, 'A device which gave a clearer pass/fail indication would be easier to use than the ALCOTEST, and reduce the number of false positive results', and recommended that those currently available should be evaluated.

Only devices approved by the Secretary of State can be used by the police. The usual responsibility for testing devices and recommending their approval falls upon the Forensic Science Service. All the devices listed in Table 13.2 were approved in this way.

Following the report of the Blennerhassett Committee in 1976, manufacturers were invited to submit their screening devices for evaluation. This resulted in the first two electronic devices being

Table 13.2 *Alcohol testing devices approved for use by the police in England, Scotland and Wales (Great Britain)*

Type of Device	Name	Manufacturer	Date Approved
Screening			
Tube and Bag	ALCOTEST R80	Dräger Ltd	July 1967
	ALCOTEST R80A	Dräger Ltd	May 1975
	ALCOLYSER	Lion Ltd	June 1979
Electronic	ALCOLMETER SL-2A	Lion Ltd	August 1979
	ALCOLMETER SL-400	Lion Ltd	September 1987
	ALCOLMETER SL-400A	Lion Ltd	
	ALCOTEST 7410	Dräger Ltd	December 1993
	ALCO-SENSOR IV UK	Intoximeter UK Ltd	March 1980
Evidential			
	LION INTOXILYZER 6000 UK	Lion Ltd	February 1993
	INTOXIMETER EC/IR	Intoximeter UK Ltd	May 1983
	CAMIC DATAMASTER	Camic Ltd	May 1983

approved for use, the Lion ALCOLMETER S-L2 in 1979 and the Dräger ALERT in 1980. Both of these have now been phased out and replaced by those listed in Table 13.2.

The ALCOLMETER S-L2 utilised an electrochemical fuel cell as its detector. This contained two platinum electrodes, one in contact with air and the other with the alcohol in the breath sample. This resulted in the generation of a small voltage which was directly proportional to the BrAC (Breath Alcohol Concentration) and hence BAC. Because the fuel cell sampling system required only 1.5 ml of breath, it was essential to ensure that top lung air was discarded first. To achieve this the instrument incorporated a pressure switch and a timing unit, and the subject had to blow hard enough (3 inches of water pressure) indicated by light 'A' becoming illuminated and long enough (2.5 seconds) to illuminate light 'B'. At this point the officer pressed a button which took a sample into the fuel cell. The voltage generated was amplified and displayed on a 'traffic light' system which showed the following:

Green	< 5 mg/100 ml	BAC
Amber	5–70 mg/100 ml	
Red and amber	79–80 mg/100 ml	
Red	> 80 mg/100 ml	

As the voltage built up to a maximum over a period of 40 seconds the lights came on in turn until the final reading was achieved. The stated accuracy of the instrument was ± 10 mg/100 ml at the calibration level 80 mg/100 ml.

Any electronic instrument requires calibration and regular checking to ensure it is retaining its calibration. In the case of the ALCOL-METER S-L2 a compressed gas mixture of argon and alcohol was used at concentrations equivalent to 80 mg/100 ml BAC for calibration, and 75 mg/100 ml for checking. The instrument operated on a disposable battery which was capable of about 400 tests.

The Dräger ALERT instrument utilised a solid-state semiconductor detector, the conductance of which increased in proportion to the BrAC. The detector had to be heated to 350 °C, which took about 90 seconds from the time it was switched on, at which point a 'ready' light was illuminated. Similar in operation to the ALCOLMETER, the subject had to blow hard enough (4 inches of water pressure) and long enough (6 seconds) so that both 'ready' and 'test' lights were illuminated, at which point the instrument automatically took a sample and the results were presented as follows:

Green	< 40 mg/100 ml	BAC
Amber	40–80 mg/100 ml	
Red	> 80 mg/100 ml	

The stated accuracy at the calibration level was ±10%. Similar to the ALCOLMETER, the ALERT had a calibration and testing regime but in this case the vapour was produced by a liquid simulator which held an alcohol solution at 34 °C ± 0.2 °C. The instrument was connected to the unit *via* the breath inlet tube and electrically *via* a lead to receive the output of the detector. A pump blew air through the solution and through the breath sampling system. The check simulator provided an OK indication if the result was 72–88 mg/100 ml, otherwise a calibration unit was used to adjust the instrument to read the correct calibration.

The instrument used rechargeable nickel cadmium batteries to supply the power needed to heat the sensor. These batteries provided 30 tests per battery charge and required 16 hours to recharge.

Following the introduction of electronic devices there was a marked decrease in false positive results, and in laboratory testing, a reduction in the number of such results to nearly zero for samples below 50 mg/100 ml. This was despite an increase of about 80% in the number of people tested.

Since the initial introduction of electronic breath alcohol screening devices, further devices have been approved for use under the legislation. These include the ALCOLMETER S-L2A, the ALCOL-METER SL-400 and the ALCOLMETER SL-400A all of which are improved versions of the original one approved in 1979. In addition the ALCOTEST 7410 produced by Dräger was approved in 1993 and the ALCO-SENSOR IV UK produced by Intoximeters UK Ltd was approved in 2000. These devices are all based on a fuel cell detector and some have the facility to provide results in digital form although for normal police use the display is PASS, WARN or FAIL. The more recent devices have increased sophistication in their operation and sampling with improved software to reduce operator involvement thus reducing training needs. All are checked regularly using compressed gas alcohol standards. For full up to date information on approved devices visit the Home Office Web site at http://www.homeoffice.gov.uk/ppd/oppu/orders.htm.

13.5.2 Substantive Methods and Instrumentation

The 1962 Act introduced the provision of breath as an analytical medium, but it was only after the 1967 Act that breath was used for screening purposes. However the report of the Blennerhassett Committee in 1976 recommended *inter alia* that 'blood alcohol concentrations should usually be determined for forensic purposes by the analysis of breath, using devices which would be kept at police stations. To this end, action should be put in hand to select suitable instruments and test them in realistic conditions in anticipation of the necessary change in the law'.

One of the reasons for this recommendation was the knowledge that instruments for this purpose were designed, manufactured and in use in other parts of the world, particularly the USA and Canada. The instruments available at that time operated on four principles:

13.5.2.1 Chemical Oxidation. This was the method used by Harger who pioneered the DRUNKOMETER. There have been a number of instruments designed on this principle, the best known of which is the Smith & Wesson BREATHALYZER, the model 1000 being the fully automated version. When an instrument is set up and ready for operation, a subject blows into a heated breath tube. The instrument automatically discards the first 450 ml of sample. This is achieved by the sampling system having two cylinders in series, first a small accurately known volume cylinder of about 50 ml and then the larger 450 ml cylinder. The subject's breath first fills the small cylinder, which then overflows into the larger cylinder. When both are full the contents of the small cylinder is taken as the analysis sample. The cylinders are maintained at 50 °C ± 3 °C to prevent condensation. The measuring system contains two glass ampoules of fixed concentration acid dichromate solution, one unopened to act as reference and one opened and used for the analysis. The ampoules are placed on either side of a tungsten lamp and the light passing through each ampoule passes through a blue filter and on to a photocell situated behind each ampoule. The position of the lamp is adjusted to give equal amounts of light to each photocell. When a breath sample has been taken as above, the breath sample from the small cylinder is bubbled through the contents of the opened ampoule. The orange/yellow acid dichromate is reduced to green chromic sulphate. The amount of light passing through the ampoule is increased and the system is unbalanced. The lamp is moved towards the reference ampoule until the balance is restored. The distance the lamp is moved is directly proportional to the amount of reagent reduced and hence the amount of alcohol in the sample, thus providing a reading which can be presented in BAC terms as that of the subject. This instrument, or the earlier version of it, the model 900, was used throughout the USA for many years.

13.5.2.2 Gas Chromatography. In the 1970s a commercial breath alcohol instrument (the INTOXIMETER), utilising gas chromatographic principles, was introduced. The sampling system was similar to that described for screening devices in that the subject had to blow hard enough for at least four seconds, at the end of which at least two litres of air were expired. When this had been achieved, if the pressure was maintained for a further six seconds, or if it started to drop, a small sample (0.25 ml) of the air was taken and injected into the top of the gas chromatographic column. If this was not achieved the instrument did not operate. The analytical system used a short column containing Porapak Q and 1% Carbowax as packing material held in an oven at 80–95 °C. A mixed gas supply of 60% nitrogen and 40% hydrogen provided both carrier gas for the column and fuel for the flame ionisation detector. Once calibrated with a standard sample the instrument would integrate the peak area, convert to an equivalent BAC, and display the result on the front panel.

This instrument was not used extensively in the USA, although it provided satisfactory results.

13.5.2.3 Infrared Absorption. Organic compounds absorb infrared (IR) radiation at specific wavelengths. The wavelengths at which these absorptions occur depend upon the molecular structure of the compound. Different parts of the molecular structure absorb radiation and resonate at only specific wavelengths. The precise location of these wavelengths and the intensity of their absorption provides a means for identifying a compound and determining its concentration. Several commercially available instruments have been produced for breath alcohol analysis based on this principle. The most popular in the late 1970s was the CMI INTOXILYZER, although more manufacturers have since produced instruments. There are disadvantages, which will be discussed, but these are far outweighed by the disadvantages encountered by the first two principles mentioned.

The IR based instruments all work on similar principles in that they incorporate a gas cell (the size and pathlengths used for analysis vary between instruments), which is heated to prevent condensation. On analysis the breath sample in the gas cell is compared to a reference cell of the same pathlength which contains air. Alcohol absorbs IR radiation at a number of wavelengths, but the wavelengths of choice are usually 3.39 or 3.48 micrometres. Since water vapour has a small but finite absorption at 3.48 micrometres, instruments have to be designed to compensate for this. The other disadvantage of the technique is that the wavelengths are not specific for alcohol, and acetone for example, also absorbs at these wavelengths. Acetone can be present on the breath of an alcohol-free ketonic subject or a non-hospitalised diabetic. Various instruments deal with this in different ways. The INTOXILYZER made a small mathematical deduction while the Lion INTOXIMETER 3000 incorporates a semi-conductor for the detection of the acetone.

All instruments of this type have sampling systems that attempt to sample alveolar air and overcome the problems of subjects trying to fool the system. Advantages of the method are that it gives an immediate read-out of the result and the sample cell can be purged and shown to give a zero result before analysing a breath sample.

13.5.2.4 Fuel Cell. A commercial instrument was produced by Lion Laboratories based on the principle already described and was used successfully by the police in Northern Ireland and some other countries. It has however, not received widespread use as a substantive instrument.

13.5.3 Instrument Evaluation and Introduction

Following the recommendation of the Blennerhassett Committee, the Forensic Science Service undertook a lengthy and detailed review of available instrumentation. This was followed, with full co-operation of certain police forces, by a field trial of examples of instruments operating on the first three principles mentioned above. The conclusions reached were:

'This was a practical assessment of three types of substantive breath alcohol test instruments which are available and in use in various parts of the world. In the course of the trial policemen tested over 1500 motorists who had been arrested under the Road Traffic Act 1972, in a representative selection of police premises, with a high degree of efficiency despite the minimal training received.

We found that each type of instrument produced results which were generally in good agreement with the certified blood results and therefore subject to the appropriate experimental allowances, reliable instrument maintenance and the use of such instruments by well trained personnel, substantive breath testing could be considered as a viable alternative to blood testing.'

As a result of this recommendation the Road Traffic Act 1981 was introduced, which not only introduced the use of substantive breath alcohol test equipment, but also provided a prescribed BrAC of 35 µg/100 ml.

Following legislation, the most up-to-date instrumentation available was initially laboratory tested and three instruments chosen for a second field trial. These were the Camic BREATH ANALYSER (British), the Lion INTOXIMETER 3000 (American Design, British manufacture) and the Lion AUTO-ALCOLMETER (British). The first two of these instruments both use infrared absorption while the third uses a fuel cell detector for the detection of alcohol.

Since it was realised that the public would have some concerns about the introduction of such instrumentation, two requirements were introduced to try to provide reassurance. Firstly, all instrument manufacture and maintenance, and the production and certification of the simulator solution *i.e.* a sample which simulates the breath of a subject with a BAC of 80 mg/100 ml, were overseen by the British Calibration Service. Secondly, satisfactory simulation check results had to be achieved prior to and after breath testing of a subject, and these results had to appear on the instrument printout of the analytical results.

For routine use Home Office Circular No 46/983 specifies limits of ± 3 mg/100 ml and therefore, the check readings must fall within the range 32–38 mg/100 ml. Also the police should not prosecute for readings of less than 40 mg/100 ml. Although this does not have the same scientific basis as the allowance made in blood and urine analysis the basic principle is similar.

The first two of the instruments mentioned above were selected, approved for use by the Secretary of State and introduced throughout England, Wales and Scotland in 1983. There followed some concern by the public about the use of such instrumentation, and the move from 'trial by analyst' to 'trial by machine' operated by a non-scientist. This was bolstered by the media resulting in a fully independent review of their use by Professor Sir William Paton; the result of which exonerated the use of such equipment which remained in use until phased out in 1999 and replaced by the second generation of instruments.

In 1998 the Forensic Science Service on behalf of the Home Office issued a Quality Framework Document for Evidential Breath Testing Instruments. This set out the requirements that instruments needed to meet to receive type approval under the legislation. This contained improvements such as: (i) the detection of mouth alcohol; (ii) the detection of interfering substances; (iii) the detection of significant differences between successive samples; (iv) and a whole series of checks and safeguards to ensure the correct calibration and operation of the instrumentation. All instruments and breath simulators (both liquid and compressed gases) needed to be certified and regularly checked under the UK Accreditation Service NAMAS.

Currently, there are three evidential instruments approved for use under the current legislation:

1. Lion INTOXILYZER 6000 UK.
2. INTOXIMETER EC/IR.
3. CAMIC DATAMASTER.

All these instruments operate on the principle of infrared analysis. The specificity for ethanol is ensured in the Lion INTOXILYZER 6000 and the CAMIC DATAMASTER by multi-wavelength measurement, the Lion instrument using five wavelengths and the Camic using three. The INTOXIMETER combines both infrared and fuel cell analysis.

All the instruments involve a higher degree of sophistication than the first generation particularly with regard to their operation and in their production and storage of results.

Considering that the use of the original instrumentation was exonerated by the report under the independent chairman Professor Sir William Paton nearly 20 years ago, it is hardly surprising that the advances in technology incorporated in the second generation ensure their continued effective use and acceptance.

13.6 TECHNICAL DEFENCE

In conclusion of this chapter, there are three aspects where the courts may wish to hear expert testimony in relation to interpretation of results obtained and the reader might be interested in these aspects. The first two relate to back-calculations either in the case of alleged laced drinks or in the case of the 'hip flask' defence. The third is in relation to interfering substances.

The results of back-calculations must always be considered suspect, since they involve so many approximations and often unverified assumptions. However, if the appropriate precautions are taken and the circumstances sufficiently extenuating, then some estimate may be made which may be useful.

As far as prescribed-level offences are concerned, as redefined following the Transport Act 1981, there is a specific provision requiring an assumption to be made (except in limited circumstances). This is that the breath, blood or urine alcohol level at the time of the offence shall not be less than in the analysed specimen, in which case the certification of analysis if unchallenged, proves this level.

In the case where an offence has been committed and the subject claims that they had been misled in what they had been drinking, then it can be claimed that these are special reasons for not disqualifying the person from driving.

When a person is misled by a third party as to what they had been drinking, such that more alcohol had been consumed than realised or intended, then the subject has to prove a 'special reason' for non-disqualification for a Section 6 offence. This requires the subject not only to prove that they had been misled, but also that the extra alcohol they had been misled or deceived into drinking could account for the fact that the alcohol in their breath, blood or urine exceeded the limit. Unless the evidence is so clear-cut that a layman can decide this, then scientific testimony must be called to show this fact.

The 'laced drink' mitigation is obviously open to abuse especially if there is a 'helpful friend'. Because of this the courts view the 'laced drink' defence with some suspicion and insist upon rigorous standards of proof. The law is now clear that if this 'helpful friend' admits participation in the actual lacing of the drink then they expose themself to a charge of procuring the offence.

The other instance is the case of the 'hip flask' defence. In this case the certified body fluid alcohol level is over the legal limit but it is alleged that drink had been taken after ceasing to drive and before the specimen for analysis was provided. This sometimes occurs when someone, who is not over the limit, has an accident and has a drink from a 'hip flask', or in a home or public house, to steady their nerves before the police arrive to investigate the incident. In this case the subject will claim that the drink consumed after the accident was sufficient to take them over the prescribed limit. If such defence is offered it is beholden on the subject to prove their case and not just suggest it. If such a defence is offered, an estimation will have to be made of the contribution of the additional drink to the body alcohol level. In that case it will be necessary to consider the nature and size of drinks consumed, the time of drinking, the time at which the body fluid sample was taken and its alcohol concentration, and the assumed rate of elimination by the drinker together with the best information of the body weight of the subject.

The complexity of the absorption and elimination process was dealt with earlier in this Chapter and the problems associated with it were expressed. Thus, due care must be taken in any attempt to provide an interpretation. When called upon to do so, forensic scientists will perform the necessary calculations and advise the courts not only of the results but also of the assumptions made in obtaing them.

Finally, it is sometimes claimed that a person has been in contact with alcohol vapour to such an extent that it caused an elevated BAC. It can however be clearly stated that BACs which are significant will never be produced by inhalation of alcohol vapour since it is important to remember that alcohol in the breath is in equilibrium with alcohol in the blood. Thus, if it is claimed that alcohol was responsible for raising the BAC from 80 mg/100 ml to a higher level, the atmosphere must exceed the concentration of the higher level claimed. Since high concentration of alcohol vapours are unbearably irritating to breathe, the highest BAC ever likely to be achieved under these circumstances would be about 10–20 mg/100 ml.

Occasionally, subjects claim that their exposure to other organic substances, such as solvents, have been responsible for elevated results. Whilst it must be remembered that infrared analysis has never been claimed to be specific for alcohol, and other substances do absorb at the same wavelength, the likelihood of such substances being present on the breath is very limited. Also certain factors need to be taken into account. Firstly, provided a person in a normal workplace is not subject to vapour levels above the maximum permitted levels under Health and Safety recommendation, then significant breath levels will not be achieved. Secondly, if maximum permissible levels are exceeded, the vapour may be absorbed into the blood stream but are also eliminated quite rapidly. Subjects not only have to leave the work environment and drive away before being stopped but must also provide a second positive breath test at a police station some 30 minutes later, by which time the likelihood of significant levels being present is remote. It is fair also to point out that for subjects with a BrAC close to the limit, the option of having the alcohol level determined in their blood is available. Despite all these points, the second generation of instruments are required to detect and indicate the presence of any interfering substance thus negating this possible defence.

13.7   CONCLUSION

This chapter has attempted to provide the salient features of the Road Traffic legislation, case law and body fluid analysis, all of which the forensic scientist must have knowledge of if required to perform alcohol analyses. The interested reader is recommended to read the articles in the bibliography and many other treatises that are available on this subject.

13.8   BIBLIOGRAPHY

F. Blennerhassett, *Drinking & Driving* (Report of the Department Committee), HMSO, London, 1976.

P.G.W. Cobb and M. Dabbs, *Report on the performance of the Lion Intoximeter 3000 and the Camic Breath Analyser Evidential Breath Alcohol Measuring Instruments during period 16th April 1984–15th October 1984*, HMSO, London, 1985.

V.J. Emerson, R. Holleyhead, M.D.J. Isaacs, N.A. Fuller, D.J. Hunt, *J. Forensic Sci. Soc.*, 1980, **20** (1), 3–70.

M.D.J. Icaacs, J.M. Jacobs, V.J. Emerson, G.C. Broster and D.J. Hunt, *Field Trial of Three Substantive Breath Alcohol Testing Instruments*, HMSO, London, 1982.

H.J. Walls and A.R. Brownlie, *Drink, Drugs and Driving*, 2nd edn, Sweet & Maxwell, 1985.

Approved Devices, http://www.homeoffice.gov.uk/ppd/oppu/orders.htm.

CHAPTER 14

The Analysis of Body Fluids

NIGEL WATSON

14.1 INTRODUCTION

The first suspect to have been convicted largely on the basis of deoxyribonucleic acid (DNA) analysis of blood samples was sentenced at the Crown Court at Leicester on January 22nd 1988. This case marks an important milestone in the science of forensic body fluid analysis. Since this case, DNA technology has become commonplace in forensic laboratories around the world and has been instrumental in establishing both guilt and innocence in many court cases. The terrorist attack on the World Trade Centre resulted in a mass disaster victim identification problem on an unprecedented scale and was compounded by the crash of American Airlines Flight 587 on Monday September 12th 2001 with 251 passengers and nine crew in a New York neighbourhood. The huge number of samples for testing required what has been described as 'industrial-scale DNA identification' and use was made of conventional DNA profiling and the latest analytical techniques available at the time. Other applications of particular interest have included the analysis of human remains alleged to be those of the Tsar and Tsarina of Russia, murdered on or around July 16th 1918, and the verification of the death of Josef Mengele, the Nazi doctor at Auschwitz.

Mention of forensic scientific analyses of blood have been recorded in 13th century Chinese texts but the modern science of blood-typing began with a discovery by Karl Landsteiner of the different types of blood, what we now call the ABO blood-typing system. The first account, published in 1901, reported that the blood types present in two week old serum stains on linen could be determined. By 1902 the four blood types, A, B, O and AB had been described.

This chapter describes the chemistry, biochemistry and biology utilised by the forensic scientist in the analysis of body fluid samples

377

(blood, semen and saliva) and hair. Greater emphasis has been placed on the DNA aspects of the discipline to reflect the advances and the interest in this area since 1985. However, consideration has been given to the biochemical tests used to search for blood, to identify stains that might be blood and to identify the species of origin of the blood. Another aspect is the interpretation of the blood splash or splatter patterns observed at many crime scenes to establish the events of the crime.

Consideration has also been made of the immunological and protein blood-grouping methodologies. In many laboratories around the world these techniques are being superseded by DNA analyses. However, the older techniques are still used in many laboratories and they will continue to be of interest because of their use in many important and historic cases.

14.2   BIOLOGICAL EVIDENCE

Blood and other body fluids, and dried stains arising from these, may contribute important physical evidence in three ways. The occurrence of a blood or body fluid stain in a certain position can be of value as evidence. For example, finding a seminal stain on a blanket or bed sheet can support the account given of a crime by a rape victim, or the occurrence of blood on a weapon can substantiate an account of a crime.

The shape, position, size or intensity of a body fluid stain may support a particular account of events concerned with a crime rather than a number of alternative accounts. When a blood drop strikes the surface the shape of the resulting stain will depend upon the angle of impact, ranging from a perfectly circular mark, arising from a perpendicular impact, to an elongated mark where the length of the mark is proportional to the angle between the trajectory and the normal. Therefore, it is often possible, from a collection of blood marks, to determine their point of origination and if there was more than one point source of the blood. This kind of information is clearly of great importance in testing the validity of conflicting accounts of the sequence of events of a crime. The examination of the blood staining patterns as illustrated earlier in Chapter 5 may often yield more useful information than just the biochemical analysis of the blood.

Finally, blood-typing analysis of body fluids and their stains can eliminate whole groups of people as suspects. Forensic blood-typing analysis is comparative so if a stain does not match a suspect, then that person cannot have been the source of the stain material. If there is a match of the blood types then the person is one of a group of people who could be the source of the stain. In any population of people there will be a group who share the blood types or combination of blood types found. The implication that the stain did come from a particular person is stronger if the type or combination of types is rare hence the size of the group is small. If the type or combination of types is common, the group is large, so the possibility that the stain came from some other person, who, by coincidence possessed the same type or combination of types, is more believable.

14.2.1 Blood

Blood constitutes about 7.7% of the body weight of a person and it acts as a transportation system. It is composed of a fluid, plasma, which accounts for 55% of the total volume, with the balance being made up of cells. Plasma is 90% water and is a straw-coloured, almost clear liquid. It contains dissolved plasma proteins, metal ions and organic substances. The plasma proteins are composed of different types including serum albumin, serum globulins and serum enzymes. The globulins are a group of proteins that include the antibodies.

Antibodies recognise and bind to specific chemical groups. They mark foreign materials as being 'non-self' to the host organism and the cells of the immune system assimilate and destroy such marked non-self objects. It is possible to collect and purify antibodies that react to specific chemical configurations. The production of antibodies can be induced in an animal by injecting small quantities of a foreign material An animal may therefore be 'immunised' against specific proteins. The materials to which the antibodies bind specifically are called antigens. Analytical tests, called immunological tests, are used in body fluid analyses and exploit the specificity and binding characteristics of antibodies.

Serum is the fluid left after blood clotting and so differs from plasma in that it lacks the clotting agents. Plasma is prevented from clotting by the addition of an anticoagulation agent, heparin or ethylenediamine tetra-acetic acid (EDTA), to a blood sample as it is collected. The term 'anti-serum' refers to serum used in medical treatments or tests because it contains antibodies with a useful specificity.

As indicated above, the remaining 45% of the volume of the blood is composed of cells. The most common type is the red blood cell, or erythrocyte, which is circular in mammals but oval in other vertebrates. There are about five million cells in one microlitre of blood. The red cells have no nucleus and are the only cells to contain the pigment, haemoglobin, which gives blood its red colour. The outer cell membrane, in common with all other body cells, carries biochemical markers that may be utilised by the forensic scientist.

Another blood cell type is the white blood cell, or leucocyte, and this is concerned with the immunological defence of the body. In contrast to the red cells, the leucocytes are present at much lower cell densities: around five to ten thousand per microlitre. White cells have a nucleus and contain DNA in the form of a protein-DNA complex called chromatin. The chromatin is composed of units called chromosomes.

All human cells, with the exception of erythrocytes and spermatozoon cells, have forty-six chromosomes that include one pair of sex chromosomes plus twenty-two pairs of analogous chromosomes. The sex chromosomes are called X and Y. Males have one of each type and females have two X chromosomes. The chromosomes carry genetic information encoded in the DNA which determines the characteristics of the host. The chromosomes of each pair carry information relating to the same characteristics. One member of each pair is from the host's mother and the other from the father. Although the information relates to the same characteristic, such as eye colour, the two members of each pair of chromsomes may carry different versions of the genetic message. The inheritance of one chromosome out of each pair is random, so that each new baby possesses a shuffled collection of half of the chromosomes available from each parent.

The blood also contains particles called 'platelets'. These are fragments of larger cells in the bone marrow, megakaryocytes, contain no nucleus but have an important role in the process of blood clotting. Clotting, or coagulation, is the process of localised solidification of blood at an injury site and the term 'agglutination' refers to the binding of blood cells together by antibodies.

14.2.2 Semen

Semen is a suspension of cells called spermatozoa, or sperm, in seminal fluid. The sperm cell is the male sex cell, or male gamete, and its biological function is to pass genetic information from the father to the female gamete, the egg or ovum. Human gametes carry a set of 23 chromosomes, one from each of the 23 pairs making up the normal human complement of 46. One millilitre of semen from a normal healthy adult male contains around sixty to one hundred million sperm cells. The average ejaculate is three millilitres within a range of between one and six millilitres. Most of the volume of the semen, 75 to 90%, is made up of material from a group of organs known collectively as the accessory glands. These are the prostate gland, the seminal vesicles, Cowper's glands and the glands of Littre. The prostrate gland produces a slightly acid fluid (~pH 6.4) rich in calcium with also zinc, sodium, citric acid and the enzymes fibrinolysin and acid phosphatase. The seminal vesicles are two glands between the scrotum and the rectum which secrete a fluid rich in fructose, phosphoryl choline, citric acid and ascorbic acid.

Spermatozoa were first discovered by the Dutch microscopist, Antonie van Leeuwenhoek, and reported in a letter written in November 1677, published in the Philosophical Transactions of The Royal Society of London. Normal sperm cells have a head, a midpiece and a tail. Spermatozoon cells can swim, the midpiece acting as a motor causing the tail to move so that the cell can move of its own accord. The head is oval shaped, three to five micrometers in length and one and a half micrometers thick. The midpiece is about seven to eight micrometers in length with the tail being at least forty-five micrometers long. Smaller headed spermatozoa less than four micrometers in length also exist.

Each sperm cell has only half of the usual complement of DNA of a standard cell, however the nucleus represents a much larger proportion of the mass of the cell than standard cells and the density of the cell suspension in semen means that semen is in fact an excellent source of DNA.

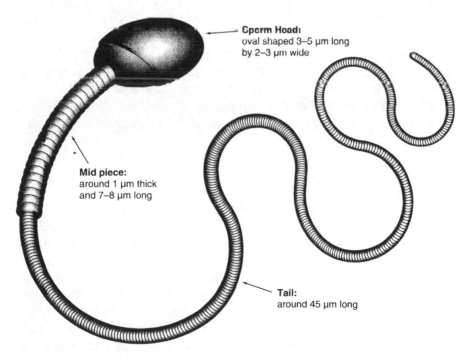

Figure 14.1 *A human sperm cell*

14.2.3   Saliva

Saliva is a watery fluid secreted by the salivary glands of the mouth. It contains mucin that assists the passage of food into the oesophagus, and salivary amylase that digests starches. The saliva also contains cells and hence DNA. These are mainly bacterial cells but saliva also contains cells shed from the inside surfaces of the cheeks.

14.3   TESTS FOR BLOOD AND BODY FLUIDS

The location of blood or other body fluids often requires a 'search test'. For example, blood stains on a dark coloured garment, or dried saliva or semen can often be difficult to detect. To find blood stains a piece of filter paper can be rubbed over an area of the garment and the paper tested by chemical tests for haemoglobin and hence the presence of blood, (see Section *14.3.1*). To find semen stains a dampened piece of blotting paper is applied to an area of a suspected stain. The moist paper will pick up some of the dried seminal fluid from the stain. On spraying with a solution of reactants (see Section *14.3.2*), the paper will change colour in any area where seminal material has been transferred. The colour change is caused by an enzyme in the semen called acid phosphatase. In an adaptation of the test, the presence of saliva may be detected by means of an enzyme component of saliva namely, salivary amylase.

The tests are not entirely specific because other substances, generally of plant origin, may undergo the same reactions as the haemoglobin and salivary or seminal enzymes. These tests are used, even when the stain is visibly quite obvious, to eliminate possible alternatives such as paint or ink or food stains which might be mistaken for body fluid stains.

14.3.1   Tests for Blood

The haemoglobin in the blood of mammals is characteristic of that tissue and occurs in no other. It has the capability of behaving as an enzyme in the presence of hydrogen peroxide, when it can catalyse the oxidation of materials. This property of haemoglobin is used as a biochemical test for the detection of blood.

Certain dyes, *e.g.* leuco malachite green (LMG), have the characteristic of existing in two different states, a 'reduced' form and an 'oxidised' form. Some of these are colourless in the reduced state but brightly coloured in the oxidised state. If a drop of a solution of such a dye, kept in its reduced state, is applied to a test blood stain no colour should be observed. However, if a drop of hydrogen peroxide is added, the colour of the oxidised dye develops. This is a two-step test and is applied to testing stains by rubbing the stained area of an item with a filter paper. A drop of the reduced dye is added to the paper, followed by a drop of hydrogen peroxide. An indication that the stain may be blood can be assumed only if the colour of the oxidised dye is observed after the addition of the hydrogen peroxide. Note that the blood of any individual of any species will produce a positive result and that no information regarding the origin of the blood can be gained. Such tests are termed presumptive tests.

Another example of a dye used in these tests is phenolphthalein. This dye is colourless in its reduced state and pink in its oxidised form. This test is often called the Kastle Meyer, or KM, test. Luminol (3-amino phthalhydrazide) can also be used for presumptive testing. When oxidised, Luminol undergoes a chemiluminescent reaction, *i.e.* a reaction in which light is emitted. If combined with an oxidising agent, sprayed onto a surface and then viewed in darkness, any luminescence will betray the presence of a bloodstain.

14.3.2 Tests for Semen

14.3.2.1 Microscopy. The most definitive test for semen is the microscopical identification of spermatozoa. Due to the relatively large quantities of DNA present in sperm heads and the fact that DNA can be detected if treated with a staining reagent, sperm cells can be readily identified by their distinctive shape when viewed through a high power microscope. Semen is the only body fluid that possesses sperm cells.

14.3.2.2 The Acid Phosphatase Test for Semen. It is possible for semen not to contain any spermatozoa; this is a condition called azospermia, which may be deliberate in the case of a vasectomised male, or may arise due to a medical condition. In the absence of spermatazoa, identification of a sample as semen can be made by testing for the presence of the enzyme acid phosphatase. This enzyme is present in semen at very high levels and can be detected by a reaction with α-naphthyl phosphate in the presence of diazotised ortho-dianisidine. If present, the enzyme produces a purple colour and the degree of the colouration produced is proportional to the quantity of enzyme present. This reaction can be used as a presumptive test for stains suspected of being semen, or as a search test.

The reaction may be used quantitatively in circumstances where a known volume of fluid is to be analysed. An example of this is a vaginal swab where the volume of fluid collected by the swab can be estimated. A portion of the swab can be added to the solution of the test reagents and the intensity of the colour generated can be related to the quantity of acid phosphatase present. Vaginal fluid possesses acid phosphatase activity in the absence of semen but at much lower levels. Elevated acid phosphatase levels are therefore indicative of the presence of semen.

14.3.2.3   Choline and p30.   The choline originating from the seminal vesicles may be detected by microcrystal tests producing characteristic crystals. A more specific test is for a protein called Prostrate Specific Antigen (PSA or p30) that is semen-specific. This protein is made in the prostrate gland and may be detected by a variety of immunological tests. It is useful for confirming the presence of semen in the absence of sperm cells. It should be noted that p30 is also present in male urine and a very simple test for p30 is available commercially and is used for forensic applications.

14.3.3   Tests for Saliva

The salivary amylase present in saliva may be used to identify this body fluid. A specimen of a stain, suspected of being saliva, is removed and placed in a solution of soluble starch. If, after the addition of an iodine solution, a deep blue colour is produced, the stain does not contain any amylase. If amylase is present it breaks down (hydrolyses) the starch with no blue colouration of the solution. Therefore, a positive test for saliva is marked by an absence of colour. A search test can be made by impregnating an absorbent paper with an insoluble starch-dye complex, dampening it and applying to the area being searched on an item. If any saliva stain is present the amylase will release the dye from the complex and this area of the paper will remain colourless.

14.3.4   Determination of the Species of Origin

A question that may well be asked is 'Is the blood human?' To answer this question two different types of test can be employed. One approach is to test for serum proteins that are species-specific and the other is to test for particular DNA sequences that are unique to humans. Most of the former tests rely upon detecting the species-specific proteins by immunological tests that use antibodies produced against the species-specific marker proteins. In the latter case, DNA probes can be made that respond only to regions of the DNA that are unique to humans or certain primates.

In practice, unless it is important to determine which species a non-human blood or bloodstain came from, such tests will not be routinely carried out. The DNA-typing tests, which have become the method of choice, will not yield a result if used on non-human DNA except, perhaps, for certain primates.

14.4 BLOOD-TYPING

In many countries conventional blood-typing has been largely super-seded as a means of human identification in forensic science. However, it is still an important aspect of forensic analyses for several reasons. Firstly, it has been and continues to be used in many countries and secondly, it is very useful in certain instances where a large number of samples need to be eliminated by quick and inexpensive screening.

The identification of blood types has been the subject of continuous development. Major developments have included the discovery of more typing systems that are inherited independently of each other and of the ABO system. Some of these are immunological systems and use specific antisera that can discriminate between the different types of marker belonging to a typing system. Another development has been the dis-covery of protein variants. These can be detected by a technique called electrophoresis. In this technique, a solution containing the mixture of proteins is subjected to an electric field that causes the different proteins to separate. The protein variants are identified by observing differences in their separation.

Once distinct types of blood have been identified, the population can be divided into groups of people who share the same blood types. Many of the blood types can be determined from blood stains and certain blood types can be determined in other body fluids. In the forensic context, a comparison of the blood types of the suspect and the stains can eliminate the suspect as a source of the blood if blood types do not match.

Several independent blood-typing systems can be co-analysed to improve discrimination. As the number of typing systems increases, the number of people who share the same combination of blood types will shrink. Blood-typing therefore, becomes more discriminatory and a coincidental match of a stain with a suspect becomes less likely. The ultimate goal would be to use a large number of typing systems such that no two people would have exactly the same combination of results, thus providing discrimination between the bloods of individual people. However, the quantity of blood stain available may be limited and restrict the number of blood-typing systems that can be used.

14.4.1   Genetics

The study of the inheritance of biological traits is called genetics and one important part of genetics is the study of variations in inherited traits. These variations, called polymorphisms (literally 'many forms'), are where several different versions of a trait co-exist simultaneously within a population. This means that we can classify people who share the same type of trait into groups.

The polymorphic traits which are of most interest to the forensic scientist faced with examining a blood or body fluid stain are those of a biochemical nature. Features such as the sugar level or the concentrations of various metal ions or other components are not usually considered because they are potentially capable of showing variation over time for a given individual. In certain instances, these may indeed be useful but in most cases the analyst seeks to identify and classify features which are determined by the inherited genetic make-up of the individual in question. These inherited features are likely to remain constant and consistent throughout the lifetime of the individual and are independent of the state of health of the donor. Fortunately, many of the inherited biochemical features that can be detected in blood or body fluids are those which are under the influence of a single inherited piece of information, or gene.

The gene is the unit of information which determines a single inherited trait. The location at which a gene is present in the genetic content of a cell of a person is called the 'genetic locus'. Although the biochemical features are generally controlled by a single gene, different versions of that gene giving rise to different versions of the feature may co-exist within a population. The different versions of the gene are called 'alleles'.

Every person inherits one set of genetic information from their mother and another set, analogous to the maternal set, from their father. The term 'genotype' is used to describe the combination of alleles a person has inherited for any given genetic locus. Therefore, all people possess a dual set of inherited genetic information. It is entirely possible that the allele of a particular gene inherited from one's mother is exactly the same as the corresponding allele inherited from one's father in which case the genotype is said to be 'homozygous'. However, it is also possible that two different allelic versions of a gene are inherited. In this case the genotype is 'heterozygous' for that trait.

14.4.2   Immunological Markers

Following its discovery the ABO system was soon put to use for the analysis of blood stains found in crimes and also in cases of disputed paternity. Since that time other immunological systems have been described including Rhesus, MNS, Kell, Duffy, Lewis, Km and Gm. Many have been employed in forensic examinations. The ABO system was the first blood-typing system to be characterised. It was also found that the ABO type could be determined from other non-blood body fluids in the majority of individuals. It has probably been the most widely used of the biochemical polymorphisms. There are four types of blood in the ABO system. These variations occur because of the inheritance of different alleles for a gene, which specifies an enzyme that converts a precursor into either substance A or substance B. There is a third allele, O, which fails to modify the precursor. The pair of alleles which can exist and are their corresponding blood types are as follows:

Alleles	AA	AO	BB	BO	AB	OO
Blood Type	A	A	B	B	AB	O

The precursor occurs on the surface of the red blood cells. It is modified according to the types of enzymes present as dictated by the alleles present. Notice that AO and AA allele combinations both give the same blood type as do BO and BB combinations. Another feature of the system is that type A people have anti-B antibodies in their blood. These antibodies are able to bind to type B red blood cells and cause the cells to clump (agglutinate). Similarly, type B people have anti-A antibodies that will agglutinate type A cells. Type O cells are not affected by either antibody type. It is this agglutination reaction that renders blood transfusion between incompatible blood types dangerous, if not fatal.

Antisera as described above are used for blood type testing. A variety of test methods can be employed with blood or body fluid stains and they all exploit the specificity of binding of the relevant antisera with the ABO markers on red cells.

14.4.3 Protein Markers

Many polymorphic protein systems have been described. Not all of them are suitable for forensic analyses and, of those that are, some do not occur in non-blood body fluids. The most widely employed techniques for discriminating between the variant types are electrophoretic or a closely related technique called isoelectric focusing (IEF). Phosphoglucomutase (PGM) is an enzyme that is present in blood and other body fluids and polymorphism of this enzyme can be detected by using IEF.

Electrophoresis is a technique used very extensively in the analyses of biomolecules. It is based upon the principle that charged particles will move when placed in the electric field between two electrodes in a conducting solution. The rate of motion is called the electrophoretic mobility and it is influenced by the strength of the field (*i.e.* the voltage applied), the charge on the particle and the mass of the particle.

Electrophoretic techniques are usually carried out on a sheet of an open pore gel. The gel is similar to a sponge in its microscopic structure and it is filled or impregnated with an aqueous salt solution. These gels are usually either made from a refined type of starch called agarose or from a polyacrylamide matrix.

The samples to be tested in many forensic cases are dried blood, semen stains or liquid blood samples obtained from people who may be connected with a case being invesigated. Liquid bloods are used to prepare stains on fresh pieces of cotton cloth. This preserves the sample and ensures comparable treatment of the control material from a known donor and any stain from a crime scene.

Extracts of the stains are placed at one end of the gel. All of the samples to be tested, including those from the crime scene, control bloods from possible donors and reference standard bloods of known type kept in the laboratory, are treated similarly. On applying an electric current the components in the samples will migrate along the gel. After a period these will focus into bands. This is because the enzymes are ionic, *i.e.* they carry positive and negative charges, and stop moving when the positive and negative charges cancel out each other. The point at which this ocurs is called the isoelectric point (pI) and is unique for a given enzyme.

After separation the PGM molecules are detected by a reaction mixture applied to the gel. The PGM converts a colourless substrate in the reaction mixture into a coloured product. This detection system is therefore specific to the PGM and although other serum proteins will be present in the electrophoresis gel it is only the PGM that is visualised.

When electrophoresis is used to analyse the PGM content in a sample, three types of the enzyme may be found namely type 1, type 2 or type 2-1. A PGM type 1 individual will produce a single band in the gel. This band is composed of molecules that have migrated together at a relatively high rate. Type 2 PGM individuals are distinguished by a single band which has moved relatively slowly compared with the type 1 band, and a type 2-1 individual produces both of these bands.

A modified form of electrophoresis is iso-electric focusing (IEF). When the PGM types are run on an IEF system the type 1 and 2 bands can be split into '+' and '−' versions. Therefore, there are in fact ten PGM types which can be identified by this IEF technique, *viz.* 1+, 1−, 1+1−, 1+2+, 1+2−, 1−2+, 1-2−, 2+, 2− and 2+2−.

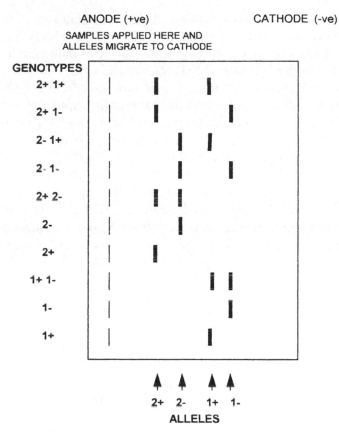

Figure 14.2 *Isoelectric focusing (IEF) results obtained for the ten PGM genotypes*

14.5 DNA AND ITS ANALYSIS

Forensic serology has always made use of techniques developed in other disciplines, from immunology to electrophoresis, genetics and more recently molecular biology. The following is an outline of the nature of DNA and some key techniques that are used in its analysis.

14.5.1 Deoxyribonucleic Acid (DNA)

This is the material that carries genetic information. The DNA exists within a cell in structures called chromosomes. Human cells possess 23 pairs of chromosomes. Human individuals, as do all other animals, originate from the combination of a male gamete (sperm) and a female gamete (ovum). Gametes carry only a half complement of DNA, *i.e.* 23 chromosomes, so the result of the combination of the gametes is a zygote that has a complete complement of DNA, *i.e.* 46 chromosomes.

All the cells of an animal arise from the subsequent cell divisions the zygote undergoes. Therefore all the cells in a person's body have identical DNA. Each one of the 46 chromosomes in human cells includes a single piece of double stranded DNA in which the two strands are wound around each other in the now famous 'double helix'. These double strands are themselves wound around a supporting structure of protein. The DNA is coiled and supercoiled enabling a vast length to be present within the microscopic space of a cell nucleus.

The DNA is composed of a backbone of alternating sugar and phosphate molecules. The sugar is a five carbon ring deoxyribose which is linked *via* the phosphate group from the 5' carbon on one sugar to the 3' carbon on the following sugar. The feature that endows the molecule with its biological significance is the attachment of a chemical group, called a base, to every sugar in the chain. Two classes of base occur in

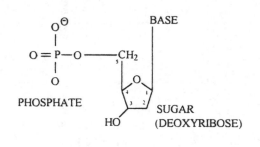

PHOSPHATE

SUGAR
(DEOXYRIBOSE)

THE PURINE BASES

ADENINE

GUANINE

THE PYRIMIDINE BASES

CYTOSINE

THYMINE

Figure 14.3   *The chemical building blocks of DNA*

DNA. One, a pyrimidine, has a six-member ring of carbon and nitrogen. The other, a purine, is composed of fused five- and six-member rings. There are four types of base in DNA: adenine (A) and guanine (G), which are purines, and cytosine (C) and thymine (T), which are pyrimidines. The DNA strand is assembled from individual units, called nucleotides, which comprise a deoxyribose sugar, a phosphate and a base. There are therefore, four types of nucleotide corresponding to the four types of base and these nucleotides can be linked to form a chain (polynucleotide).

The strands of the double helix are held together by hydrogen bonding between the bases on adjacent strands. The cytosine and guanine bases, C and G, form three hydrogen bonds and the T and A bases form two hydrogen bonds. Therefore, the two DNA strands of a pair do not carry the same base sequence but complementary sequences. Another

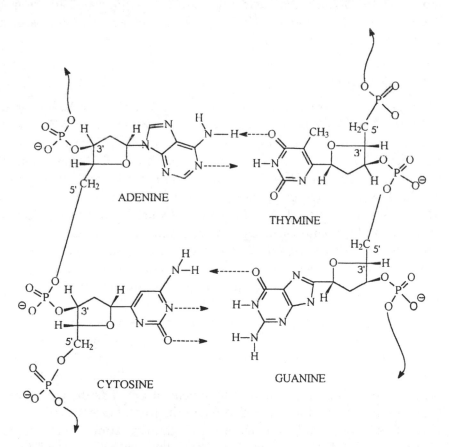

Figure 14.4 *A segment of double stranded DNA showing the links between the thymine (T) and adenine (A) bases, and the cytosine (C) and guanine (G) bases*

feature is that the direction of the backbone of the DNA strands run in opposite directions as defined by the 5′→3′ carbon to carbon linkage. For example the sequence 5′-GGCTATAAT-3′ on one strand would be matched by the sequence 3′-CCGATATTA-5′ on the corresponding strand given the A-T and G-C pairing.

The sequence of the nucleotides on the DNA strand encodes genetic information. This is called the genetic code and it describes the composition of the peptide chains and hence the proteins produced by the organism. As all of the chemical reactions of the body are controlled by enzymes, which are composed of peptide chains, it follows that the entire biochemistry of a person is dictated by the enzymes prescribed by the genetic code. Each amino acid of every peptide chain is determined by a three-base code. The sequence of the codes therefore determines the sequence of amino acids and hence the characteristics of the peptide. When the cell divides, the nucleotide sequence is replicated in the new DNA so that the daughter cells have the same sequence, and so the same code, as the parent cell.

14.5.2 DNA Analysis

DNA has to be extracted from the specimen before it can be analysed. DNA is robust and has been recovered from bones some 5,500 years old. In forensic applications two different strategies of analysis may be identified. The first of these to be used employed a combination of enzymes and DNA probes. The enzymes are restriction enzymes that cut the DNA strands in specific places, giving rise to collections of DNA fragments. These fragments can be sorted by size using electrophoresis and fragments of specific interest detected using DNA probes. The probes are DNA composed of a predetermined sequence of bases in a single strand so they will bind only to their complementary sequences. This analytical approach can be used either for human identification or to detect specific genes.

The second methodology, and one that has been almost universally adopted by the forensic community, makes use of enzymes that replicate the DNA. These enzymes are called polymerases and their function in nature is to replicate the DNA prior to cell division, so that the daughter cells possess the same complement of DNA as the parent cell. They can also be used in controlled reactions to replicate specific pieces of the DNA, over and over again, in a chain reaction. Many copies of that piece of DNA are produced and the process is called DNA amplification.

In some procedures a combination of the two analytical approaches are used. Both approaches require knowledge of the composition of the DNA at the genetic locus to be examined prior to the analysis, but the DNA amplification technique has one particular advantage since significantly smaller quantities of DNA are required for analysis.

Various extraction methods can be used to collect the DNA. For analyses by hybridisation probe it is important that the DNA is relatively undegraded and of high molecular weight. This is assessed by analysing a small quantity in a short electrophoresis gel. Within certain limits, DNA has the property of its electrophoretic mobility being proportional to its length. As a consequence small DNA fragments will move rapidly through the gel and high molecular weight DNA will migrate as a single slow moving band.

An alternative method which is suitable, but only applicable for use with amplification techniques, is the use of a commercially prepared chelating resin, *e.g.* 'Chelex'. If the specimen of stained material is boiled up in the presence of Chelex the DNA is released from the remains of the cells in the stain but remains undegraded in the solution. The Chelex removes metal ions which are required by degregative enzymes.

14.5.3 DNA Probes

These consist of short single strands of DNA, which carry a label, either an isotope or a protein molecule, to facilitate their detection. They are used to detect specific sequences of bases on fragments of DNA created by a controlled enzymic cutting of the DNA. Restriction enzymes, or restriction endonucleases, are enzymes that cut DNA strands at specific base sequences. The recognition sequences are usually four or six bases long. A wide range of enzymes with different recognition sites are available commercially. It is therefore possible to select an enzyme to give a large number of fragments from the human nuclear genome. The human genome is so large that many different sizes of fragments will be produced with most restriction enzymes. In order to identify specific fragments that carry DNA sequences of interest the collection of fragments can be separated by electrophoresis in a gel submerged in a buffer solution, *i.e.* submarine gel electrophoresis. The smaller DNA fragments will be less impeded by the matrix of the gel and so will move more quickly than larger fragments. The DNA is therefore size separated in the gel. However, when the current is turned off, the separation will be lost because the fragments will diffuse through the gel. In order to preserve the separation, the fragments are transferred from the gel to a nylon membrane immediately after switching off the current.

The single stranded DNA probe is used to identify those fragments on the membrane which are of interest. The fragments are also made single stranded so the probe will only bind by hydrogen bonding between complementary bases. The probe can be made of any desired sequence and will only bind where there are a number of bases in sequence in the DNA which are complementary to that of the probe. The probe is called a hybridisation probe because the new double stranded piece of DNA formed is a hybrid between the fragment and the probe. The process of binding between the probe and the DNA restriction fragment is called hybridisation.

Another feature of probes is that they are labelled so that they can be detected. Originally the radioisotope of phosphorous (^{32}P) was used because of the sensitivity required to detect the small quantities of DNA present. This has now largely been superseded by the use of non-isotopic detection systems and these overcome the radiation hazards of using isotopes. The non-isotopic systems utilise a dye or a protein bound to the probe.

14.5.4 DNA Amplification

The replication of a selected DNA sequence present in a sample of DNA is carried out by the Polymerase Chain Reaction (PCR). It results in an exponential increase in the number of short DNA fragments which carry the sequence of interest. The increased copy number makes many analyses possible from small samples. The PCR is carried out in a reaction mixture composed of the various components in an aqueous solution in a small plastic centrifuge tube, typically in 50 μl volume. The reaction is controlled by changing the temperature and there are three stages: denaturation, annealing and extension as illustrated in Figure 14.5. The temperature is controlled by means of a programmable thermal block called a thermal cycler. The reaction tubes are placed in the block which is heated for the first stage to around 94 °C to denature the double stranded DNA isolated from the specimen. This denaturation consists of breaking the hydrogen bonds, which hold the two strands together, so rendering the DNA single stranded. The temperature is then lowered, typically to 55 °C for the annealing stage and raised to around 75 °C for the extension phase.

The reaction components consist of an enzyme, *Taq* polymerase, which builds the DNA strand, the nucleotides which are the building blocks of the DNA and the primers. The reaction mixture also contains DNA which has been recovered from the specimen. The specimen may be a blood sample taken under clinical conditions from a suspect or victim, a blood or body fluid stain, or it may be any kind of body tissue. Hair, especially the hair roots, and mouth swabs which collect a few cells from the inside of the cheeks are good sources of DNA and involve non-invasive sampling techniques.

The amplification process operates only on a small portion of the DNA at a time. The specific target which is to be amplified is marked out or defined by a pair of short sequences of DNA which have been made synthetically and which are usually about 20 base pairs long. They are called DNA primers or oligonucleotides. Two different primers are used. The base sequence of one is selected to bind to one side of the target DNA sequence and the other primer binds to the other side of the target sequence. The target sequence of DNA bases which lies in-between the two priming sites is used as a template for the amplification and is referred to as the template DNA.

The reaction mixture contains each of the four types of nucleotide and the action of the polymerase enzyme is to add nucleotides to the 3′ end of the primers when they are bound to the template strand. The type of nucleotide added to the primer is determined by the base on the template. Note that it is the complimentary base which is added so that base paring can occur between the extended primer and the template. The polymerase will continue to add bases to extend the primers until it is interrupted. When the primers are extended this happens by the addition of bases to the 3′ site and the extension will proceed past the target site and the binding site of the partnered primer.

The process of primers binding to their respective binding site is called annealing and the process of extending the primers along the template is called the extension phase. After one complete cycle of denaturation, annealing and extension the processes may be repeated. For the second cycle there should now be double the number of templates available because the new strands created from the last cycle will themselves act as templates for the subsequent cycle.

Notice that when the new strand acts as a template the 5′ end terminates with the primer and does not continue on as the DNA isolated from the specimen would. As a consequence the extension will stop at the end and so the new DNA produced in this case will be bound on one side by the primer which has been extended and at the other end by a base sequence complimentary to the partner primer. The analogous (but 'opposite') case will be true for the other alternative product. These new PCR products are called short products and are defined by either a primer sequence or a sequence complementary to a primer at either end. They are amplified preferentially in the reaction because they are short.

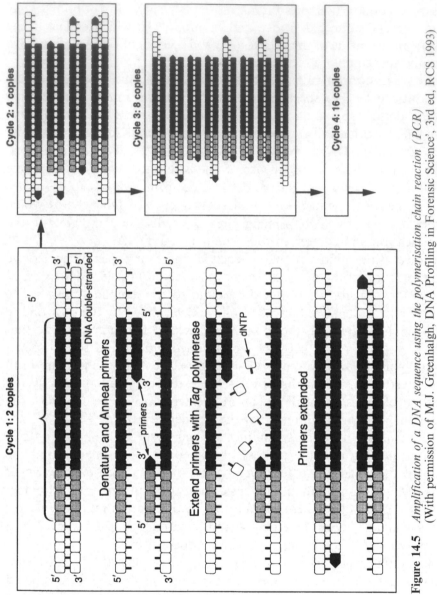

Figure 14.5 *Amplification of a DNA sequence using the polymerisation chain reaction (PCR)*
(With permission of M.J. Greenhalgh, DNA Profiling in Forensic Science', 3rd ed, RCS 1993)

Every time the reaction cycle is repeated the number of template strands should be doubled. In theory the quantity of the target site should therefore increase geometrically. However, in practice the full potential is not realised because the activity of the enzyme reduces a little with repeated temperature cycling and the number of templates for amplification increases. There are also a number of other factors which affect the number of templates produced but overall a considerable increase in the number of copies of the target site is actually achieved. The base sequence at the target site may be regarded as a signal, hence the term 'amplification', since an increase of the signal has been achieved.

14.6 FORENSIC DNA ANALYSIS

Results from the analyses of DNA in forensic samples are providing considerable evidence to help the courts and valuable information for the police in conducting their enquiries. The successes recorded have prompted major research and development programs to introduce or improve DNA analysis techniques. Whilst many types of DNA variation, or polymorphism, have been discovered, only four, the mini-satellite, microsatellite, DQA and mitochondrial DNA systems will be described here, since these are of most interest to the forensic scientist.

The blood-typing systems discussed so far, cell marker or protein based, are manifestations of polymorphisms in the DNA. However, the range of the variation possible at each of the genetic sites is limited by the need for the gene product to be a viable working molecule. Another factor limiting the power of discrimination is the need to conduct a separate analysis for each polymorphic system. This requires a different procedure for each system, often different apparatus, a separate set of reference controls to be maintained and a good deal of labour and time to conduct a comprehensive analysis. However, the greatest limiting factor is sample size and with many forensic samples there is usually insufficient material to perform a large number of blood-typing tests. This limitation has been largely overcome by the DNA analyses currently in use.

The four strategies developed and mentioned earlier may be divided between site or length polymorphisms. The length polymorphisms used vary in terms of the number of times a sequence of bases is repeated at a defined site in the genome. One category of length polymorphisms, the minisatellites, was the first ever to be employed in a criminal case. The other type of length polymorphism is called a microsatellite and this may be analysed using DNA amplification. Minisatellites and microsatellites are comparable in the sense that they are composed of repeating units of base sequence, but they are quite separate entities.

Site polymorphisms are those where the different alleles are all about the same size or length but differ in the base sequence at a defined site in the genome. DNA amplification is central to the analysis of site polymorphisms in forensic cases.

14.6.1   Minisatellites

The first ever use of the direct analysis of DNA in a criminal case employed minisatellite technology. The case required the screening of blood samples from several thousand males from the village of Enderby in Leicestershire and its locality, in the hunt for the murderer of two schoolgirls who had been assaulted and killed. The conventional police investigation had yielded a single suspect who had admitted to the crimes but was not a convincing suspect. This resulted in the police approaching Professor Alec Jeffreys and his team at Leicester University to employ their newly reported technique of 'DNA Fingerprinting' to help solve this crime. The man eventually convicted, Colin Pitchfork, had attempted to avoid supplying a blood sample by offering to pay a workmate to impersonate him on the pretence that Pitchfork had committed some other crime. The workmate had discussed this with his girlfriend who clearly saw the true situation and informed the police. Subsequent analysis of Pitchfork's DNA confirmed that it was he, and not the previous suspect, who was overwhelmingly the most likely donor of the seminal material recovered from one of the victims.

The minisatellite technology employed in the above case made use of length polymorphism – a Fragment Length Polymorphism (FLP). FLP can arise from the substitution of one base with another either creating or destroying a restriction enzyme recognition site. The creation of a new restriction site will cause the formation of two smaller fragments where there had previously been a single larger fragment. The destruction of a restriction site would have the opposite effect. Such variation may be described as a Restriction Fragment Length Polymorphism or RFLP. The differences in the length of the DNA fragments can be used as the means of separating and identifying the different alleles.

However, in complex genomes such as human nuclear DNA, these would be very difficult to detect amongst the many restriction fragments typically generated. In addition, such a polymorphism can only consist of two alleles, so the restriction site either exists or does not exist. In order to achieve useful discrimination for the forensic scientist many such polymorphisms would have to be analysed, adding to the complexity of the test.

An example of an RFLP is a tandem array of repeated DNA sequences between restriction sites. For example, consider a sentence as a sequence of letters that is analogous to a DNA sequence. If the word 'base' is repeated in tandem the sentence could be:

'The length polymorphism takes the form of a section of the **base base** sequence which is repeated in tandem.'
or
'The length polymorphism takes the form of a section of the **base base base** sequence which is repeated in tandem.'

The word 'base' in the example above represents a 'repeat sequence' or a 'motif'. The motif may consist of a few bases or over a hundred bases. The number of times it is repeated can be from one copy to hundreds of copies. Thus the length of a fragment of DNA which possesses such a variable site and which lies between identifiable fixed sites can vary enormously between chromosomes.

The name 'minisatellite' is used for these arrays of tandemly repeated sequences. These types of length polymorphisms are also called Varaiable Length Tandem Repeats (VNTR). Since there are a large number of copies of the repeat sequence it is possible to achieve a high power of discrimination from this type DNA polymorphism analysis.

A DNA probe developed by Professor Jeffreys and his group recognised sequences belonging to length polymorphisms at a number of different genetic loci. It was a Multi Locus Probe (MLP), which responded to several minisatellites simultaneously. The probes produced a complex pattern of bands. The pattern was described as being a DNA 'fingerprint' because the chance that some other person's DNA giving the same pattern is very small.

MLPs were the type of probe used initially in forensic science. However, probes that are specific for only one of the minisatellites, *i.e.* Single Locus Probes (SLP) can also be used. The use of SLPs eventually became widespread because they work with smaller quantities of DNA, and generate simpler and more easily interpreted results – either two bands in the case of heterozygotes or a single band with homozygotes. Good discrimination can be achieved when combinations of several different SLP's are used. The process for using restriction enzymes and SLP hybridisation for DNA-typing is shown in Figure 14.6.

The band patterns generated from the DNA in different forensic specimens can be compared with each other and the control or reference specimens obtained from suspects or complainers involved in a specific case. If the complex band patterns generated by an MLP match between two DNA samples then it is highly likely that the two samples

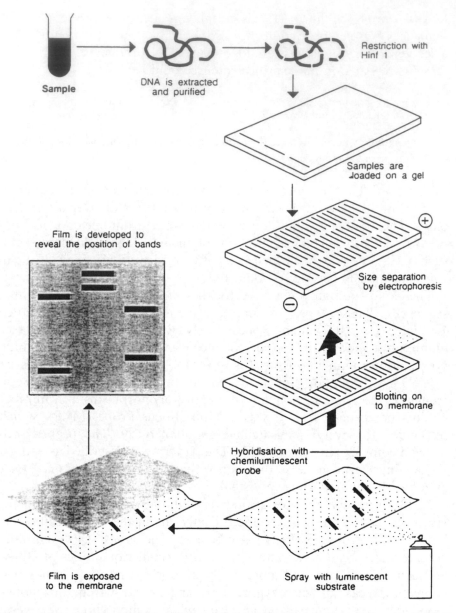

Figure 14.6 *The process for DNA-typing by use of restriction enzymes and SLP hybridisation*

originated from the same person. The chance that such a match has occurred by coincidence can be calculated, but the calculation may be complex and may make use of certain assumptions and simplifications.

Many minisatellite alleles are too long to amplify easily. They also contain a large number of repeat sequences which differ considerably in length and if amplified, the PCR technique will amplify the shorter repeat sequences preferentially – a phenomenon called 'allele drop out'. They have therefore been superseded by PCR-based techniques that offer simpler, more readily automated procedures and which can be used with much smaller samples.

14.6.2 Microsatellites

Microsatellites, or Short Tandem Repeats (STRs) consist of tandem repeats of sequences of two to five base pairs. This type of variable DNA is very suitable for analysis by a combination of DNA amplification by the PCR and the analysis of the amplification products by automated instruments. Thousands of STR regions have been discovered in the human genome but only certain ones have been selected for use in forensic human identification. Depending upon the STR locus, the number of repeat units in the STRs used for forensic work range from eight to almost 30. Different alleles are distinguished by the size of the PCR amplification products generated by a given set of primers. The following example illustrates an early application of STR analysis and concerns the DNA recovered from the bones of a man buried under the name of Wolfgang Gerhard. The bones had been recovered from a cemetery at Embu in Brazil in 1985. Gerhard had drowned in a swimming accident in February 1979 and it was alleged that the deceased was in fact Dr Joseph Mengele, the 'Angel of Death' of Auschwitz who had escaped to South America at the end of the Second World War. DNA from the bones and DNA from Mengele's wife and son were compared by analysing the PCR fragments generated from ten STR genetic loci. The results were consistent with the deceased having been the father of Mengele's son. The chance of another unrelated person possessing genotypes consistent with this parentage was estimated at 1 in 1800. In the light of this, and other evidence, the German authorities have closed the case.

The use of the amplification technique ensures the sensitivity of the analysis. In addition it is possible to label the products of the DNA amplification. This amplification and labelling procedure is incorporated in automated DNA analysis instruments. The label used is a fluorescent dye that is chemically bound to one member of the primer pair used for the STR amplification. The dye labelled PCR allele is detected in a gel and since alleles contain different numbers of repeat sequences, they display different electrophoretic mobilities. The migration distances of the fluorescently labelled PCR products in the gel therefore reveal which alleles are present in the DNA sample and are hence used for the comparison of samples.

The introduction of automated instrumentation now provides rapid, high volume throughput analyses of PCR products. The automated instruments are of two types and the first type to be produced was an automated gel reader that detects fluorescent dye label PCR products seperated by electrophoresis in gels. As a further refinement, it is possible to use a combination of fluorescent dye labels, each possessing a characteristic fluorescence emission wavelength. This enables a mixture of products from different PCR reactions to be monitored. By using a computer to control and record the operation, the relative positions of each PCR product can detected across each track on the gel. An example of an STR profile obtained after electrophoresis of DNA samples is shown in Figure 14.7.

The need for a compound that exhibits the desired fluorescent behaviour and also possesses a chemistry compatable with conjugation to a DNA strand restricts the number fluorescent dyes that are available. One of the dyes, called Rox, is reserved to label a size marker calibrant. This is a collection of DNA fragments of known sizes and is used to calibrate the system so that the size of the PCR products may be determined accurately. This leaves other dyes available for the labelling of the PCR products. However, if the STR loci and the primers used are selected so that the range of allele sizes of one STR locus never overlaps that of another locus, the same dye may be used for two or more STR loci in the same lane. By means of such combinations of size range differences and fluorescence properties of the dyes it is possible to accommodate many different STR loci in a single electrophoretic separation.

A number of different configurations of genetic loci and dye label allocations have been developed. In the United Kingdom three configurations, using four (the Quad), six (the Second Generation Matrix (SGM)) and ten (SGM+) different STR genetic loci have been used. The Quad has been superseded by the SGM and SGM+ configurations since the latter give superior discrimination. These tests also include the amplification of a marker that indicates the sex of the donor of the biological stain. This marker, called Amelogenin, AMG, is located on both the X and Y sex chromosomes. The copy of the AMG marker on the X chromsome is six base pairs shorter than the AMG marker on the Y chromsome. In humans the sex of an individual is determined by the sex chromosomes. Females have two X chromosomes and no Y chromosome. Males have one X chromsome and one Y chromosome. Thus a PCR product generated from the AMG site of a person's DNA will reveal the sex of the donor: males yield two peaks, representing two different sizes of product, and females give a single peak because both of their AMG sites are the same size.

GEL ELECTROPHORESIS OF SAMPLES

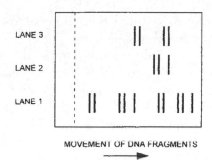

MOVEMENT OF DNA FRAGMENTS

ELECTROPHEROGRAM

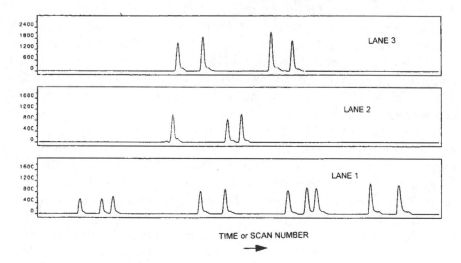

TIME or SCAN NUMBER

Figure 14.7 *Examples of STR profiles (electropherograms) generated from the electro-phoresis of DNA samples. Lane 1 – size marker calibrant; Lanes 2 and 3 – DNA samples*

The ability to be able to carry out amplification on more than one DNA sequence in a single reaction has been an important development because it permits several different regions of the sample DNA to be tested at the same time. This type of reaction is called a multiplex PCR and it exploits the fact that the primers all bind independently of one another and that the final PCR product incorporates both primers for a specific amplification site. Thus by careful selection of the dye types used and the range of lengths of the PCR products, the mixture of

products can be resolved and the DNA locus of each identified. The
result obtained is called a DNA, or STR Profile.

As mentioned above the United Kingdom forensic science laboratories use one multiplex reaction called the SGM+. The STR loci tested
by the SGM+ multiplex are always the same so that results can be compared between laboratories and hence the frequencies of the various
alleles can be compared between populations. Not all jurisdictions use
the same multiplex. The international police organisation, Interpol,
has recommended a standard set of seven STR loci, all of which are
included in the SGM+, plus the AMG locus as the minimum for a
profile to be included in a collection of STR profiles maintained by
Interpol. In the United States the FBI have specified a fourteen
loci multiplex called the Combined DNA Index System, CODIS, that
consists of thirteen STR loci and the AMG locus.

These standard multiplex systems are used by national police and
justice agencies to establish databases of DNA profiles. In essence the
profiles generated from biological evidence found in connection with
crimes, including unsolved crimes, are collected. When a suspect for a
crime has been apprehended the STR profile of that suspect can be
compared to those on the database to check for links with other crimes.
The establishment of standard multiplexes has led to the production of
commercially available kits.

In addition to the STR regions used in the multiplex systems such
as those described above there also exist STR regions that occur on
the Y chromosome. These YSTRs enable a DNA result to be obtained
from only the male contribution to a mixture and thus can have a
particularly useful role to play in, for example, sexual assualt cases.

A third strand in the evolution of the present DNA-typing technologies has been the refinement of the combination of the amplifications
of different loci of the DNA as demonstrated by the increasing numbers
of loci addressed by the multiplex kits. The loci used in any collection
should be sufficiently variable that they contribute to the discriminatory
power of the combination of loci. Their inheritance from parent to off-
spring should be independent of each other so that a person possessing
a particular combination of alleles at one microsatellite locus is not
more likely or less likely to possess a particular combination at another
loci included in the collection. As the degree of variability and the DNA
base sequences in the DNA flanking the loci of more microsatellite loci
have been characterised the number of suitable loci available for inclusion has increased. There are many advantages to using profiles of multiple loci. These include reducing the chance that two people share the
same profile, improving the discrimination between the profiles of close

blood relatives and obtaining useful information from profiles that are incomplete because only very small or degraded samples were available.

14.6.2.1 Instrumentation for the Automated Analysis of STR Multiplex Kits. The dye label chemistry is in a process of constant improvement with the consequence of requiring certain changes in the detection systems and the introduction of more advanced instrumentation. The new instruments also have incorporated improved electrophoretic separation technology. New dyes have resolved certain technical limitations and have enabled a greater range of distinct dye labels to be available. It is presently possible to use five different types of dye label, each possessing distinctive emission spectra so that the PCR products can be readily detected and distinguished from each other.

In parallel with the above a key development in separation technology has been the introduction of a range of automated electrophoresis instruments that use narrow glass or silica tubes. This type of separation, called capillary electrophoresis (CE), has many advantages but one is due to the greater heat dissipation possible from within a narrow column of separation media compared with that of the flat slab gels used in previous instruments. This allows much greater voltages to be used across the capillary than the gel resulting in much more rapid separations without overheating. CE is appropriate for certain requirements, especially where a rapid result is required from a small number of samples. On the other hand slab gels are more suitable where there is more time available for results and large numbers of samples need to be tested as cheaply as possible.

14.6.3 DQA and Polymarker

The letters 'DQA' designate a collection of alleles that form a part of a large and complicated polymorphic system called the Human Lymphocyte Antigenicity or HLA system. Unlike other DNA polymorphisms used in forensic analyses, the HLA system is determined by genetic loci that are part of the active coding position of the DNA. Prior to the widespread adoption of the STR-based DNA analyses, a test kit that combined the PCR and hybridisation probe technologies had been produced by a commercial company. The process involved the amplification of specific regions of the DNA in a sample that were located within the HLA gene complex. These amplification products were made using primers that carried a protein label and were applied to a membrane strip on which were a series of spots. Each spot contained DNA probes to different varieties of the HLA alleles. The probes were fixed to the strip and would bind and hence capture amplification products that possessed the complimentary DNA sequence. Thus the allelic variants present in the sample could be determined by detecting which spots carried the protein label and hence an amplification product.

The kit was a very effective means of testing forensic samples but it lacked the discriminatory power of the single locus probe approach especially where a battery of probes were used. The kits were supplemented by a second kit called the Polymarker Kit that tested more genetic loci and incorporated a simple minisatellite locus called D1S80. The kits proved to be a very useful means for many forensic science laboratories to introduce DNA analysis but have been superseded by the STR approach.

14.6.4   Mitochondrial DNA

The techniques described so far have been for the analysis of the genomic DNA found in the nucleus of a cell. Within a cell there are other components (cell organelles) and one of these is the mitochondrion which is associated with the generation of energy within the cell. Between 100 and 1000 mitochondria may be found in one human cell and each one contains two or three copies of its mitochondrial genome, its own DNA. Each cell will therefore contain hundreds of copies of the mitochondrial DNA, compared to only one copy of the nuclear DNA.

Within the cells of a human all the copies of the mitochondrial DNA have the same sequence. The mitochondrial DNA consists of a loop of double stranded DNA of 16,596 base pairs. The entire sequence of this loop was reported in 1981 in a paper in *Nature* by Anderson *et al.*, and hence it is called the 'Anderson sequence'. In contrast to the nuclear DNA, almost all of the mitochondrial DNA sequence is used for coding and only a relatively small portion called the 'D' (diversity) loop, or the 'control region' has no specific coding function. The control sequence includes a point at which mitochondrial DNA replication always starts. This is called the origin of replication and the Anderson sequence of bases is numbered starting at number 1 at the origin and onwards up to 16,596. Two regions have been found to exhibit a high degree of sequence variation between individuals. These regions, called hypervariable regions (HVR) 1 and 2, occur within the control region.

The mitochondrial genome has been shown to mutate at a rate of five to ten times faster than the nuclear genome. This has been attributed to a combination of the absence of a 'proof-reading' of repair mechanism in mitochondrial DNA replication and a poor fidelity of the mitochondrial polymerase in its action. Another factor that must be taken into account in the use of mitochondrial DNA for forensic analysis is the fact that the mitochondrial genome is maternally inherited, *i.e.* we all inherit mitochondrial DNA from our mothers.

The analysis of mitochondrial DNA is therefore of greatest benefit where the sample size is greatly restricted so that the occurrence of multiple copies of mitochondrial genome per cell is clearly advantageous, and also where family relationships through the maternal line are important. As the mitochondrial genome effectively consists of a single chromosome, and there appears to be no mechanism for crossing over between different mitochondrial genomes, HVR 1 and HVR 2 cannot be regarded as separate unlinked genetic loci. The entire genome must be considered as a single genetic locus. As such it dose not have the discriminating power of a combination of nuclear genetic loci.

The usefulness of mitochondrial DNA analysis has been demonstrated by its application to the resolution of the authenticity of a collection of bones, alleged to be the remains of Tsar Nicholas II, the Tsarina, their three daughters, the family doctor and two servants. These individuals were reputedly murdered in 1917 during the Russian Revolution and this act has remained a contentious issue in Russia to this day. Various accounts of the disposal of the remains had been recorded and there was also the question of the authenticity of various individuals claiming to have escaped or survived the murders. Two amateur historians made a discovery of a collection of bones and other remains at a site around 20 miles from Ekatrinberg. The nature of some of the artefacts recovered with the buried remains, and the number and nature of the skeletons in terms of age and gender was consistent with them being the remains of the murdered Tsar, his family and entourage. In a collaborative exercise between the Engelhardt Institute of Molecular Biology, Moscow, and scientists of the British Forensic Science Service, a DNA analysis was made of the skeletal remains.

Both STR analyses and mitochondrial genome sequencing were employed. The STR analyses were conducted over five loci and it enabled a tentative distinction of a family group of blood relatives, those with genotypes consistent with a mother, a father and three children. The mitochondrial DNA analysis was of particular importance because individual skeletons could be linked to the maternal lineage of the Russian royal family through the mitochondrial DNA. This was possible because the authenticated living descendants of the Russian royal family maternal lineage, including HRH Prince Philip Duke of Edinburgh, could be traced and the genealogy of the European royal families is well documented. Paradoxically, the skeleton identified tentatively as being that of the Tsar matched the mitochondrial HVR sequences of contemporary living maternal descendants perfectly except for a single base mismatch. Such a high degree of match, when taken in the context of the other supporting evidence, strongly supports the authenticity of the claims for the remains but yet is not the perfect match. It is clear that the true significance of the single base mismatch can only be interpreted by taking mutation rates into account.

Many forensic laboratories now take advantage in refinements of DNA sequencing technology to test selected portions of the mitochondrial genome by directly sequencing the DNA. Whilst mitochondrial analysis is not a replacement for STR analysis it has an important application where the samples are highly degraded and where the maternal inheritance can be used in an investigation.

14.7 BIOLOGICAL EVIDENCE IN COURT

Part of the forensic scientist's role is to present the results of an analysis to a court and, where need be, interpret the significance of the findings. The interpretation of the results obtained from the chemical analyses of body fluid stains are largely self-evident and mostly non-controversial. The area that has been the subject of the greatest controversy has been the interpretation of the results obtained from the genetic analyses of body fluids.

When the best methods available were those of the immunological and protein polymorphisms, the diversity of each grouping system was sufficiently limited that in relatively short periods an operational forensic laboratory would have observed all or most of the genotypes. The scientist could estimate the genotype frequencies by counting them. With most of the DNA-typing systems, however, the larger numbers of allelic alternatives at the polymorphic loci create the situation where a laboratory will not have observed all the possible allele combinations, or genotypes, until extremely large numbers of samples have been tested. Therefore, the analyst is unable to quote precisely a genotype frequency and inform a court as to how great is the chance that there might be a match in the DNA for two unrelated people.

The current practice is to estimate the frequency of the genotype by calculations based upon the frequencies of alleles that have been observed. Using this approach, the analyst can quote a value of perhaps one chance in millions of a coincidental match, even when millions of people have not been tested. Problems could possibly occur in making this estimate because certain assumptions concerning the behaviour of human populations have to be made, primarily in that a given human population has developed by random mating, and that there is no 'structure' in the distribution of alleles throughout the population.

Structure in this context would be groups of people who work and live within a larger population but who marry and have children only within their own group. An allowance is therefore made for these and other factors to ensure the reliability of the estimates for the uniqueness of the genetic-typing being presented to a court.

In most court cases, where DNA evidence is introduced, a questioned sample, *i.e.* one whose origin is unknown, yields a DNA profile that matches that of the suspect. The question therefore arises that this match might not be due to the suspect being the origin of the questioned sample, but that the suspect possesses the same profile of that of the questioned sample by chance. It is part of the responsibility of the expert witness to guide the court as to how likely and hence how credible this explanation of the match is. Clearly an estimate of how common the profile in question is among the relevant population can serve to indicate what the chance of a coincidental match is. The more frequent the profile is, the more credible a coincidental match. If only a single location in the DNA is considered then the genotype found will be shared by a group of people in the population. If a combination of genotypes is considered the frequency of the combination will be smaller, with the proviso that the genotypes are inherited independently of each other, that is the occurrence of a given genotype at one locus does not predispose a genotype found at an other locus.

The discriminatory power of the combination of ten STR loci of the SGM+ is such that, except in special circumstances, the frequency of any STR profile will not be greater than one in one billion and indeed will often be very much smaller. Many scientists will now simply report the frequency as being not greater than one in a billion. In some instances the evidence may be reported in terms of a 'likelihood ratio'. Very simply this is a comparison of two competing hypotheses. The prosecution hypothesis is that the observed match between the DNA profiles is because the accused is the origin of the questioned sample. If this is so then the chance that questioned and reference samples have matching profiles can be said to be one (on a scale from zero, *i.e.* absolutely no chance, to one, *i.e.* absolute certainty).

The rival hypothesis is that the accused possesses the matching profile by coincidence. The chance of this being the case can be estimated as the frequency of the occurrence of the profile in question. For simplicity suppose the profile had a frequency of one in ten, *i.e.* 0.1. If we take the ratio of the chance of the match occurring because the accused is the origin of the questioned sample, *i.e.* one, to the chance that the match arose because the suspect coincidentally possessed the profile, *i.e.* 0.1, then 1/0.1 is 10. This value of ten is the likelihood ratio. It means that it is ten times more likely that the match between the questioned sample and the reference sample from the accused occurs because the accused is the origin of the questioned sample than if it came from some other person. Notice that no account has been taken of any other evidence than the DNA tests. The DNA-testing should be considered within the context of the rest of the case. Nonetheless, the development of these very highly refined and powerful technologies has resulted in an extremely powerful tool for justice.

Certain difficulties do, however, exist. One of the most difficult circumstances is where a piece of biological evidence is composed of a mixture of body fluids from different people. Where there are only two contributors, and with one having a greater proportion of the DNA than the other, then it may be easy to allocate big peaks in the electropherogram to the dominant contributor and the smaller peaks to the subordinate with some confidence. However, there are many complicating circumstances, for example, the profiles of one or other of the contributors may be incomplete or there may be more than two contributors. Another difficulty is the existence of artefacts in the test results. These artefacts arise from the technical aspects of the PCR process and produce very small peaks in addition to the alleles detected in the electrophoresis. In unmixed samples these are very readily recognised for what they are and can be easily discounted. However, where mixtures are possible it can be a difficult task to discriminate between an artifactual result and a contribution from a very weak component of a mix. The sophisticated instrumentation in combination with the highly refined design of the amplification kits has reduced the number of instances where mixtures are truly problematical but there will always be difficult examples.

14.8   DEVELOPMENTS IN DNA TESTING

DNA testing procedures continue to be refined and developed and some of the areas of interest are highlighted below.

14.8.1   Low Copy Number PCR

This describes a combination of refinements of the existing test matrices to improve the sensitivity of the test to the point where the DNA from a single cell can give a result. These high sensitive analyses are known collectively as Low Copy Number (LCN) amplification and the term arises from the low number of copies of the genome present in such small samples. The use of CE for the amplification product analysis has also improved sensitivity.

The interpretation of the results of LCN requires great care. Firstly the danger of contamination is much greater. The sensitivities of these techniques is so great that very tiny amounts of contaminating DNA from investigators, or from DNA in cellular debris present on a substrate before any stain has been deposited can yield alleles. As a consequence of the amplification being taken to its limit, the major and minor components may not be readily distinguishable.

At the limits of the sensitivity of the test, alleles may be detected in some replicates of the test and not in others. This phenomenon is called allele drop-in and arises when very few potential template molecules are present in a PCR reaction tube. If one allele is successfully amplified prior to another over the very initial cycles then that allele's amplification will prevail over the other to the extent that only the prevalent one is detected. Upon a repeat of the test it is not necessarily the same allele that prevails and is hence detected. This lack of reproducibility introduces a greater degree of uncertainty than in conventional test interpretations and therefore requires the alleles to be observed in more than one amplification. However, the number of times an allele needs to be observed in separate amplifications will depend on the type of sample and may be open to debate.

14.8.2 Mass Spectrometry

Techniques have been developed whereby PCR products can be embedded within a crystalline matrix consisting of small organic compounds. The products of different PCRs can be thus embedded in an array. A short pulse of laser light is then used to volatilise and ionise the DNA and the matrix. This process is called 'matrix assisted laser desorption/ionization' or 'MALDI'. The ionised DNA molecules are therefore in the gas-phase and can be analysed by mass spectrometry (MS). The size of the DNA amplification products of STR loci will depend upon the number of tandem repeats and consequently so will the atomic mass of the fragments. The molecular weight will influence the time-of-flight of the ions and therefore the alleles may be inferred. The technique is called matrix assisted laser desorption/ionisation time-of-flight mass spectrometry or MALDI-TOF-MS.

The advantages of this technique include the very rapid analysis time, seconds for each sample, and the highly accurate measurement of mass thus eliminating the need for allelic ladders. Furthermore, no fluorescent labels on the primers are required.

14.8.3 Trait Identification

The accumulation of knowledge of gene function has provided, and will continue to yield, genetic information that might be used actively in the investigative process to predict the appearance, or mental proclivities of individuals, rather than retrospectively to link evidence with suspects once they have been located and apprehended. To some extent this is what criminal intelligence databases set out to achieve. At present only the sex of the donor is predicted. As understanding of the interaction of the relevant genes progresses we may expect traits such as eye, hair and skin colour, colour blindness and other less common traits to become predictable from genetic predispositions that are inferred from the DNA in evidence samples.

14.8.4 DNA Microarray Technology

This is a very flexible technology, developed primarily to enable rapid co-testing for a large number of sequence variants. It can take a number of forms but broadly speaking most of them follow the following general pattern. The tests are carried out on glass plates, commonly about the size of a glass microscope slide onto which DNA sequences, called 'targets' are immobilised. These immobilised strands are arranged into groups of like strands immobilised into specific locations making up an array of spots in predetermined patterns. Independently, samples of DNA, called probes, which come from the specimen being tested, are labelled with fluorescent dyes. The immobilised targets are exposed to a solution of the probes and where there is complementarity between these, hybridisation occurs. The array is scanned and those spots that fluoresce, *i.e.* have hybridised a fluorescently labelled probe, are detected and their positions within the array recorded.

A large part of the effort to develop these devices is to conduct gene expression analysis in the course of drug discovery. However, large numbers of nucleotide site polymorphisms are known to exist. These consist of a site in the DNA at which the type of base present can vary. This is called a Single Nucleotide Polymorphism (SNP). These can be readily amplified and the amplification products detected on DNA microarrays. Potentially many thousands of SNPs can be tested simultaneously by microarray devices.

14.9 CONCLUSION

The continued refinement of the existing DNA technologies is certain to continue. A more daunting prospect is the possibility in future to identify traits related to mental stability or the predisposition of individuals to violent acts. The increasing understanding of the links between a person's genetic inheritance and the influences this may have on their behaviour, coupled with the new technologies currently under development, promise to open the door to new avenues for the forensic application of DNA analysis.

14.10 BIBLIOGRAPHY

J.M. Butler, *Forensic DNA Typing*, Academic Press, 2001.

W.G. Eckert and S.H. James, *Interpretation of Bloodstain Evidence at Crime Scenes*, Elsevier, New York, 1989.

R.E. Gaensslen, *Sourcebook in Forensic Serology, Immunology, and Biochemistry*, The National Institute of Justice of the US Department of Justice, Washington DC, 1983.

M. Krawczak and J. Schmidtke, *DNA Fingerprinting*, Bios Scientific Publishers, Oxford, 1994.

J.M. Walker and E.B. Gingold, (ed), *Molecular Biology and Biotechnology*, The Royal Society of Chemistry, Cambridge, 1993.

FBI Combined DNA Index System (CODIS) Home Page, http://www.fbi.gov/hq/lab/codis/index1.htm

CHAPTER 15

Presentation of Expert Forensic Evidence

TREVOR ROTHWELL

15.1 INTRODUCTION

The forensic scientist may be skilled at his or her particular branch of science and may indeed be a world expert on the subject. Such expertise is of very little value if the expert concerned is unable to communicate adequately both on paper and in the witness box.

The end product of almost every forensic scientific investigation consists of a report which may be used by police officers, prosecuting authorities, defence lawyers and the judiciary, and ultimately by those members of the general public who will comprise the jury. It is essential, therefore, that the forensic scientist is able to put together a report encapsulating the results of the scientific tests that have been undertaken in such a fashion that the information is readily accessible to a non-scientist. On occasion the scientist will have to appear in person in the witness box to explain and, if necessary, defend the conclusions reached in the laboratory, and in order to do this effectively the scientist will need to develop yet another set of skills.

Various aspects important in this presentation of expert forensic evidence will be discussed. The culmination of any criminal investigation is likely to be a trial within the criminal justice system. Accordingly, the first part of this chapter is devoted to a brief outline of the legal processes operating in the United Kingdom, so that the scientist's contribution to the criminal trial can be put into context. The duties and responsibilities of the expert witness are outlined and then the respective roles of prosecution and defence will be explored in more detail, to demonstrate the similarities and differences for the forensic scientist working for one side or the other.

The need for adequate quality control measures and the importance of proper training for the expert witness are also considered, together with the content and format of the forensic scientist's report. Disclosure of expert evidence is an important issue, therefore the obligations on the prosecution and the defence to disclose the nature and substance of the expert evidence prior to trial will be mentioned.

Finally, some practical advice is offered to those about to embark on the ultimate stage of the expert's work – appearing in court and giving evidence.

15.2 THE LEGAL SYSTEM AND THE COURTS

The legal system to be outlined is broadly that which exists in England, Wales and Northern Ireland. In Scotland the structure of the courts is somewhat different, although the principles of justice   and the responsibilities of the forensic scientist – remain essentially the same. Some of the differences between Scotland and the rest of the United Kingdom will, however, be highlighted.

15.2.1 The Lawyers

The Crown Prosecution Service (CPS) is responsible in England and Wales for the compilation of the prosecution evidence and its presentation in court. Either branch of the legal profession – solicitors or barristers – may appear in the magistrates' courts. At present only barristers may appear as counsel in the higher courts. This is not the case in Scotland, and there are moves afoot in England to open up certain of these proceedings to solicitors as well. The Crown Prosecution Service will brief counsel where necessary; members of its staff will often present the case directly in the lower courts.

Prosecution in the Scottish courts is undertaken by the Procurator Fiscal Service, operating through local offices in a manner which is similar to that of the Crown Prosecution Service. A Procurator Fiscal (solicitor) can demand further investigative work to be carried out and may take a more active part in an investigation than the equivalent officer of the Crown Prosecution Service. The Crown Court Office administers the Procurator Fiscal Service and also, where necessary, instructs counsel to appear in the High Court.

The defendant is usually represented by a solicitor who will appear for him or her at the magistrates' court. Where the case is to be tried at a higher court in England, the solicitor will brief a barrister to act as counsel for the defendant at the trial. For a trial by jury in Scotland, a solicitor will brief a barrister (advocate) to act for the defendant.

15.2.2   Magistrates' Courts

Most criminal cases are tried in magistrates' courts. Lay magistrates, who form the bulk of the magistracy, do not possess specialist legal training. They come from all walks of life, act in a voluntary capacity and commonly sit as a bench of three, advised by a legally qualified clerk. There are a number of stipendiary magistrates; such individuals are legally qualified, are paid for the work which they do, and usually sit alone. Stipendiary magistrates are often employed in urban areas where caseloads are heavy. Magistrates are limited both in the types of case which they may try and in their powers of sentencing.

Every criminal case must pass through the magistrates' courts. The vast majority are tried in these courts; the defendant is sentenced appropriately and no other court is involved. The most serious and complicated cases are committed by the magistrates to the next tier of court, the Crown Court, to be tried by judge and jury. Defendants who are to be tried for certain less serious offences in England and Wales may also elect for jury trial.

15.2.3   Crown Courts

The Crown Court is presided over by a judge who sits with a 12 member jury selected from members of the public. While the judge is there to advise on matters of law, it is the responsibility of the jury to decide upon matters of fact and to give their verdict at the end of the proceedings. The full range of sentences available within the criminal justice system may be used for a defendant found guilty at the Crown Court.

Because of their serious nature, it is Crown Court cases in which the forensic scientist is most likely to become involved. The Royal Commission on Criminal Justice, which considered, among other topics, the use of scientific evidence in criminal trials and reported in July 1993, indicated that about a third of all contested cases in the Crown Court involved scientific evidence.

15.2.4   Appeals

Following conviction, a defendant may lodge an appeal against sentence and/or conviction on various well defined grounds. An individual who has been sentenced by magistrates may appeal to the Crown Court, while an individual sentenced at that court may go to the Court of Appeal, Criminal Division, where the case will be heard by senior members of the judiciary. Under relatively recent legislation it is also possible for an application to be made that the sentence of the Crown Court was too lenient.

The Divisional Court of the Queen's Bench Division of the High Court considers appeals on points of law or procedure in cases dealt with in the magistrates' courts; such appeals may originate from either prosecution or defence. For all types of case, and provided the circumstances are appropriate, the final court of appeal is the House of Lords.

15.2.5   Coroners' Courts

The coroner does not have a responsibility to try individuals, but to inquire into the causes of deaths which may have occurred in suspicious or unusual circumstances. Following an examination of the circumstances, the coroners' inquest will return a verdict of, say, accidental death. Coroners may sit by themselves or, in appropriate circumstances, may empanel a jury. The office of coroner is a very ancient one, and those who hold the office must be qualified either legally or medically – in London, coroners often possess both qualifications.

15.2.6   Scottish Courts

The District Courts form the lowest tier of the courts in Scotland, dealing with the least serious criminal matters. The Sheriff Court is the principal local court, exercising a wide jurisdiction in both criminal and civil matters. A sheriff, who will be legally qualified and in some ways may be likened to a stipendiary magistrate, may sit alone (Summary Procedure), or with a jury of 15 members (Solemn Procedure) depending on the seriousness of the crime to be tried. A defendant cannot elect for a trial by jury, and only cases involving serious crimes are heard before a jury in a High Court. The High Court of Judiciary sits in Edinburgh and other major towns, and these courts will also hear appeals from the lower courts. There is no right of final appeal to the House of Lords from the Scottish Courts. The equivalent of the coroner's inquest is a fatal accident enquiry, which is presided over by a sheriff.

15.2.7   Civil Courts

Criminal cases form only a part of the workload of the courts in the United Kingdom; other matters such as insurance or personal injury claims are usually part of the work of the civil courts. The burden of proof may be different in the civil courts. Conviction in a criminal trial requires proof beyond reasonable doubt. In the civil courts a case may be proved on a balance of probabilities.

The forensic scientist may well become involved in civil matters, being requested to provide advice or to undertake scientific testing in a parallel fashion to that of the criminal investigation. Although the comments in this chapter are directed specifically to the criminal trial procedure, the forensic scientist concerned with civil casework will need to adhere to the same basic principles.

15.2.8   The Course of the Criminal Trial

The forensic scientist's work is commonly orientated to the production of information which may be used as evidence in a criminal trial. In order that the contribution of the forensic scientist can be seen in the context of the whole proceedings, the basic structure of such a trial will be described.

At the commencement of the trial in England and Wales, and following such initial steps as the swearing in of the jury, the prosecution will open its case with a brief outline of the evidence to be adduced. In Scottish courts there are no opening speeches and the procedure starts by going straight into the evidence. Witnesses for the prosecution will then be called one by one to give their evidence. Each will be questioned by the lawyer acting for the prosecution about those aspects of the case of which that individual has knowledge; this is referred to as the 'evidence-in-chief'. Having given their evidence, the witness may be questioned by the lawyer acting for the defence. In this 'cross-examination' the witness may be asked, for instance, to expand on answers previously given to the prosecution's lawyer and, at the end, the prosecution lawyer may wish to briefly re-examine the witness on points raised during the cross-examination.

When the prosecution has presented its case the defence lawyer will open the case for the other side, again calling such witnesses as are necessary to make the various points which the defence wishes to raise. The procedure is similar to that employed with the prosecution witnesses, except that of course the defence lawyer will conduct the examination-in-chief, and the prosecuting lawyer the cross-examination. The defence witnesses may or may not include the defendant, who is under no obligation to say anything at trial. If they choose to do so they are usually the first witness.

When all the evidence has been adduced the closing speeches commence, and here the procedure differs between magistrates' and Crown courts. In the magistrates' court only the defence lawyer speaks, summing up the various points of that side's case. In the Crown Court the procedure is more complex and there are three speeches: firstly from the prosecution's lawyer; secondly from the defence; and finally from the judge who, while remaining neutral, must ensure that the jury is properly briefed for its task. In England, Wales and Northern Ireland the jury of 12 peolple may return verdicts of 'guilty' or 'not guilty'. For a verdict of guilty, 10 of the 12 jurors must agree. In Scotland the additional verdict of 'not proven' is available for use in appropriate instances for the jury of 15 people, and eight of the jurors must agree for a verdict.

15.2.9 The Role of the Witness

The role of the witness is to place before the court such information as may be relevant to the case in question. For the majority of witnesses this is limited to evidence of facts that have formed a direct part of their experience. Witnesses are generally not permitted to give 'hearsay' evidence; in other words, the witness cannot repeat second-hand information that has been told to them. One or other lawyer will rapidly intervene if they feel that a witness is overstepping the mark in this respect.

15.3 THE EXPERT WITNESS

Forensic scientists are among those with specialist expertise who may be called upon to give evidence in the courts, providing advice on matters as diverse as the causes of fire, medical negligence, and stresses and strains in bridges. The role of such a specialist, however, is different from that of the ordinary witness, in that the expert is not present simply to repeat the facts adduced by the scientific tests that have been undertaken, but to offer an interpretation of the findings in the context of the case. Accordingly, expert witnesses may give, indeed are encouraged to give, opinions as well as factual evidence.

The expert may be expected to give the court information which falls outside the general knowledge of the judge or jury. If asked, the expert may legitimately give an opinion on any issue which falls within his or her competence, although there may be reasons why the judge may subsequently advise the jury to disregard such information. This opinion may include, for example, relevant probability estimates to illustrate that a particular blood group has been found to occur in one person in every 20 in the general population of the United Kingdom.

There are circumstances, for instance, with the results of certain DNA profiling tests, where the evidence may appear to be so powerful that the forensic scientist may believe it to be beyond any reasonable doubt that another individual might be responsible for the incident. However, the expert must not phrase his or her evidence in those terms. The question of the guilt or innocence of the accused does not fall within the remit of the expert, or any other, witness. This issue is one which must be ultimately addressed only by those who will decide the case, that is, the magistrates or jury.

15.3.1 The Duty of the Expert

The responsibility of the prosecuting lawyer is to present the case against the defendant; to marshal and present such facts as may serve to show that without any reasonable doubt the accused individual perpetrated the crime in question. The responsibility of the defence lawyer is to counter the arguments adduced by the prosecution and to demonstrate that the prosecution has not made a convincing case. Witnesses called by the prosecution or defence, although they are simply presenting evidence of fact, may also be expected to have a view of the guilt or otherwise of the subject of the trial. In that sense both sides are partial and biased to the particular party whom they represent. *It is vital that the expert stand apart from this partiality. The role of the expert witness is to use his or her experience and skill to provide impartial and unbiased evidence to the court.*

15.4 PROSECUTION AND DEFENCE

15.4.1 Equality of Arms

The system of criminal justice which applies in the United Kingdom is essentially adversarial. The case against the accused will consist of all those facts which the prosecution considers necessary to convince the court of the accused person's guilt; the defence will put together all such information as it considers necessary to nullify the prosecution case. A basic tenet of the criminal justice system is to ensure 'equality of arms' and it is important that appropriate advice and assistance is available equally to both prosecution and defence. Such assistance includes adequate facilities for undertaking any necessary scientific examination.

In the UK either or both prosecution and defence may introduce scientific evidence into the criminal proceedings. The introduction of scientific evidence by one side does not necessarily mean that the other side will also automatically wish to employ a scientist, although where the prosecution seeks to rely on detailed or involved science it is usual for the defence to brief their own expert. The extent of any investigation is entirely the responsibility of whoever commissions the scientific work.

The situation in the UK thus differs from that in some other jurisdictions where an accusatorial system of criminal justice is practiced. In this system an examining magistrate or *juge d'instruction* may commission and set the parameters for scientific work to be done on behalf of the court.

15.4.2 The Forensic Scientist and the Prosecution

For many years the police service in this country has had available to it a comprehensive service for the investigation of crimes using a wide range of scientific techniques. In England and Wales this service is currently provided by the Forensic Science Service, a Home Office agency. The Scottish police service operates its own laboratories, and the Northern Ireland Office operates a laboratory in Belfast. Operating broadly within the public sector, these laboratories are large and well equipped, and offer a comprehensive service and an extensive range of facilities.

It is quite usual for the police officer to have direct contact with the forensic scientist during the investigation of a crime. This contact may be close and ongoing; in a major case the scientist may become a part of the investigating team. The scientist is likely to be involved in the investigation from near its beginning and will almost always be the first forensic scientist to undertake work on any particular item. In forensic science the sequence of events which form part of any investigation may assume some importance, as will demonstrated later.

The forensic scientist working for the prosecution will usually con duct an in-depth examination of all the items and materials relevant to an investigation, often undertaking complex scientific tests which together may take many weeks to complete. All data produced will be recorded in a case file from which the scientist will in due course compile a report detailing the salient facts. This report will be passed to the police investigating officer and hence to the prosecuting authority.

15.4.3 The Scientist Working for the Defence

Scientific work for the prosecution is usually commissioned by the police as a part of the investigative process, while work for the defence is commissioned by that side's legal team, and the position of a forensic scientist working for the defence is both similar to and different from that of a scientist employed by the prosecution.

It is likely that the individuals employed to advise the prosecution and defence will possess similar qualifications and experience; both will be engaged upon a search for independently verifiable facts; both will be looking at similar aspects of the case. The scientist engaged by the defence, however, may have no involvement with the case until much or all of the initial work for the prosecution has been completed. Accordingly, it is seldom that the scientist working for the defence sees items in the condition which they were in during the commission of the incident. Instead, the defence scientist will often need to work from information and results provided by the prosecution scientist and any conclusions which the defence scientist draws must take these limitations into account.

Unlike the prosecution scientist, the expert engaged by the defence will frequently be asked to prepare a report within a very short timescale. The work done on behalf of the prosecution, because it tends to be comprehensive and wide-ranging, may take weeks, if not months, to complete and may well not be delivered to that side's legal team until after the case has been committed for trial at Crown Court. Lawyers acting for the defence will not generally know whether they wish to brief their own experts until they have been made aware of the content of the prosecution scientist's report. This may result in precious little time being available before the trial is set down for hearing for any work to be undertaken on behalf of the defence team.

The provision of adequate forensic science facilities for the defence is a relatively new concept. It has developed through the growth of small private independent laboratories, often founded by experienced forensic scientists who, having previously practised in the large public sector 'prosecution' laboratories, have seen the need for suitable facilities to be made available to all those involved in the criminal justice system. More recently, disparities between the provision available to the two sides in a criminal trial have become blurred as the public sector laboratories have opened their doors to work from defence lawyers, while the private practitioners bid for work from organisations such as the police.

15.4.4   The Sequence of Events in a Forensic Examination

The fact that only one scientist is likely to be in a position to see an item in its original condition may have a significant effect on the subsequent course of events.

Let us take the example of an attempt to demonstrate contact between two people through an examination of their clothing. One way in which this may be achieved is by looking at the range of textile fibres transferred between one set of clothing and the other, removing for microscopic examination the superficial fibres adhering to the surface of the garments. Such fibres are extremely tiny and are easily lost from the surface, and other fibres may be readily transferred on to a garment through accidental contact. Accordingly the forensic scientist will accord the highest priority to removal of the superficial fibres very early in the examination in order to obviate the possibility of accidental contamination occurring during subsequent handling of the items.

Having removed the superficial fibres from, say, a jacket during its examination it is not possible to repeat this task and any other expert looking at the garment subsequently will not see the item in its original state. For any subsequent examination the scientist will be presented with a jacket together with a completely separate set of tapings bearing the superficial fibres which had at one time been present on the jacket and this may limit the options available to the second examiner.

This situation has its analogies in many other fields. For example, following an initial autopsy by a forensic pathologist working on the instructions of the coroner, further examinations may be requested by other interested parties such as lawyers acting for an accused or for the relatives of the victim. Only the pathologist who conducts the original autopsy will be able to see the body in its original state, and while other doctors may subsequently conduct their own dissection of the material they will be constrained by the nature of the original autopsy.

In many circumstances the use of photographs and video recordings may be of immense value in showing, for example, the original disposition of the various items found at a crime scene or the appearance of injuries on a body. In other instances, however, there is at present no way of recording the appearance of an item in such a manner as to demonstrate its original condition to anyone who may subsequently express an interest. The removal of superficial fibres is a case in point and the efficiency and effectiveness of this procedure has essentially to be taken on trust.

15.4.5 The Role of the Second Examiner

The role of the second or subsequent scientist is thus to examine such items as are available and to make an assessment of the way in which the initial examination was carried out. This will usually involve detailed scrutiny of the documentary material – test results, chromatographic charts, *etc.* – produced by the original scientist to check that appropriate procedures were carried out to the proper standards. Relatively less often will there be any need to repeat actual scientific tests. Although there are occasions on which the defence takes the initiative and produces items for scientific examination that have not already been looked at by the prosecution, in the majority of cases it is the prosecution's scientist who takes first bite at the cherry and the defence's expert who has to take the role of the second examiner.

15.4.6   The Need for Both Prosecution and Defence Experts

The responsibility of the scientist is to search for verifiable scientific facts and to provide impartial evidence to the court. It is, therefore, legitimate to question the necessity for both prosecution and defence to employ their own expert, and if they do, why the two experts may come up with different conclusions. The most obvious justification for the use of independent expert witnesses is the need to ensure the equality of arms referred to earlier. Thus, to ensure the maximum probability of a fair trial, whatever facility is available to one side should be made available to the other. If the prosecution intends to rely on scientific findings as one plank of its evidence at trial, it is right that the defence should be able to test independently the strength of that evidence.

While equality of arms may justify the need to employ independent forensic science experts, it does not explain why, after an examination of the same materials and results, they may come up with different answers. In fact, and not surprisingly, it is very rare for the experts to disagree over the scientific facts adduced in any particular case. Where differences may arise it is more likely to be in the interpretation placed upon the findings, and for these differences there may be a variety of reasons.

In some areas of science there is room for genuine ambiguity about the meaning of particular scientific findings. Certain autopsy findings can, in the view of one forensic pathologist, point to one conclusion about the origin of particular injuries; another practitioner may consider a different conclusion more probable in the circumstances. It is often not possible to determine that one pathologist is more 'right' than the other and the court will be left to decide between the two experts.

There will be other circumstances in which different conclusions can be drawn because the scientists involved are working on different assumptions. Forensic science is seldom an absolute science and the scientist has to be given a framework within which to design his or her programme of work. The scientist working for the prosecution, although usually having the advantage of time and of conducting the initial examination of the items, will be working to the information provided by the police. While this information may be detailed, it is unlikely to include the defendant's own view of the course of events.

The defence scientist, on the other hand, although hampered by time and other pressures, almost always has the undeniable advantage of having heard both the rationale behind the prosecution expert's work and the defendant's own explanation of events. Using this broader spread of knowledge, the defence scientist may be able to posit other possibilities or explanations which provide an alternative, and sometimes better, fit with the observed facts. These other explanations may assist the defence in pursuing its case, although in many instances the information may also be of value to the prosecution. Where such information is discovered it should not be withheld.

From time to time the suggestion is made that the forensic scientist should be more precisely a servant of the court itself rather than of one side or the other. Attractive though such a suggestion may be in emphasising the impartiality of science, it would appear that within the adversarial system of justice which operates in the United Kingdom the current arrangements are satisfactory. In most cases it is expected that the scientists concerned will come up with very similar conclusions. In those cases where conflicts exist it remains with the court to decide between the two strands of scientific evidence.

15.5 THE IMPORTANCE OF QUALITY

The evidence adduced by the forensic scientist may be crucial in securing the conviction of an accused person, who may then be given a heavy fine or a long period of imprisonment. On the other hand, scientific findings may assist the lawyer to obtain an acquittal. The responsibility which rests upon the shoulders of the forensic scientist is thus high and no effort must be spared to ensure the veracity and accuracy of any evidence which is provided. Each and every individual who provides expert evidence to the courts bears a similar responsibility. The need for quality and the maintenance of standards has been obvious since the inception of forensic science, and for many years comprehensive quality management schemes have been in operation, certainly in the public sector laboratories.

15.5.1 The Individual

The quest for quality involves the individual as well as the work process. The qualities necessary to perform well as a forensic scientist are not necessarily the same as those required, say, to conduct a research programme. Although forensic work is sometimes at the frontiers of science, particularly, for example, with DNA profiling, it often necessitates common sense rather than a deep scientific understanding. It always requires an interest and ability in putting over scientific concepts to non-specialists in a clear and unambiguous fashion.

Individuals seeking a career in forensic science require a basic background in an appropriate subject. This will often consist of a first degree in some aspect of chemistry, biology or biochemistry. There are currently two higher degree courses devoted specifically to the specialism and such courses can provide a valuable background for the potential forensic scientist. Detailed training in forensic science is provided on the job and, along with developing technical skills, such training will include considerable practice in the compilation of reports. Specific training for those scientists who give evidence in court, utilising real lawyers and a mock court, is undertaken by the Forensic Science Service. Similar training has for many years been provided as part of the MSc course at the Forensic Science Unit of the University of Strathclyde and there are now commercial organisations who can offer such training.

Aside from the specialist academic courses there were originally no formal qualifications in forensic science. The Forensic Science Society has introduced a series of diplomas in particular aspects of the profession, for instance, in the examination of firearms, examination of documents, fire investigation, and scene of crime investigation. A scheme of qualifications now being developed under the aegis of National Vocational Qualifications (and its Scottish equivalent) should in time cover most of the skills and competencies needed by the practising forensic scientist at all stages of their career. In Scotland an expert witness can only give evidence if the person has authority from the Secretary of State for Scotland.

Up to now the usual course of a scientist's career has been for the individual to join one of the public sector laboratories, to be trained in some appropriate discipline and to remain within the sector for the duration of that person's career. As has been explained, these laboratories have traditionally carried out the examination of work submitted by the police and have thus been associated with the prosecution. Over the years some members of staff at these establishments have seen the need for equivalent facilities to be made available to the defence and have initiated laboratories within the private sector for this purpose. Thus far these latter laboratories have not generally been in a position to recruit scientists direct from university and to offer them basic training. Transfer of staff has thus effectively been limited to movement from the public into the private sector, although in a truly free market it might be preferable for personnel to be able to interchange freely in either direction between the two sectors. However, the large laboratories are able to offer well designed programmes of training and the opportunity to obtain excellent all-round experience which can only be to the advantage of the profession as a whole.

15.5.2 Setting Standards

To be assured that work standards are adequate it is necessary to define the appropriate standard, to set up work processes which will achieve that standard and to introduce a monitoring system to ensure compliance. Proper quality management must therefore be integral to every aspect of the work of a laboratory; it is not something which can be added on at the end of the production process.

15.5.3 Case Documentation

Underpinning the quality of work performed by the forensic scientist must be the basic documentation which comprises the case file. This should record every action undertaken by the scientist and by every assistant or other person working on their behalf. In this respect such 'actions' are likely to include:

1. Telephone calls and written correspondence between the scientist and those for whom the work is being carried out. Such material may involve lists of items for examination, statements from other witnesses, invoices, requests for attendance at court, *etc*.
2. Notes to assistants or other colleagues regarding work carried out.
3. Detailed records of all the examinations carried out and the tests undertaken, together with the interpretation of the results – these records will constitute the bulk of the file.
4. Drafts of the final report, and a copy of the report itself.

All this material should be indexed, for instance through the use of a minute sheet attached to the file.

As far as possible the layout of the case file should be in logical sequence, firstly because it enables the material to be checked through more easily, but also because it will prove to be of enormous value should the need arise for the scientist to appear in court. It is all too easy to lose one's place during intensive cross-examination in the witness box!

15.5.4 Assuring the Quality of the Work

Test procedures to be employed must be reliable and capable of yielding accurate and precise results. One way of assuring this is to rely on published procedures that have been subjected to peer review. Although there may be occasions when the individual scientist will have to design a new procedure for a specific investigation, in the main there is no place for any expert to use tests which have not been generally accepted by the scientific community as a whole.

Another basic tenet of quality assurance is that significant decisions should be checked by a colleague. Thus, for instance, the rationale behind a blood-grouping test or the results of a chromatographic separation of fibre dyestuffs would be noted by the worker performing the test and confirmed by a colleague. The object of such peer review is not to belittle the work of any individual, but to obviate the commission of errors which may easily occur when, say, a sample tube becomes misplaced.

One method of assessing the efficiency and effectiveness of a forensic science laboratory is through the submission of simulated case material for examination. The larger laboratories have developed a system in which such cases can be worked on without the examiners being aware of the provenance of the material and this has proved to be a useful performance monitor.

It is particularly important that the report itself, the product of the scientist's activities, should be subject to thorough quality assurance procedures. This is likely to include independent scrutiny of the report by another scientist with a similar background, followed by a discussion of the salient points in which the writer is invited to justify the conclusions reached. More generally, such discussion promotes the sharing of experience, which can only result in the overall improvement of quality. Sharing experience through checking each other's work and discussing reports may be easier to organise in large laboratories than in small independent units; the need, however, is universal.

The scrutiny should not be limited to scientific matters, because it is vital not simply that the conclusions are correct but that they may be understood by the reader. In this respect it is also valuable to have the material read by a non-specialist; that, after all, will be its fate when the report leaves the laboratory.

15.5.5 Time Limits

While the time taken for laboratory examination may not generally be considered as an aspect of quality it is nevertheless important. In

Scotland a time limit is imposed on the prosecution, because a case must be committed for trial within 110 days of a person being taken into custody. Time limits have now been introduced south of the border. Although not a major consideration, therefore, the existence of custody time limits may have to influence the conduct of the investigation.

15.6 THE FORENSIC SCIENTIST'S REPORT

The forensic scientist's report is the product of all the effort which has gone into the investigation. It must contain all relevant detail and must be laid out in such a way that its content is immediately accessible to the non-scientist, whether police officer, lawyer or lay jury member. Although the image of the forensic scientist is of the expert in the witness box, in fact the vast majority of cases do not involve the scientist's appearance in court. In these circumstances, therefore, the report has to stand on its own feet without the possibility of explanation or clarification from its writer. Accordingly, it is absolutely essential that the forensic scientist's report is clear and unambiguous, and contains all the information necessary to explain the scientific findings.

There is another factor which should be taken into account. In most instances the only scientific evidence to see the light of day in any particular case will be the report prepared by the prosecution's scientist. According to the 1993 Royal Commission's Report, in about three-quarters of all cases (and even in two-thirds of those cases rated as 'very important') the scientific evidence was not contested because there was no basis for any such challenge.

As well as being clear and unambiguous, the scientist has an overwhelming responsibility to be frank and fair in whatever statement they make, because there is unlikely to be another scientist – or anyone else in either legal team – in the position to question any aspect of the material. That thought need not create paranoia in the mind of any potential or practising forensic scientist; any one individual can only proceed to the best of their ability. Nevertheless, it should always be at the back – or perhaps at the forefront – of the scientist's mind when the report is being written.

15.6.1 Format

No formula can be given for the format of the forensic scientist's report; the 'shape' of a report is dependent on the nature of the investigation and the predilections of the writer. The end result of working for the prosecution will usually be a report which uses the layout specifically designed for witness statements in England and Wales or court reports in Scotland. This may also be the case where the work has been carried out on behalf of the defence, although a straightforward technical report format may be acceptable. In every instance it is important that the report is signed by the scientist concerned and dated.

In recent years much effort has been expended by forensic scientists from all types of background in devising formats which are user-friendly and, because in the majority of cases the scientist will not be present in person (either at conferences or in court) to interpret what is written, the importance of this cannot be stressed too highly. The report will usually contain the following information, although neither the sequence of headings nor the mode of presentation will always be the same.

1. *Name and address.* The report should open with the relevant details of the scientist who has complied it: name, laboratory and qualifications. The standardised format employed for witness statements has prenominated spaces for these details. If appropriate, details of the expert's experience in the subject may also be included in the introduction.

2. *Outline of circumstances.* There is logic in then outlining the course of events of the incident that is under investigation, giving just sufficient detail to put into context the scientific tests and examinations which have been carried out. This information is hearsay evidence in that it has been given to, rather than generated by, the scientist and the report must make quite clear that this is the situation.

3. *Outline of scientific work carried out.* An outline of the scientific work carried out will show the scope and depth of the expert examination. It should indicate the questions that have been addressed and may clarify those which have not been, or cannot be, addressed. A common expectation of those without an understanding of the subject and its limitations is that examination of an item by a forensic scientist will reveal everything there is to know about that item. This is, of course, very far from the truth. The scientist can only offer answers to such questions as have been asked; sometimes even this cannot be achieved. It is important that the effective parameters of the examination are made clear from the start.

4. *List of exhibits examined.* It is usual to list the items that have been examined and from whom these items were obtained in order to

maintain the chain of continuity of the relevant exhibits in the case. In Scotland these exhibits are usually referred to as labelled productions. The importance of continuity is to demonstrate that proper track has been kept of the progress of each and every item from the moment of recovery at the scene of the incident, or elsewhere, to the time of production in court during the trial and that nothing untoward has been allowed to befall them during any intervening period of time.

5. *Description of the work carried out.* The work undertaken by the forensic scientist should then be described, relating the examination and test procedures to the various items in the most logical order possible. The scientist should explain the justification for each element of work carried out and provide a comprehensive summary of the results achieved. Sometimes it may be appropriate to go into the detail of the test procedure itself, although the level of scientific understanding of the report's potential readership must never be forgotten. It is most important that every procedure should be placed in context in order that the reader is made aware both of the rationale behind it and of the meaning and limitations of the results obtained.

6. *Interpretation of the findings.* The results of every procedure carried out need to be interpreted in the context of the case. On occasion there may be more than one interpretation of a particular set of test results. It may be necessary to detail the various possibilities, suggesting which explanation may be more likely within the conditions that pertain. It will always be important to explain the tests and procedures in a full and clear manner.

There will be occasions on which the information adduced from the scientific tests appears to be of more value to the 'other' side. For instance, a forensic scientist briefed by the prosecution may find that the results of certain tests that have been carried out might assist in the defence of the person accused of the crime. Conversely, and probably more frequently, a defence scientist may turn up information of potential value to the prosecution. The finding of trace evidence, such as textile fibre transfer, which may help to demonstrate contact between two individuals, is after all more likely to be of positive value to the prosecution than to the defence. In order to retain integrity, the forensic scientist has a duty to be open about such findings and not to withhold information because it does not appear to suit the demands of the legal team on behalf of which the scientist's skill has been sought.

7. *Conclusions*. The results of the various tests should be drawn together to form a series of conclusions. These should be clear and unambiguous; it is well that the scientist compiling them realises that they may be the only part of the report to be read by at least some of the readers.

8. *The use of assistants*. Forensic science today is very specialised and the scientist who compiles the report may in fact be putting together the work of several colleagues who have assisted during the examination. It is usual to list the names of these colleagues and to indicate which aspects of the work they have carried out. On occasion these assistants will be called to give evidence along with the scientist who has signed the report. When that happens, the assistants will only be expected to give factual evidence of their role in the case examination. They will not be expected to provide any opinions about the tests or the results obtained or to act as expert witnesses in their own right.

9. *Appendices to the report*. Certain facets of the scientific work, for instance blood-grouping tests, may result in information that is more readily documented in tabular form than described in words. The scientist's report should be in a form suitable to be read in court and it may well be difficult to read out a table of information in any way that makes sense. For this reason it is important to describe the information to the best of the scientist's ability in the body of the text, but it may be useful in addition to incorporate the data into a table which can be appended to the report. Copies of the tabulated information can be supplied to those requiring them.

15.6.1.1   Corroborated Evidence.   The requirement in Scotland is that an expert's work should be corroborated by a colleague and that both scientists should put their names to what then becomes a joint report. The individual who corroborates the evidence must have taken part in the work, seen it done or be able to verify that it has been done.

15.6.2   Disclosure of Expert Evidence

The forensic scientist working for the police must ensure that officers are made aware of all the scientific evidence relevant to the case. Such information must be passed to the prosecuting authorities. In turn it will be disclosed to the defence who must be made aware of anything on which the prosecuting scientist may have relied in forming the scientific conclusions. The disclosure rules are wide ranging:

'... if expert witnesses are aware of experiments or tests, even if they have not carried them out personally, which tend to disprove or cast doubt upon the opinions they are expressing, they are under an obligation to bring the records of them to the attention of the police and prosecution.'

(*Royal Commission on Criminal Justice 1993 (9.46)*)

Following disclosure, the defence is entitled to access to case files and other documentation, and to any other information used by the prosecution's scientist. There is thus a duty upon this scientist to retain pertinent documents and materials in order that they may be made available to the defence if required.

Under the Crown Court (Advance Notice of Expert Evidence) Rules 1987, the defence must disclose to the prosecution any expert evidence on which it may wish to rely at trial and this parallels the situation which applies to the prosecution.

Sometimes the defence may simply wish to use the advice of an expert in order to discredit the prosecution evidence, rather than to call the expert to court to give evidence directly on their behalf. In this situation there is no obligation upon the defence to disclose the results of any tests that, although carried out on their instruction, tend to support the prosecution's case. The scientist should note, however, that any such decision to withhold information in this situation will be taken by lawyers; it does not affect the obligation on the scientist to disclose all the facts to those who have commissioned investigation.

15.7 GIVING EVIDENCE IN COURT

Scientists may be called to court to give evidence on behalf of either prosecution or defence. The advice which follows is based on the assumption that it is lawyers working for the former who have requested the expert to attend and it will thus be the prosecution which conducts the evidence-in-chief. The procedure will be similar where it is the defence which has called for the expert to be present, except that the initial examination will be conducted by the defending lawyer.

15.7.1 Preparation

The need to give evidence in court for the first time is a daunting prospect; indeed, it often remains so throughout the forensic scientist's career. There is a sense of stepping into the unknown, of entering a world with rules different from those of the laboratory. In order to give a good 'performance', and to assist the court in the best way, adequate preparation is vital and the expert should go through all of the relevant material in good time before the trial. Reviewing the material may necessitate looking again at items such as microscope slides and chromatographic results, together with appropriate papers or other items in the literature which provide a background to the scientific findings.

The expert should ensure that the case file is suitably annotated and indexed so that important details can be located quickly while in the witness box. Relevant supporting papers should be taken to court so that they can be consulted should the need arise. It is not necessary to memorise a mass of detail, as the expert will normally be allowed to refer to their report and other papers in the witness box. The potential witness should, however, try to think through the type of question that might be asked concerning the work which has been undertaken.

In most important and complex cases the scientist may be invited to attend a pre-trial conference with the counsel who will conduct the case in court. Such a meeting can be useful to both lawyers and scientists, providing, for instance, an opportunity to discuss potential problems such as aspects of the evidence which may be difficult to explain in simple terms.

In Scotland expert witnesses may be precognosed prior to trial. Precognition may be undertaken by prosecution or defence and involves discussion of the expert's findings with either the Procurator Fiscal or the defence counsel, or agents acting on their behalf. During precognition, notes are taken which will form the basis of a statement of what the witness will say during the subsequent trial. The statement is not admissible as evidence in court and should not be confused with the expert's own signed report.

15.7.2   Practical Details

The expert is likely to have to explain complex issues to non-scientists. It would be surprising if some of these issues did not pass over the heads of at least some members of the jury and in such circumstances it may be that the demeanour of the expert leaves as much of an impression on the jury as what was actually said. For these reasons an expert who is to appear in court must take particular care with such fundamental items as dress and appearance, unfashionable though it may be draw attention to such issues.

While Crown Courts are usually well signposted and easy to find, the same may not be true of magistrates' courts. The potential expert witness should accordingly arrive early enough to locate the court with time to spare. Most large court buildings will have a number of trials going on at any one time and the court list notice board will indicate the relevant courtroom. Having found the courtroom it is important to let the legal team know that the expert has arrived. That is simple if the session is about to commence, as the witness will be able to locate some member of the team either inside or outside the courtroom. Arrival while the court is in session may present more difficulties and the witness may have to wait for an usher to emerge from the court in order to give a message to the lawyer concerned.

Because the lawyers are trying to paint a picture of the events that occurred during the course of an incident, it is usually important to them that witnesses give their evidence in a logical and predetermined sequence. Lawyers are often unable or unwilling to hear an expert's evidence out of sequence and for this reason the witness may well find that they have to wait around for hours, and sometimes days, before going into the witness box. This waiting creates frustration and may generate a feeling that the lawyers are not giving proper value to the presence of the expert. However, the lawyers are in charge of the case; they have a responsibility to adduce the various elements of the evidence in the manner that they consider is going to have the maximum impact on judge and jury.

15.7.3 The Witness Box

The witness who is eventually called to give evidence will be shown to the witness box by a court usher and then 'sworn'. This involves repeating the words of the oath while holding a copy of the New Testament. Appropriate variations of the oath are available to those of other faiths; alternatively it is entirely in order to 'affirm' instead of swearing on the Bible or any other religious text. Copies of the relevant oaths and affirmations will be provided by the court clerk and there is no need for the potential witness to memorise these prior to their first appearance in court.

New expert witnesses are occasionally confused as to the correct form of address for the magistrate or judge. Magistrates should be addressed as 'Sir' or 'Madam' with 'Your Honour' or 'My Lord' (depending on the particular individual involved) being appropriate for Crown Court judges.

15.7.4 Evidence-in-Chief

The lawyer who has called the witness will commence the evidence-in-chief. The first task will be for the witness to give their name and address; this latter will be the address from where the report emanated and experts should normally be careful not to give their home addresses in open court. The expert's qualifications will then be elucidated and witnesses are occasionally asked how long they have been practising their specialised science. More detailed examination of an expert's qualifications and experience in the topics to be discussed is rare in this country, although it may be more common in, for instance, the United States.

There is an assumption that the expert witness, like the police officer, will need to refer to appropriate notes before answering a question. This is often clarified at the start of the evidence-in-chief by the lawyer confirming with the judge or magistrate that the scientist's notes may be consulted in the witness box. On rare occasions either the judge or the opposing lawyer may object to such reference being made. In my view, at such times the expert has a duty to make it clear that, because scientific issues may be complicated, without consulting the relevant notes there is a danger that incorrect information may be given to the court.

The lawyer will then return to the report which the scientist has prepared. Salient points may be picked out and the witness asked to expand on, say, the nature or significance of a particular test.

15.7.5   Giving Expert Evidence

In order to answer questions, whether during the evidence-in-chief or later in cross-examination, the expert should take care to refer to the relevant part of the report, for it is important that the evidence given is accurate. Answers should be as brief as practicable, directly addressing the question and not proffering superfluous information. On the other hand, care should be taken that sufficient information is provided to enable the court to understand the significance of the scientific findings. In this respect answering 'yes' or 'no' to a question may not be adequate, even though a lawyer may on occasion ask the expert to respond in this manner. If the response to a question is not immediately obvious the witness should say so; there is no disgrace in not being able to answer a question. An answer must not be concocted on the spot in the hope that it will prove correct.

There may also be occasions when the expert feels that there are points which the lawyers should have picked up, but have not done so. This can be a difficult situation, as the lawyers must be left to organise the flow of the evidence in whatever ways they consider best suit their respective cases. Nevertheless, if the scientist considers that vital information which could have a direct bearing on the course of the case has not been disclosed during the course of their evidence-in-chief or cross-examination, then perhaps the scientist should seek advice from the judge before leaving the witness box.

The witness box is no place for the flippant or off-hand remark, nor for a condescending manner. The expert witness is there to be a servant of the court, and the proceedings should be treated with proper seriousness.

15.7.6 Cross-Examination

Once evidence-in-chief has been given, cross-examination may begin. The lawyer conducting this aspect of the trial may wish to revisit various points of the evidence, perhaps to test whether other explanations of the scientific findings might be sought. Popular drama would give the impression that cross-examination is always aggressive and conducted so as to destroy the credibility of the witness, but, while this is sometimes the situation, it is more usually carried out in a civilised manner. Nevertheless, it is vital that the witness listens carefully to the questions and takes sufficient time to give a considered answer.

During preparation of the report, the forensic scientist will have considered the manner in which the results of the various scientific tests should be interpreted. Decisions will have been taken as to which of the possible explanations appears to be the most likely in the circumstances of the case. These decisions will have been subjected to appropriate quality assurance procedures through peer review and the like. Accordingly, the forensic scientist should have no difficulty in explaining and defending the reasoning behind the conclusions given in their report. The lawyer who has called the expert will be expecting that expert to support their conclusions even when different potential explanations for the findings are posited. Nevertheless, the scientist who considers these alternative explanations to have some validity has a duty to say so. The expert must tread a fine line, adhering to his or her conclusions while admitting, in appropriate circumstances, that a different explanation for the observed facts may exist.

Forensic scientists may, of course, be called by either prosecution of defence. Where the prosecution intends to rely on scientific evidence it is common for the defence legal team to brief another scientist to act on their behalf. Accordingly it may be that the scientist who appears for the prosecution finds that a scientist acting for the defence is present in the court while the former is giving evidence, and *vice versa*. Where another scientist is present it is common for them to 'feed' questions to the lawyer who is undertaking the cross-examination. Such questions are likely to be relevant and designed to illuminate any differences in the way in which the two scientists have interpreted the data. The presence of an expert from the opposing side should not lead to any foreboding. It usually enables the relevant information to be drawn out in the most useful way for the court.

15.7.7   Re-Examination

Following cross-examination the prosecuting lawyer has the opportunity to clarify any points which have been raised by the defence during that examination. However, fresh evidence cannot be introduced. It may also be that the judge or the chairman of the magistrates will wish to question the witness.

15.7.8   Releasing the Witness

When the evidence has been given and the cross-examination completed the lawyer will usually ask the magistrate or judge to release the individual, who is then free to go unless for some reason one side or the other wishes them to remain in the vicinity of the court. If the witness is not clear that permission to leave the court has been given, he or she should confirm the situation before leaving the witness box.

15.7.9   And Afterwards

Often the expert will step down from the witness box with a feeling of a job well done. On other occasions the expert will experience an anticlimax, with the feeling that points could perhaps have been explained in a better way. There will even be times when the witness will wish the floor to open in order that they can disappear without trace! Whatever the reaction, there is advantage in a debriefing session with colleagues in order to discuss the course of the examination while it is still fresh in the mind.

15.8   CONCLUSIONS

The presentation of evidence in both written and oral form is the culmination of all the effort expended by the forensic scientist in carrying out their investigation. Accordingly, the scientist must ensure that their evidence is accurate, reliable and is delivered in such a manner that it can be clearly and easily understood by all those who may need access to the information.

15.9   BIBLIOGRAPHY

D. Corker, *Disclosure in Criminal Proceedings*, Sweet and Maxwell, London, 1996.
S. Leadbeatter, (ed), *Limitations of Expert Evidence*, Royal College of Physicians of London, 1996.

Report of the Royal Commission on Criminal Justice, Cm 22632, HMSO, London, 1993.

P. Murphy, *A Practical Approach to Evidence*, Blackstone Press, London, 2001.

P. Roberts and C. Willmore, *The Role of Forensic Science Evidence in Criminal Proceedings. (Research Study No. 11 prepared for the 1993 Royal Commission on Criminal Justice)*, HMSO, London, 1993.

B. Robertson and G.A. Vignaux, *Interpreting Evidence; Evaluating Forensic Science in the Court Room*, Wiley, Chichester, 1995.

J. Smith, *Cases and Materials on Criminal Law*, Butterworths, London, 1999.

Subject Index